Multi-armed Bandit
Allocation Indices

WILEY-INTERSCIENCE SERIES IN SYSTEMS AND OPTIMIZATION

Advisory Editor

Peter Whittle

Statistical Laboratory, University of Cambridge, 16 Mill Lane, Cambridge CB2 1SB, UK

GITTINS—Multi-Armed Bandit Allocation Indices

Multi-armed Bandit Allocation Indices

J.C. GITTINS

Mathematical Institute, University of Oxford

With a Foreword by
Peter Whittle

JOHN WILEY & SONS

Chichester · New York · Brisbane · Toronto · Singapore

Library of Congress Cataloging-in-Publication Data:

Gittins, John C.,

   (Wiley series in probability and mathematical
statistics.   Applied probability and statistics)
   Bibliography: p.
   1. Experimental design.   I. Title.   II. Series.
QA279.G55   1989   519.5   88-17411
ISBN 0 471 92059 2

British Library Cataloguing in Publication Data available

Gittins, J.C.

   Multi-armed bandit Allocation Indices
   1. Indices
   I. Title
   512'.72
   ISBN 0 471 92059 2

Typeset by Interprint Malta Limited
Printed and bound in Great Britain by Anchor Press Ltd, Tiptree, Essex

Contents

Foreword

The term 'Gittins index' now has firm currency in the literature, denoting the concept which first proved so crucial in the solution of the long-standing multi-armed bandit problem and since then has provided a guide for the deeper understanding of all such problems. The author is, nevertheless, too modest to use the term so I regard it as my sole role to reassure the potential reader that the author is indeed the Gittins of the index, and that this book sets forth his pioneering work on the solution of the multi-armed bandit problem and his subsequent investigation of a wide class of sequential allocation problems.

Such allocation problems are concerned with the optimal division of resources between projects which need resources in order to develop and which yield benefit at a rate depending upon their degree of development. They embody in essential form a conflict evident in all human action. This is the conflict between taking those actions which yield immediate reward and those (such as acquiring information or skill, or preparing the ground) whose benefit will come only later.

The multi-armed bandit is a prototype of this class of problems, propounded during the Second World War, and soon recognised as so difficult that it quickly became a classic, and a by-word for intransigence. In fact, John Gittins had solved the problem by the late sixties, although the fact that he had done so was not generally recognised until the early eighties. I can illustrate the mode of propagation of this news, when it began to propagate, by telling of an American friend of mine, a colleague of high repute, who asked an equally well-known colleague 'What would you say if you were told that the multi-armed bandit problem had been solved?' The reply was somewhat in the Johnsonian form: 'Sir, the multi-armed bandit problem is not of such a nature that it *can* be solved'. My friend then undertook to convince the doubter in a quarter of an hour. This is indeed a feature of John's solution: that, once explained, it carries conviction even before it is proved.

John Gittins gives here an account which unifies his original pioneering contributions with the considerable development achieved by both himself and other workers subsequently. I recommend the book as the authentic and authoritative source-work.

PETER WHITTLE

ix

Preface

A prospector looking for gold in the Australian outback has to decide where to look, and if he prospects for any length of time must repeatedly take further such decisions in the light of his success to date. His decision problem is how to allocate his time, in a sequential manner, so as to maximise his likely reward. A similar problem faces a student about to graduate from university who is looking for employment, or the manager of a commercial laboratory with several research projects competing for the attention of the scientists at his or her disposal.

It is to problems like these that the indices described in this book may be applied. The problems are characterised by alternative independent ways in which time or effort may be consumed, for each of which the outcome is uncertain and may be determined only in the course of applying the effort. The choice at each stage is therefore determined partly on the basis of maximising the expected immediate rate of return, and partly by the need to reduce uncertainty, and thereby provide a basis for better choices later on. It is the tension between these two requirements that makes the decision problem both interesting and difficult.

The classic problem of this type is the multi-armed bandit problem, in which several Bernoulli processes with different unknown success probabilities are available, each success yielding the same reward. There are already two good books on this subject: by Presman and Sonin (1982), and by Berry and Fristedt (1985). Both books go well beyond the simple version of the problem just stated: most notably to include, respectively, a negative exponentially distributed rather than constant interval between trials, and a study of the consequences of discounting rewards in different ways.

The aims of this book are to expound the theory of dynamic allocation indices, and to explore the class of problems for which they define optimal policies. These include sampling problems, like the multi-armed bandit problem, and stochastic scheduling problems, such as the research manager's problem. Tables of index values are given for a number of sampling problems, including the classical Bayesian multi-armed bandit problem. For the most part, and except where otherwise indicated, the book is an account of original work, though much of it has appeared before in the publications bearing the author's name which are listed in the references. The mainstream of the book flows through Chapters 1 to 4. It breaks into independent sub-streams represented by Chapter 5, Chapters 6 and 7, and Chapter 8, then reforms briefly for the review of Chapter 9 as it swirls into the ocean of the future.

Readers should have some prior knowledge of university-level probability theory, including Markov processes, but not necessarily measure theory. Measurable functions may be regarded for our purposes simply as reasonably well-behaved functions, without serious loss of understanding. I hope the book will be of interest to researchers in chemometrics, combinatorics, economics, numerical analysis, operational research, probability theory and statistics. Few readers will want to read from cover to cover. For example, both chemometricians and numerical analysts are likely to be interested mainly in Chapter 7 and the tables which it describes: chemometricians as potential users of the tables, and numerical analysts for the sake of the methods of calculation. As another example, Chapters 1, 2, 4 and 8 could form the basis for a specialised lecture course in applied probability for graduate students. The exercises at the ends of the chapters have been included as an aid to the use of the book in teaching, and are not particularly difficult. In Chapter 3 and §§5.3 and 5.4 the mathematics becomes rather more intricate, and to some extent more sophisticated, than elsewhere; almost everyone will be well advised to avoid most of the details at these points, at least at first.

My own interest in the theory of stochastic allocation arose through efforts to provide aids to research managers in choosing between competing projects, and to research chemists in deciding which compounds to test in the hope of finding a new drug to treat a particular condition. These aids now exist in the form of computer programs. These are mentioned in Chapters 5 and 6 respectively, and more fully described in Gittins and Bergman (1985), which complements this book by giving a general review of statistical methods for pharmaceutical research planning.

It is a pleasure to acknowledge the stimulus and encouragement which this work has received over the years by conversations, and in some cases collaboration, with Joao Amaral, Tony Baker, John Bather, Sten Bergman, Owen Davies, Kevin Glazebrook, David Jones, Frank Kelly, Aamer Khan, Dennis Lindley, Peter Nash, David Roberts and Peter Whittle. I should particularly like to thank Joao Amaral, Frank Geisler and David Jones for the substantial effort involved in calculating index values, and Brenda Willoughby for her painstaking work in typing the manuscript. The book is dedicated to Hugh and Anna, who have grown up during its preparation.

JOHN GITTINS,
Mathematical Institute,
Oxford University,
March 1988.

CHAPTER 1

Introduction

This book is about mathematical models for optimizing in a sequential manner the allocation of effort between a number of competing projects. The effort and the projects may take a variety of forms. Examples are: an industrial processor and jobs waiting to be processed; a server with a queue of customers; an industrial laboratory with research projects; any busy person with jobs to do; a stream of patients and alternative treatments (yes, the patients do correspond to effort—see Problem 5 later in this chapter); a searcher who may look in different places. In every case effort is treated as being homogeneous, and the problem is to allocate it between the different projects so as to maximize the expected total reward which they yield. It is a sequential problem, as effort is allowed to be reallocated in a feedback manner, taking account of the pattern of rewards so far achieved. The reallocations are assumed to be costless, and to take a negligible time, since the alternative is to impose a travelling salesman-like feature, thereby seriously complicating an already difficult problem.

The techniques which come under the heading of dynamic programming have been devised for sequential optimization problems. The key idea is a recurrence equation relating the expected total reward (call this the *payoff*) at a given decision time to the distribution of its possible values at the next decision time (see equation (2.1)). Sometimes this equation may be solved analytically. Otherwise a recursive numerical solution may, at any rate in principle, be carried out. This involves making an initial approximation to the payoff function, and then successive further approximations by substituting in the right-hand side of the recurrence equation. As Bellman (1957), for many years the chief protagonist of this methodology, pointed out, using the recurrence equation involves less computing than a complete enumeration of all policies and their corresponding payoffs, but none the less soon runs into the sands of intractable storage and processing requirements as the number of variables on which the payoff function depends increases.

For the problem of allocating effort to projects the number of variables is at least equal to the number of projects. An attractive idea, therefore, is to establish priority indices for each project, depending on its past history but not that of any other project, and to allocate effort at each decision time only to the project with the highest current index value. To calculate these indices it should be possible to calibrate a project in a given state against some set of standard projects with simple properties. If this could be done we should have a reasonable policy

1

without having to deal with any function of the states of more than one project.

That optimal policies for some problems of effort allocation are expressible in terms of such indices is well known. The first three problems described in this chapter are cases in point. Chapter 3 sets out a general theory, including several theorems asserting the optimality of index policies under different conditions, which puts these results into context. In fact five of the six problems described in this chapter may be solved fairly rapidly by using Theorem 3.6 and its Corollary 3.10. Problem 4 requires Theorem 5.1, which is a continuous-time version of Theorem 3.6.

Since they may change as more effort is allocated, these priority indices may aptly be, and often are, termed *dynamic allocation indices*. The main aim of this chapter is to exhibit their range of application by describing a number of particular instances. A second aim is to give an informal introduction to the main methods available for determining indices. These are by (i) interchange arguments, (ii) exploiting any special features of the bandit processes concerned, in particular those which lead to the optimality of myopic policies, (iii) calibration by reference to standard bandit processes, often involving iteration using the dynamic programming recurrence equation, and (iv) using the fact that a dynamic allocation index may be regarded as a maximized equivalent constant reward rate.

Problem 1. Single-machine scheduling (see also §§2.5 and 3.6). There are n jobs ready to be processed on a single machine. A cost c_i is incurred per unit of time until job i has been completed, and the service time for job i is s_i. In what order should the jobs be processed so as to minimize the total cost?

Problem 2. Goldmining (see also §2.9). A man owns n goldmines and a goldmining machine. Each day he must assign the machine to one of the mines. When the machine is assigned to mine i there is a probability p_i that it extracts a proportion q_i of the gold left in the mine, and a probability $1 - p_i$ that it extracts no gold and breaks down permanently. To what sequence of mines on successive days should the machine be assigned so as to maximize the expected amount of gold mined before it breaks down?

Problem 3. Search (see also Chapter 8). A stationary object is hidden in one of n boxes. The probability that a search of box i finds the object if it is in box i is q_i. The probability that the object is in box i is p_i, and changes by Bayes' theorem as successive boxes are searched. The cost of a single search of box i is c_i. How should the boxes be sequenced for search so as to minimise the expected cost of finding the object?

For Problem 1 an optimal schedule is one which processes the jobs in decreasing order of c_i/s_i, as may be shown by a simple interchange argument.

For Problem 2 an optimal policy is one which, when x_i is the amount of gold remaining in mine i on a particular day, allocates the machine to a mine j such that

$$\frac{q_j x_j}{1 - p_j} = \max_i \frac{q_i x_i}{1 - p_i}.$$

For Problem 3, if box j is searched next under an optimal policy then

$$p_j q_j / c_j = \max_i p_i q_i / c_i.$$

For Problems 2 and 3 a policy is specified by a sequence of numbers corresponding to the order of the mines to which the machine is allocated until it breaks down, or the order in which the boxes are to be searched until the object is found. The policies just quoted may therefore be shown to be optimal in the class C of all those policies specified by sequences in which each $i(= 1, 2, \ldots, n)$ occurs an infinite number of times, by arguments involving interchanges of adjacent numbers in such a sequence. Since any policy may be matched by a policy in C for any arbitrarily large initial portion of the specifying number sequence, it follows that the above policies are optimal in the class of all policies.

The optimal policies for Problems 1, 2 and 3 may be described as follows. At each stage where a job, mine or box is to be selected there is a real-valued function defined on the set of available alternatives, and the one selected is such as to maximize this function. These functions, then, are dynamic allocation indices for the three problems.

We have described these indices as functions on the sets of available alternatives. More precisely, the index for Problem 2, for example, is a function of the three variables, x_i, q_i and p_i, which constitute the information available for mine i when the amount of gold which it contains is x_i.

The point of this observation is that our indices do not simply tell us which alternative to choose at a given stage in a particular problem with n alternatives. Rather, they specify the optimal policy for a problem with any set of alternatives of a given type. This means that if we know that such an index exists for a given class of problems, without as yet having an explicit expression for it, it is sufficient, in order to find an explicit expression, to consider problems within the class for which $n = 2$, and indeed even this restricted class of problems need not be solved in full generality, as our next example shows.

Problem 4. Industrial research (see also §2.5 and Chapter 5). The manager of a team of industrial scientists has n research projects which may be carried out in any order. Switches from project to project cause negligible loss of time, whether these occur when a project has been successfully completed or not. The value to her employer of completing project i at time t is $V_i e^{-\gamma t} (i = 1, 2, \ldots, n; \gamma > 0)$. The time which the team would need to spend on project i in order to complete it has distribution function $F_i(t)$ and density function $f_i(t)$. What policy should the manager follow in order to maximize the expected total value generated by the n projects?

There is no obvious elementary means of answering this question, so let us take advantage of the fact that the problem may be shown (Theorem 5.1) to have an index, and proceed to find an expression for the index by considering the case $n = 2$. Project 1 will be an arbitrary project of the type described. Project 2 we suppose to require a non-random time Δ to complete, and $V_2 = \lambda\Delta$. Let C_Δ denote the class of projects of this type for a given Δ and different positive values of λ.

For a set of projects all of which are in C_Δ it is best to work on them in order of decreasing λ-values. This is almost obvious, as such a policy results in the larger undiscounted values $\lambda\Delta$ being discounted least by the factors $e^{-\gamma t}$, and is easily checked by an interchange argument. It is also not difficult to convince oneself that in this case work on a project should not be interrupted before the project has been completed. Thus for C_Δ the λ-values define an index.

Consider then the case $n = 2$, where project 1 is arbitrary and project 2 is in C_Δ. If we can find a λ for which it is optimal to work initially either on project 1 or on project 2 then, since an index is known to exist for the entire set of projects, it follows that one such index function may be defined by assigning the index value λ to project 1 as well as to project 2. In effect we may calibrate the entire set of projects by using C_Δ as a measuring device. Any monotone function of λ would of course serve equally well as an index.

By the optimality property of the index rule it follows that if in a given problem it is optimal to work on a member of C_Δ then, if the problem is modified by adding further members of C_Δ all with the same λ-value, it remains optimal to work on these in succession until they have all been completed. Thus for calibration purposes we may suppose that several replicates of project 2 are available. For good measure suppose there is an infinite number of replicates. (Discounting ensures that the maximum expected reward over all policies remains finite.) Note, too, that since the value of Δ is arbitrary we may as well work with the limiting case as Δ tends to zero, so that our infinity of replicates of project 2 reduce to a continuous reward stream at the undiscounted rate λ, an object which will appear later, and termed a *standard bandit process* with parameter λ, or simply $S(\lambda)$.

We should like, then, a criterion for deciding whether it is best to work on project 1 first, with the option of switching later to $S(\lambda)$, or to start on $S(\lambda)$. Denote these two alternatives by 1λ and λ. Note that the index rule never requires a switch back to project 1 in the case of positive Δ, and such a switch is therefore not required for optimality either for positive Δ or in the limiting case of a standard bandit process.

Under 1λ work proceeds on project 1 up to some time t (possibly ∞), or until project 1 terminates, whichever occurs first, and then switches to $S(\lambda)$, from which point 1λ and λ yield identical reward streams. Thus 1λ is preferable iff

$$\int_0^t Vf(s)e^{-\gamma s}\,\mathrm{d}s > \int_0^t \lambda[1-F(s)]e^{-\gamma s}\,\mathrm{d}s, \tag{1.1}$$

these two expressions being, respectively, the expected reward from 1λ up to the switch to $S(\lambda)$, and the expected reward from λ up to that point. The uninformative suffix 1 has been dropped in (1.1). The right-hand side of (1.1) takes the given form since $\lambda e^{-\gamma s}\delta s + o(\delta s) = $ return from $S(\lambda)$ in the interval $(s, s+\delta s)$, and $1 - F(s) = \mathbf{P}\{\text{switch to } S(\lambda) \text{ under } 1\lambda \text{ takes place after } s\}$.

It is best to start work on project 1 iff (1.1) holds for some $t > 0$, i.e. iff

$$\lambda < \sup_{t>0} \frac{V\displaystyle\int_0^t f(s)e^{-\gamma s}\,\mathrm{d}s}{\displaystyle\int_0^t [1-F(s)]e^{-\gamma s}\,\mathrm{d}s}. \tag{1.2}$$

It follows that the right-hand side of (1.2) is the required index for a project on which work has not yet started. For a project on which work has already continued for a time x without successful completion the density $f(s)$ and distribution function $F(s)$ in (1.2) must be replaced by the conditional density $f(s)/[1-F(x)]$ and conditional distribution function $[F(s)-F(x)]/[1-F(x)]$, giving the general expression for the index

$$\sup_{t>x} \frac{V\displaystyle\int_x^t f(s)e^{-\gamma s}\,\mathrm{d}s}{\displaystyle\int_x^t [1-F(s)]e^{-\gamma s}\,\mathrm{d}s}. \tag{1.3}$$

For a project for which $\rho(s) = f(s)/[1-F(s)]$ is decreasing, (1.3) reduces to $V\rho(x)$. Since $\rho(x)$ is the probability density for completion at time x conditional on no completion before x, using $V\rho(x)$ as an index corresponds to maximizing the expected reward during the next infinitesimal time interval, a policy sometimes termed a *myopic* policy because of its neglect of what might happen later, or disregard of long-range vision, to continue the optic metaphor. What we have shown is that for Problem 4 there is no advantage to be gained from looking ahead if all the completion rates $\rho(s)$ are decreasing, but otherwise there may well be an advantage.

The indices already derived for Problems 1, 2 and 3 are all myopic. However, we shall meet later more general versions of each of these problems for which short-sighted optimization is often not optimal. This leads to more complex expressions for the indices, which were obtained in full generality only as a result of the index theorem. In each case the need to look ahead arises from the possibility that the index for the alternative selected may increase at subsequent stages. This conflict between short-term and long-term rewards is neatly shown by the following deceptively simple sounding problem.

Problem 5. Multi-armed bandits (see also §§6.4 and 7.4). There are n arms (treatments) which may be pulled (allocated to patients) repeatedly in any order. Each pull may result in either a success or a failure. The sequence of successes and failures which result from pulling arm i forms a Bernoulli process with unknown success probability θ_i. A success at the tth pull yields a reward $a^{t-1}(0<a<1)$, whilst an unsuccessful pull yields a zero reward. At time zero each θ_i has a beta prior distribution and these distributions are independent for different arms. These prior distributions are converted by Bayes' theorem to successive posterior distributions as arms are pulled. Since the class of beta distributions is closed under Bernoulli sampling these posterior distributions are all beta distributions. How should the arm to pull next at each stage be chosen so as to maximize the total expected reward from an infinite sequence of pulls?

For this problem, in contrast to the previous one, there is an obvious solution. This is at each stage to pull the arm for which the current expected value of θ_i is largest. Thus if the current distribution for θ_i is Beta(α_i,β_i) $(i=1,2,\ldots,n)$ the suggestion is that the next arm j to be pulled should be such that $\alpha_j/(\alpha_j+\beta_j)=\max_i[\alpha_i/(\alpha_i+\beta_i)]$.

This policy is simply stated, intuitively plausible, and in most cases very good. However, it is unfortunately not optimal. This is because it is myopic, and in this problem there is some conflict, though not usually a strong one, between short-term and long-term rewards. To see this, suppose $n=2$ and $\alpha_1/(\alpha_1+\beta_1)=\alpha_2/(\alpha_2+\beta_2)$. The suggested index rule is indifferent between the two arms, though if, for example, $\alpha_2+\beta_2\gg\alpha_1+\beta_1$, so that the variance of θ_1 is much greater than for θ_2, it is much more likely that further pulls on arm 1 will lead to appreciable changes in $\alpha_1/(\alpha_1+\beta_1)$ than it is for comparable changes to occur when arm 2 is pulled. This strongly suggests that it must be better to pull arm 1, as the expected immediate rewards are the same in both cases, and there is more information to be gained by pulling arm 1, information which may be used to achieve higher expected rewards later on. This hunch turns out to be well-founded as we shall see in §7.4.

In fact there seems to be no simple way of expressing the optimal policy, and a good deal of effort has been invested in working it out iteratively by means of the dynamic programming recurrence equation for the expected total reward under an optimal policy. When $n=2$, and with beta prior distributions with parameters (α_i,β_i) $(i=1,2,)$, this takes the form

$$R(\alpha_1,\beta_1,\alpha_2,\beta_2)=\max\left\{\frac{\alpha_1}{\alpha_1+\beta_1}[1+aR(\alpha_1+1,\beta_1,\alpha_2,\beta_2)]\right.$$

$$+ \frac{\beta_1}{\alpha_1 + \beta_1} a R(\alpha_1, \beta_1 + 1, \alpha_2, \beta_2),$$

$$\frac{\alpha_2}{\alpha_2 + \beta_2} [1 + a R(\alpha_1, \beta_1, \alpha_2 + 1, \beta_2)]$$

$$\left. + \frac{\beta_2}{\alpha_2 + \beta_2} a R(\alpha_1, \beta_1, \alpha_2, \beta_2 + 1) \right\}. \qquad (1.4)$$

The reasoning behind this is as follows. If arm 1 is pulled the probability of a success is the expectation of θ_1 with respect to a Beta(α_1, β_1) distribution, namely $\alpha_1/(\alpha_1 + \beta_1)$. This quantity is the multiplier of the first expression in square brackets on the right-hand side of (1.4), which is the expected total reward under an optimal policy after the first pull, conditional on this resulting in a success. To see this, note that the success yields an immediate reward of 1 and results, by applying Bayes' theorem, in a posterior distribution for θ_1 which is Beta$(\alpha_1 + 1, \beta_1)$. The prior distribution for θ_2 remains the current distribution, as a pull on arm 1 yields no relevant information. Thus the expected total reward under an optimal policy from this point onwards is exactly the same as the expected total reward from the outset with priors which are Beta$(\alpha_1 + 1, \beta_1)$ and Beta(α_2, β_2), except for an additional factor a, namely $a R(\alpha_1 + 1, \beta_1, \alpha_2, \beta_2)$. Similarly the probability of a failure if arm 1 is pulled is $\beta_1/(\alpha_1 + \beta_1)$, and after this the conditional expected total reward under an optimal policy is $a R(\alpha_1, \beta_1 + 1, \alpha_2, \beta_2)$. Thus the first expression in the curly brackets is the expected total reward under an optimal policy after the first pull if this is on arm 1, and, of course, the second expression is the similar quantity when arm 2 is pulled first.

The sum of the four parameters on which the function R depends is one greater where R appears on the right-hand side of (1.4) than for the left-hand side. This means that given an approximation to R for all parameter values such that $\alpha_1 + \beta_1 + \alpha_2 + \beta_2 = N$, an approximation to R for $\alpha_1 + \beta_1 + \alpha_2 + \beta_2 = N - 1$ may be calculated using (1.4). Recall that for a beta distribution the parameter values are positive. For a given positive integer N the number of positive integer solutions of the equation $\alpha_1 + \beta_1 + \alpha_2 + \beta_2 = N - 1$ is $(N-2)(N-3)(N-4)/3!$, so this is the number of calculations using (1.4) that are required. This approximation for $\alpha_1 + \beta_1 + \alpha_2 + \beta_2 = N - 1$ may now be substituted in the right-hand side of (1.4) to give an approximation for $\alpha_1 + \beta_1 + \alpha_2 + \beta_2 = N - 2$. Successive substitutions yield an approximation to R for all integer parameter values totalling N or less, and the corresponding approximation to the optimal policy. The total number of individual calculations required is

$$\frac{1}{3!} \sum_{r=4}^{N-1} (r-1)(r-2)(r-3) = \frac{1}{4!} (N-1)(N-2)(N-3)(N-4). \qquad (1.5)$$

One of the main theorems for discounted Markov decision processes asserts that repeated iteration using an equation of the form (1.4) gives convergence to the true optimal reward function. Since large parameter values imply tight distributions for θ_1 and θ_2 it is also possible to give increasingly good initial approximations for R for large values of N. Thus we have a method which may be used to give increasingly accurate approximations to the entire function R by increasing the value of N from which the iterative calculations begin.

For $a \leqslant 0.9$ a good approximation may be obtained with $N = 100$, From (1.5) we note that this requires around 10^6 individual calculations. The storage requirement is determined by the largest array of R values required at any stage of the calculation, which is to say by the set of values corresponding to parameter values which sum to N. This set has $(N-1)(N-2)(N-3)/3!$ members—around 10^5 for $N = 100$. Thus the computing requirements are appreciable even when $n = 2$. A similar analysis shows that for general $n(>1)$

$$\text{number of individual calculations required} = \frac{(N-1)!}{(2n)!(N-2n-1)!},$$

$$\text{storage requirement} = \frac{(N-1)!}{(2n-1)!(N-2n)!}.$$

This means that for $N = 100$ the calculations are quite impracticable for $n > 3$.

These computational difficulties may be drastically reduced by taking advantage of the index theorem. For this problem it is convenient to use as a calibrating set those arms for which the probability of success is known. These are in effect discrete-time standard bandit processes.

Suppose then that $n = 2$, arm 1 has a success probability θ which has a prior Beta(α, β) distribution, and arm 2 has a known success probability p. The dynamic programming recurrence equation for this problem may be written as

$$R(\alpha, \beta, p) = \max \left\{ \frac{p}{1-a}, \frac{\alpha}{\alpha+\beta}[1 + aR(\alpha+1, \beta, p)] + \frac{\beta}{\alpha+\beta}aR(\alpha, \beta+1, p) \right\}. \quad (1.6)$$

The expression $p/(1-a)$ is the expected total return from a policy which always pulls arm 2, since such a policy yields an expected reward of $a^{t-1}p$ at the tth pull. The reason for its appearance in (1.6) is that if it is optimal to pull arm 2 once, then after this pull our information about the two success probabilities is precisely the same as it was before, and it must therefore be optimal to pull it again, repeatedly. This may be checked by an argument based on the exact analogue of (1.4) for the present problem, which is the same as (1.6) except that $p/(1-a)$ is replaced by $p + aR(\alpha, \beta, p)$.

Equation (1.6) may be solved iteratively for $R(\alpha, \beta, p)$ starting with an approximation for values of α and β such that $\alpha + \beta = N$, just like equation (1.4). The set of (α, β) values for which the two expressions in curly brackets in (1.6) are equal defines those arms with beta priors whose index value is equal to p. Thus index values may be calculated as accurately as required for all beta priors by solving (1.6) for a sufficiently fine grid of values of p, thereby defining optimal policies for all n.

An analysis like that carried out for (1.4) shows that to solve (1.6) for a single value of p

number of individual calculations required $= \frac{1}{2}(N-1)(N-2)$,
storage requirement $= N - 1$.

For a grid of p-values the first of these numbers is multiplied by the number of values in the grid, and the second is unchanged. Thus for $n = 2$ the index method requires fewer calculations for $N = 100$ for grids of up to a few hundred p-values than the direct method, does very much better in terms of storage requirement, and at the same time solves the problem for $n > 2$.

In brief, Problem 5 has been brought within the bounds of computational feasibility by the index theorem. In this respect it is typical of a large number of problems to which the theorem applies, and for which it does not lead to a simple explicit solution.

Because of its status as the archetypal problem in sequential resource allocation, the multi-armed bandit has been used as the source of the generic term *bandit process* to describe one of the alternative reward-(or cost)-generating processes between which resources are allocated. Thus each of the jobs, mines, boxes, projects and arms of Problems 1 to 5, respectively, defines a bandit process.

Problem 6. Choosing a job. A man is faced with a number of opportunities for employment which he can investigate at a rate of one per day. For a typical job this investigation has a cost c and informs the man of the daily wage rate w, which has a prior probability distribution $F(w)$. The job has probability $\frac{1}{2}$ of proving congenial or, on the other hand, uncongenial, and the man rates these outcomes as worth g or $-g$ per day, respectively. The congeniality becomes apparent on the Nth day in the job, where N is a random variable with a Geometric (p) distribution. The discount factor for any reward or cost which occurs on day t is a^{t-1}; unlike the other parameters, a is the same for all the jobs. Decisions to take or leave a job may be taken at any time, and with effect on the day following the decision. How should the man decide whether to take a job, and then whether to leave it if he does take it, when he discovers its congeniality?

The bandit processes for this problem are, of course, the various jobs. A (discrete-time) standard bandit process is simply a job for which the wage rate and congeniality are both known. The sum of these two quantities we term the reward rate.

The index for a bandit process is, as always, equal to the reward rate of the standard bandit process which leads to indifference between the two when these are the only bandit processes available. In the economics literature, and in the present context, this equivalent reward rate is termed the *reservation wage*. Consider first a job J with a known wage rate w, but for which the congeniality is as yet unknown, together with an alternative standard job $S(x)$ with a known reward rate x. Suppose that initially there is indifference between the two jobs. It follows that indifference must persist until the congeniality for job J is known. This is because after $n(<N)$ days working on job J the distribution of $N-n$, now conditional on $N>n$, is Geometric (p), the same as the initial unconditional distribution of N, and the rewards available from that time onwards therefore have precisely the same structure as at the outset. This is except for the discounting, which scales all future rewards downwards by the same factor when we compare the reward streams with those available after n days on job J, thus leaving the optimal policy unchanged. We may suppose, therefore, that our man starts working on job J and carries on doing so for N days, at which point he becomes aware of the congeniality. If the job is congenial it must now be optimal to remain in it, since it was optimal to do so even without this knowledge. If the job is uncongenial it must now be optimal to switch to $S(x)$, since our assumption of initial indifference between the two jobs means that the switch was already optimal without this knowledge. This then is one optimal policy. The indifference assumption means that it is also optimal to work permanently on $S(x)$. (Note that we are endowing our man with an infinite working life, or—which in the present context is equivalent—one of L days, with $a^L \ll 1$.)

The expected rewards from each of these two policies may now be equated and the equation solved for x. Before doing this note that for either policy we have

$$\mathbf{E}(\text{Reward}) = \mathbf{E}(\text{Reward during first } N \text{ days})$$
$$+ \tfrac{1}{2}\mathbf{E}(\text{Reward after } N\text{th day}|J \text{ is congenial})$$
$$+ \tfrac{1}{2}\mathbf{E}(\text{Reward after } N\text{th day}|J \text{ is uncongenial}).$$

Moreover, the last of the three terms on the right-hand side is the same for both policies, since in both cases $S(x)$ is the job our man works on after the Nth day if job J is uncongenial. Thus the equation reduces to

$$w[1 + a(1-p) + a^2(1-p)^2 + \cdots] + \tfrac{1}{2}(w+g)(1+a+a^2+\cdots)$$
$$\times pa[1 + a(1-p) + a^2(1-p)^2 + \cdots]$$
$$= x[1 + a(1-p) + a^2(1-p)^2 + \cdots] + \tfrac{1}{2}x(1+a+a^2+\cdots)$$
$$\times pa[1 + a(1-p) + a^2(1-p)^2 + \cdots],$$

which on summing the geometric series becomes

$$\frac{1}{1-a(1-p)}\left[w + \frac{pa(w+g)}{2(1-a)}\right] = \frac{1}{1-a(1-p)}\left[x + \frac{pax}{2(1-a)}\right]. \tag{1.7}$$

The index for job J is therefore

$$x = w + \frac{pag}{2(1-a)+pa}. \tag{1.8}$$

Thus it is better to take job J until its congeniality is apparent than a job with a known reward rate of w, although initially w is the expected reward rate for J.

The margin of the preference for J is increasing in p, a and g. This is to be expected, for reasons to which we shall shortly return.

In this derivation of the index x we were able to show that only one policy starting with job J needed to be considered. Call this policy P_1 and write $x = x_1$. If, for example, our model had allowed for several different degrees of congeniality this would not have been so easy. A more formal and general procedure is required, and we now digress to consider this question before evaluating the index for a job with an unknown wage rate.

Equation (1.7) is an expression of the fact that the expected reward from working on job J under policy P_1 is equal to that available from $S(x_1)$ over the same random time. The factor $[2(1-a)+pa][1-a(1-p)]^{-1}[2(1-a)]^{-1}$ by which equations (1.8) and (1.7) differ may be regarded as the expected discounted time for which the man works on job J under P_1, and x_1 as the *equivalent constant reward rate* yielded by J under this policy. For any other policy P starting with job J an equation analogous to (1.7) may be written down and solved to give an equivalent constant reward rate x. The point to note is that unless P is chosen to maximize x we do not have initial indifference between jobs J and $S(x)$, as the expected total reward is higher if the maximizing policy P_1 is followed than it is if our man works throughout on $S(x)$.

We have established then that the index defined for a bandit process in terms of calibration by standard bandit processes is equal to the maximum equivalent constant reward rate obtainable from that bandit process. This is an important and general result, not restricted to Problem 6.

Returning now to the case of a job K with an as yet unknown wage rate, it is clear that the equivalent constant reward rate is maximized by a policy of the following type. Find out the wage rate w on the first day. If $w > w_1$ accept the job. If $w_2 \geqslant w$ leave the job if it turns out to be uncongenial. If $w > w_2$ keep the job even if it turns out to be uncongenial. Since the index ξ for job K is the maximum equivalent constant reward rate it follows that

$$\xi = \sup_{w_1, w_2(w_2 \geqslant w_1)} \frac{-c + \dfrac{a}{1-a(1-p)} \displaystyle\int_{w_1}^{w_2} \left[w + \dfrac{pa(w+g)}{2(1-a)} \right] dF(w) + \dfrac{a}{1-a} \displaystyle\int_{w_2}^{\infty} w \, dF(w)}{1 + \dfrac{a}{1-a(1-p)} \displaystyle\int_{w_1}^{w_2} \left[1 + \dfrac{pa}{2(1-a)} \right] dF(w) + \dfrac{a}{1-a} \displaystyle\int_{w_2}^{\infty} dF(w)}.$$

(1.9)

It is not difficult to show that the maximizing w_1 and w_2 are such that

$$\xi = w_1 = \frac{pag}{2(1-a) + p\alpha} = w_2 - g.$$

Thus the maximization in (1.9) is effectively with respect to a single variable, and readily carried out numerically.

It is clear on general grounds that, like the case of a known wage rate, ξ must increase with p, a and g, though this is not particularly obvious from (1.9). The reason is that opportunities for achieving a high equivalent constant reward rate increase with each of these variables. In the case of g this is because large values of g increase the variability of rewards between different possible outcomes, and therefore increase the possibilities of including those with high reward rates and excluding those with low reward rates. In the case of p, any increase tends to make the time at which this selection can be made earlier, and therefore more significant. Large values of a, on the other hand, reduce the tendency for delays in the selection to diminish the equivalent constant reward rate which can be achieved, since they mean that rewards which occur after the selection are not totally insignificant compared with the earlier rewards; this effect is not confined to Problem 6, as Theorem 4.1 will show.

Models along the lines of Problem 6 have appeared in the economics literature for at least ten years, but the solution in terms of the indices (1.8) and (1.9) was obtained only with the help of the index theorem. Some of the relevant papers, which give references to a large number of others, are those by Berninghaus (1984), Lippman and McCall (1981), McCall and McCall (1981), Miller (1984) and Weitzman (1979).

EXERCISES

1.1. Write down equation (1.6) with $p + aR(\alpha, \beta, p)$ in place of $p/(1-a)$ as suggested on page 8, and hence deduce the given form of equation (1.6).

1.2. Show that in equation (1.9) the maximizing w_1 and w_2 satisfy the stated condition.

CHAPTER 2

Main Ideas

2.1 INTRODUCTION

Chapter 1 in large measure set out to dazzle the reader with the diversity of the allocation problems which succumb to a solution in terms of indices. In Chapter 2, on the other hand, the emphasis is on unity rather than diversity, as we point out the common features of the problems which may be solved in this way, and why things go wrong in the absence of these features. This prepares the way for Chapter 3, which includes all the main theorems.

The key concepts are those of a *bandit process* and of a *family of alternative bandit processes* (*FABP*). Bandit process is the generic term for the projects, jobs, gold-mines, boxes and arms of the various problems in Chapter 1. A FABP is a model for the problem when choices must be made between different bandit processes.

A FABP is a special type of exponentially discounted semi-Markov decision process, a subject on which there is a substantial literature—see, for example, the book by Howard (1971) and the review by Teugels (1976). The next two sections set out the relevant general theory. Then in §2.4 the first of the index theorems is stated. It is not proved, but pointers are given as to why it should hold.

When bandit processes are *jobs* their structure is particularly simple. These are introduced in §2.5, and used in §§2.6 and 2.7 to show that if an index theorem is to hold discounting must be exponential, the time-horizon must be infinite, and bandit processes to which no effort is applied must remain in the same state. The next three sections each give an example of a problem which may be fitted into this framework, although this is not immediately obvious. Finally, in §2.11 it is shown that except in special cases index policies are not optimal when effort is divided between more than one server or processor, not more than one of which may be allocated to the same bandit process.

2.2 DECISION PROCESSES

A Markov process is a mathematical model of a system which passes through a succession of states in a manner determined by a corresponding succession of transition probability distributions which depend on the current state. This concept is one of the most powerful weapons in the probabilist's armoury. The mathematical theory is elegant, and examples abound in all branches of the physical and economic sciences.

To such a process let us now add a set of decisions available at each stage, and on which the probability distribution governing the next state of the system depends. Let us also add a set of possible rewards at each stage, so that the decision-maker receives a reward which depends on his decision and on the subsequent transition. These additions give us a Markov decision process, varieties of which are appropriate models for a wide range of situations where a sequence of decisions has to be made. Over the past thirty-five years the literature on Markov decision processes has mushroomed, starting from the work of Wald (1950) and Bellman (1957).

Various different criteria may be used to choose which decision to make at each stage. We shall consider Markov decision processes with an infinite time horizon: processes, that is to say, which go on for ever. Future rewards will be discounted, thereby ensuring that the total reward obtained is finite. The aim will be to find policies which maximize the expected value of this total discounted reward, which will be termed simply the *payoff*. Some of the most interesting applications of our theory are to non-discounted problems, whose solutions may be obtained either by letting the discount factor tend to one, or by re-interpreting the discount factor in some way. Problems 1, 2, and 3 of Chapter 1 are cases in point.

Blackwell (1965) has written an important paper on discounted Markov decision processes, and we shall need some of his results. These were proved in the first instance for a discrete time process, and we shall be interested in their extensions to the case when the times between successive decision times are themselves random variables. Processes of this type are known as semi-Markov decision process. In the following sentence, the unsophisticated mathematician may, without serious loss of understanding, ignore the words in square brackets, and regard 'class' as a synonym for 'σ-algebra'.

A *semi-Markov decision process* (or simply *decision process*) is defined on a state-space, Θ, which is a [Borel] subset of some [complete separable metric] space, together with a σ-algebra $\mathscr{X}$ of subsets of Θ which includes every subset consisting of just one element of Θ. Controls are applied to the process at a succession of *decision times*, the intervals between which are random variables. The first decision time is at time zero. Transitions between states, and rewards, occur instantaneously at these times, and at no other times. It will prove convenient to regard the control applied at a decision time as being continuously in force up to, but not including, the next decision time.

When the process is in state x at decision time t a control must be chosen from the finite set $\Omega(x)$. Application of control u at time t with the process in state x yields a reward $a^t r(x, u) = e^{-\gamma t} r(x, u)$. The quantities a $(0 < a < 1)$ and $\gamma (> 0)$ are termed the *discount factor* and *discount parameter*, respectively. $P(A|x, u)$ is the probability that the state y of the process immediately after time t belongs to $A(\in \mathscr{X})$, given that at time t the process is in state x and control $u(\in \Omega(x))$ is applied. $F(B|x, y, u)$ is the probability that the duration of the interval until the next decision time belongs to the Borel set B, given that at time t the process is in state x, and control u is applied, leading to a transition to state y. The functions

$\Omega(\cdot)$, $r(\cdot, u)$ and $P(A|\cdot, u)$ are $\mathscr{X}$-measurable, and $F(B|\cdot, \cdot, u)$ is $\mathscr{X}^2$-measurable.

In order to ensure that an infinite number of transitions do not occur in a finite time the following condition is sometimes imposed (see, for example, Ross, 1970, p. 157).

Condition A. There exist $\delta(>0)$ and $\varepsilon(>0)$ such that

$$\int_{\Theta} F((\delta, \infty)|x, y, u)P(dy|x, u) > \varepsilon \qquad (x \in \Theta, u \in \Omega(x)),$$

where (δ, ∞) denotes the unbounded open interval to the right of δ.

This condition means that the probability that the interval between decision times, and therefore between transitions, is greater than δ is always at least ε. In fact it is this conclusion which is important, rather than the condition which leads to it; for standard bandit processes every point in time is a decision time, and other arguments will be used.

A *Markov decision process* is a (semi-Markov) decision process with the property that, for some $c(>0)$,

$$F(\{c\}|x, y, u) = 1, \qquad \forall x, y, \text{ and } u.$$

Thus the interval between decision times takes the constant value c. Without loss of generality we shall assume that $c = 1$, since this is just a question of choosing the units in which time is measured.

A *policy* for a decision process is any rule satisfying the obvious measurability requirements, including randomized rules, which for each decision time t specifies the control to be applied at time t as a function of t, the state at time t, the set of previous decision times, and of the states of the process and the controls applied at those times. Thus the control applied at time t may depend on the entire previous history of the process, though not on what happens after time t, a situation which we shall describe by saying that it is *past-measurable*. A policy which maximizes the expected total reward over the set of all policies for every initial state will be termed an *optimal* policy. *Deterministic* policies are those which involve no randomization. *Stationary* policies are those which involve no explicit time-dependence. *Markov* policies are those for which the control chosen at time t depends only on t and the state at time t.

If D is a Markov decision process let $R_t(D, x)$ denote the expected total reward starting from state x at time zero under an optimal policy for the truncated process for which no further rewards accrue from time t onwards. It may easily be shown that if $\Omega(x)$ is finite for all x then, for all integers s, and x in Θ,

$$R_s(D, x) = \max_{u \in \Omega(x)} \left[r(x, u) + a \int_{\Theta} R_{s-1}(D, y)P(dy|x, u) \right], \qquad (2.1)$$

and policies which at time $t - s$ ($s = t, t-1, \ldots, 1$) and state x select a control which maximizes the expression in square brackets are optimal for the process when truncated at time t. For this problem then, there is an optimal policy which is deterministic and Markov, though it is not in general stationary.

Equation (2.1) is the justly celebrated functional equation of dynamic programming. It is a consequence of the principle that 'an optimal policy has the property that whatever the initial state and initial condition, the remaining decisions must constitute an optimal policy with regard to the state resulting from the first decision' (Bellman, 1957). By definition $R_0(D, y) = 0$, for all y, so the function $R_1(D, \cdot)$ may be calculated from (2.1). These values may then be substituted in the right-hand side and $R_2(D, \cdot)$ calculated. Repetition of this process gives us the function $R_s(D, \cdot)$ for any value of s, and optimal policies when D is truncated at values of t less than or equal to s. This procedure is often described as *backwards induction*, since, given that D is to be truncated at a certain time, it constructs the optimal policy at successively earlier points in time.

Now if an optimal policy exists for the process D without truncation, leading to an expected total reward $R(D, x)$, then clearly $R_t(D, x)$ tends to $R(D, x)$ as t tends to infinity. We should thus expect to find that under appropriate conditions the functional equation

$$X(x) = \max_{u \in \Omega(x)} \left[r(x, u) + a \int_\Theta X(y) P(\mathrm{d}y | x, u) \right] \quad (x \in \Theta), \qquad (2.2)$$

has the solution $X(\cdot) = R(D, \cdot)$. Blackwell (1965) proves, amongst other things, the following results for the case of a finite control set $\Omega(x)$.

Theorem 2.1. (i) There is at least one optimal policy which is deterministic, stationary and Markov.
(ii) $R(D, \cdot)$ is the unique bounded function which satisfies the functional equation (2.2).
(iii) A policy is optimal if and only if, for every x in Θ, the control which it chooses in state x is such as to achieve the maximum on the right-hand side of equation (2.2) when $X(\cdot) = R(D, \cdot)$.
(iv) $R(D, \cdot)$ is an $\mathcal{X}$-measurable function.
(v) The function $R(D, \cdot)$ may be determined by *value iteration* using equation (2.2).

Value iteration is the term coined by Howard (1960) to describe the following procedure, which is a natural extension of backwards induction to the situation when D is not truncated, often termed the infinite horizon case. Starting with some $\mathcal{X}$-measurable function $X_0(\cdot)$ defined on D, define and determine functions $X_n(\cdot)$ iteratively by the equation

$$X_n(x) = \max_{u \in \Omega(x)} \left[r(x, u) + a \int_\Theta X_{n-1}(y) P(\mathrm{d}y | x, u) \right] \quad (x \in \Theta, n = 1, 2, \dots). \quad (2.3)$$

If the functions $r(\cdot, \cdot)$ and $X_0(\cdot)$ are bounded, as assumed by Blackwell, it follows from Banach's fixed point theorem for contraction mappings that $X_n(\cdot)$ tends to $R(D, \cdot)$ uniformly over Θ as n tends to infinity, and it is to this convergence that part (v) of the above theorem refers.

Blackwell's results are actually for a control set which does not depend on the state, but this is a restriction whose removal requires virtually no change in the argument. A further straightforward adaptation of Blackwell's proof suffices, as remarked by Ross (1970), to establish similar results for a semi-Markov decision process D which satisfies Condition A. The functional equation now takes the form

$$X(x) = \max_{u \in \Omega(x)} \left[r(x, u) + \int_\Theta \int_{t=0}^\infty a^t X(y) F(dt \mid x, y, u) P(dy \mid x, u) \right] \quad (x \in \Theta). \quad (2.4)$$

We shall also replace the condition that $r(\cdot, \cdot)$ be bounded by Condition B. This is a weaker condition, whose statement requires some additional notation. Let $x(0)$ be the state of the semi-Markov decision process D at time zero, $x_g(t)$ the state at time t if policy g is followed, and $R_g(D, x)$ the expected total discounted reward under policy g when $x(0) = x$. Let $[t]$ denote the first decision time after time t.

Condition B. $E[a^{[t]} R_g(D, x_g([t])) \mid x(0) = x]$ tends to zero uniformly over all policies g as t tends to infinity, for all $x \in \Theta$.

This condition will be assumed always to hold. We now proceed to discuss its role, and that of the stronger condition which it replaces.

The quantity $X_m(x)$ is the supremum over all policies of the expected total discounted reward from the Markov decision process D if it is truncated at time m, with an additional reward at time m of $a^m X_0(x(m))$, where $x(m)$ is the state at time m. The supremum is attained by the policies which at time $m - n$ ($n = m, m - 1, \ldots, 1$) and state x select a control which maximizes the expression in square brackets in equation (2.3). This is simply backwards induction once again.

Since $X_0(\cdot)$ is bounded it follows that $X_m(\cdot)$ must tend at each point to $R(D, \cdot)$, provided that the contribution to the expected total discounted reward which accrues after time t tends to zero uniformly over all policies as t tends to infinity. Thus $X_m(\cdot)$ tends at each point to $R(D, \cdot)$ under Condition B. This convergence is the key point in the proof of Theorem 2.1. The stronger assumption of bounded $r(\cdot, \cdot)$ ensures that this convergence is uniform over Θ, but this is not an essential condition.

The condition that $X_0(\cdot)$ be bounded may also be replaced by the obvious analogue of Condition B. These weaker conditions lead to the following modified version of Theorem 2.1 for a semi-Markov decision process D satisfying Conditions A and B, and for which the control set $\Omega(x)$ is finite for all $x \in \Theta$.

Theorem 2.2. (i) There is at least one optimal policy which is deterministic, stationary, and Markov.
(ii) $R(D, \cdot)$ is the unique solution of the functional equation (2.4) which is such that $E[a^{[t]} X(x_g([t])) \mid x(0) = x]$ tends to zero uniformly over g as t tends to infinity.
(iii) A policy is optimal if and only if, for every x in Θ, the control which it chooses

in state x is such as to achieve the maximum on the right-hand side of equation (2.4) when $X(\cdot) = R(D, \cdot)$.

(iv) $R(D, \cdot)$ is an $\mathscr{X}$-measurable function.

(v) Value iteration using equation (2.4) converges to $R(M, \cdot)$ at each point in Θ for which $\mathbf{E}[a^{[t]} X_0(x_g([t])) | x(0) = x]$ tends to zero uniformly over g as t tends to infinity, where $X_0(\cdot)$ is the initial approximation.

This theorem incorporates those results from the established theory of dynamic programming which will be needed in our discussion of the special types of semi-Markov decision process with which this book is concerned.

2.3 BANDIT PROCESSES AND SIMPLE FAMILIES OF ALTERNATIVE BANDIT PROCESSES

A *bandit process* is a semi-Markov decision process for which the control set $\Omega(x)$ consists, for every state x, of two elements 0 and 1. The control 0 *freezes* the process in the sense that when it is applied no reward accrues and the state of the process does not change. Application of control 0 also initiates a time interval in which every point is a decision time, so that Condition A does not hold. In contrast, control 1 is termed the *continuation* control. Apart from measurability conditions, Condition A, when restricted to $u = 1$, and Condition B, the functions $r(\cdot, 1)$, $P(\cdot | \cdot, 1)$, and $F(\cdot | \cdot, \cdot, 1)$ may take any form.

At any stage, the total time during which the process has not been frozen by applying control 0 is termed the *process time*. The process time when control 1 is applied for the $(i + 1)$th time is denoted by $t_i (i = 0, 1, 2, \ldots)$. The state at process time t is denoted by $x(t)$. The sequences of times t_i and states $x(t_i)$, $i = 0, 1, \ldots$ constitute a *realization* of the bandit process, which thus does not depend on the sequence of controls applied. The reward at decision time t if control 1 has been applied at every previous decision time, so that process time coincides with real time, is $a^t r(x(t), 1)$, which we abbreviate to $a^t r(t)$. A *standard* bandit process is a bandit process with just one state, and for which every point in time is a decision time. Thus a policy for a standard bandit process is simply a random Lebesgue-measurable function $I(t)$ from the positive real line to the set $\{0, 1\}$, specifying the control to be applied at each point in time. The total reward yielded by such a policy is $\lambda \int a^t I(t) dt$, the parameter λ depending on the particular standard bandit process.

An arbitrary policy for a bandit process is termed a *freezing rule*. Given any freezing rule f for a non-standard bandit process, the random variables $f_i (i = 0, 1, 2, \ldots)$ are past-measurable, where $f_i (\geqslant f_{i-1})$ is the total time for which control 0 is applied before the $(i + 1)$th application of control 1. We also define $f_{-1} = 0$.

A policy which results in control 1 being applied up to a time τ, and control 0 being applied thereafter, will be termed a *stopping rule*, and τ the corresponding *stopping time.* In particular a policy which divides the state space into a *continuation set*, on which control 1 is applied, and a *stopping set*, on which control 0 is applied, is a deterministic stationary Markov stopping rule. If f is a stopping rule, $f_i = 0$ if $t_i < \tau$ and $f_i = \infty$ if $t_i \geq \tau$, for all i. We note that τ may take the value infinity with positive probability. Stopping rules have been extensively studied, for the most part in the context of stopping problems (e.g. see Chow, Robbins and Siegmund, 1971), which may be regarded as being defined by bandit processes for which $r(x, 0) \neq 0$.

We now come to the central concept of a *family of alternative bandit processes* (*FABP*). This is a decision process formed by bringing together a set of n bandit processes all with the same discount factor. Time zero is a decision time for every bandit process, and at time zero control 1 is applied to just one of them, bandit process B_1 say, and control 0 is applied to all the other bandit processes until the next decision time for B_1 is reached. Again, this is a decision time for every bandit process and just one of them is continued, B_2 say (which may be B_1 again), whilst the others are frozen by applying control 0 until the next decision time for B_2. Once again, this is a decision time for every bandit process and just one of them is continued.

A realization of a family of alternative bandit processes is constructed by proceeding in this way from one decision time to the next. The state space for the FABP is the product of the state spaces for the individual bandit processes. The reward which accrues at each decision time is the one yielded by the bandit process which is then continued. The control set at a decision time is given by the set of bandit processes which may then be continued, application of control i to the FABP corresponding to selection of bandit process i for continuation. In general, as discussed in §3.9, this set may not always include all n bandit processes. Families of alternative bandit processes all of which are always included in the control set will be described as *simple families of alternative bandit processes* (*SFABPs*).

The following notation will be used in conjunction with an arbitrary bandit process B. $R_f(B)$ denotes the payoff under the freezing rule f. Thus

$$R_f(B) = \mathbf{E} \sum_{i=0}^{\infty} a^{t_i + f_i} r(t_i).$$

$R(B)$ denotes the payoff under the freezing rule, termed the null freezing rule, for which $f_i = 0$ $(i = 0, 1, 2, \ldots)$. Also

$$W_f(B) = \mathbf{E} \sum_{i=0}^{\infty} a^{f_i} \int_{t_i}^{t_{i+1}} a^t \, dt, \quad v_f(B) = R_f(B)/W_f(B), \quad \text{and}$$

$$v'(B) = \sup_{\{f : f_0 = 0\}} v_f(B).$$

Similarly, for stopping rules,

$$R_\tau(B) = \mathbf{E} \sum_{t_i < \tau} a^{t_i} r(t_i), \quad W_\tau(B) = \mathbf{E} \int_0^\tau a^t \, dt = \gamma^{-1} \mathbf{E}(1 - a^\tau),$$

$$v_\tau(B) = R_\tau(B)/W_\tau(B), \quad \text{and} \quad v(B) = \sup_{\tau > 0} v_\tau(B). \tag{2.5}$$

All these quantities naturally depend on the initial state $x(0)$ of the bandit process B. When necessary $R_f(B, x)$ and $W_f(B, x)$, for example, will be used to indicate the values of $R_f(B)$ and $W_f(B)$ when $x(0) = x$.

The expected reward $R_f(B)$ when the freezing rule f is applied to B is now the expected reward from a standard bandit process with parameter $v_f(B)$ to which controls 0 and 1 are applied at the same times as they are to B under f. Thus $v_f(B)$, and similarly $v_\tau(B)$, is an equivalent constant reward rate, and $v'(B)$ and $v(B)$ are maximized equivalent constant reward rates. Clearly $v'(B) \geqslant v(B)$, and it is actually not difficult to show that $v'(B) = v(B)$. The conditions $f_0 = 0$ and $\tau > 0$ in the definitions of $v'(B)$ and $v(B)$ mean that the policies considered are all such that at time zero control 1 is applied. This restriction is required to rule out zero denominators $W_f(B)$ or $W_\tau(B)$. In the case of $v'(B)$ it also has the effect of removing a common factor from the numerator and denominator of $v_f(B)$ for those f for which $f_0 \neq 0$, and otherwise implies no loss of generality.

For a Markov bandit process it seems natural to modify the definitions of $W_f(B)$ and $W_\tau(B)$ by multiplying by the factor $\gamma(1 - a)^{-1}$. Thus

$$W_\tau(B) = \mathbf{E}(1 + a + a^2 + \cdots + a^{\tau - 1}),$$

the expected discounted time up to time τ if the tth time unit is discounted by the factor a^{t-1}. The factor $\gamma(1 - a)^{-1}$ will be termed the *discrete-time correction factor*. These modified definitions, and the corresponding ones for $v_f(B)$, $v(B)$, $v_\tau(B)$ and $v(B)$, all of which are divided by the discrete-time correction factor, will always be used in the Markov case.

As Theorem 2.3 shows, $v(B, x)$ is the dynamic allocation index for bandit process B in state x, when B belongs to a SFABP.

2.4 A FIRST INDEX THEOREM

For a SFABP $\mathscr{F}$ made up of bandit processes $\{B_j : j = 1, 2, \ldots, n\}$ let x_j be a typical element of the state space θ_j of B_j, and μ_j a real-valued function defined on θ_j. A policy for $\mathscr{F}$ will be described as an *index policy with respect to* $\mu_1, \mu_2, \ldots, \mu_n$ if at any decision time the bandit process B_k selected for continuation is such that $\mu_k(x_k) = \max_j \mu_j(x_j)$.

Theorem 2.3. The Index Theorem for a SFABP. A policy for a simple family of alternative bandit processes is optimal if it is an index policy with respect to $v(B_1, \cdot), v(B_2, \cdot), \ldots, v(B_n, \cdot)$.

Because of this theorem an index policy with respect to $v(B_1,.), v(B_2,.),\ldots,$ $v(B_n,.)$ will be referred to simply as an *index policy*. As a first step towards understanding why the theorem holds consider the following problem.

Problem 7. There are n biased coins with p_i the probability of a head when coin i is tossed. The coins may be tossed once each in any order. Each head tossed yields a reward, the reward from a head on the tth toss being $a^{t-1}(0 < a < 1)$. In what order should the coins be tossed so as to maximize the expected total reward?

For this problem the answer is obvious, the coins should be tossed in decreasing order of p_i.

Now suppose there are mn coins arranged in the form of an $m \times n$ matrix, that coin (i,j) cannot be tossed before coin $(i-1,j)(2 \leqslant i \leqslant m, 1 \leqslant j \leqslant n)$, and p_{ij} is the probability of a head when it is tossed. We may think of the coins arranged in n piles, and that a coin may be removed and tossed only from the top of a pile. Otherwise the problem is as before.

If $p_{i-1,j} \geqslant p_{ij}(2 \leqslant i \leqslant m, 1 \leqslant j \leqslant n)$ there is no difficulty. Call this the decreasing case. Under these conditions the coins should be tossed in decreasing order of p_{ij}. Since this is the optimal policy without the constraints on the order of tossing, and since it does not violate the constraints, it must remain optimal when they are imposed.

Note that the problem is essentially unchanged if the matrix $[p_{ij}]$ is extended downwards by adding an infinite number of rows of zeros. With this extension the n columns of the matrix define a simple family of n alternative bandit processes. Our argument shows then that for this SFABP the myopic policy, which always selects the bandit process yielding the highest immediate expected reward, is optimal for the decreasing case. An easy additional argument shows that this policy is an index policy in the strict sense.

In the general case things are more complicated. We should clearly try to toss coins with high p_{ij}s early in the overall sequence, but how should we proceed if these desirable coins are not at the tops of their respective piles? In some circumstances the myopic policy is still optimal, but it is not difficult to construct $[p_{ij}]$ matrices for which the optimal policy depends on the second, third, or for that matter the kth, coin from the top of a pile, where k is arbitrary.

The way forward becomes apparent when we recall that $v(B, x)$ is the maximized equivalent constant reward rate for bandit process B in state x. Given two standard bandit processes $S(\lambda_1)$ and $S(\lambda_2)$ with $\lambda_1 > \lambda_2$, the reward from selecting $S(\lambda_1)$ for a time T_1 and then selecting $S(\lambda_2)$ for a time T_2 is clearly greater than the reward if the order is reversed and $S(\lambda_2)$ is selected for T_2 followed by $S(\lambda_1)$ for T_1. This is because it is always better for high rewards to be given priority over low rewards, thereby incurring less discounting. This principle carries over immediately to the case of two arbitrary bandit processes.

Lemma 2.4. If bandit processes B_1 and B_2 have indices $v(B_1)$ and $v(B_2)$ with

$v(B_1) > v(B_2)$), and these are achieved with stopping times σ_1 and σ_2, the expected reward from selecting B_1 for time σ_1, and then B_2 for σ_2, is greater than from reversing the order of selection.

Proof. We have

$$-(\log_e a)v(B_1) = \frac{R_{\sigma_1}(B_1)}{1 - \mathbf{E}a^{\sigma_1}} > -(\log_e a)v(B_2) = \frac{R_{\sigma_2}(B_2)}{1 - \mathbf{E}a^{\sigma_2}}.$$

Thus

$$R_{\sigma_1}(B_1) + \mathbf{E}a^{\sigma_1} R_{\sigma_2}(B_2) > R_{\sigma_2}(B_2) + \mathbf{E}a^{\sigma_2} R_{\sigma_1}(B_1).$$

Since B_1 and B_2 are stochastically independent, this is the required inequality.   □

The reward stream from a SFABP is formed by splicing together portions of the reward streams of the constituent bandit processes. Lemma 2.4 shows that for any two independent consecutive portions from different bandit processes the expected reward is higher if the portion with the higher equivalent constant reward rate comes before the portion with the lower equivalent constant reward rate. It is thus a reasonable conjecture that it would be optimal to select a bandit process with an index value at least as high as those of any other bandit process, and to continue it for the stopping time corresponding to the index value. Call the policy defined by repeatedly selecting bandit processes in this way a *modified* index policy, and the intervals between successive selections the *stages* of such a policy.

There are two points to note about this conjecture. The first is that at this stage in the argument it really is only a conjecture. Lemma 2.4, although strongly suggestive, is not a complete proof. The main difficulty is the reference to *independent* consecutive portions. This does not hold if the bandit process to be selected at a given stage may depend on the realizations of those selected earlier, and it is not obvious that an optimal policy can be found without this sort of dependence. The second point is that it leaves us with the problem of relating an index policy with a modified index policy. They are in fact identical, as the following lemma shows.

Lemma 2.5. If $v(B, x(0)) = v_\sigma(B, x(0))$ and ρ is a stopping time for B then $\mathbf{P}\{\rho < \sigma \cap v_{\sigma - \rho}(B, x(\rho)) < v(B, x(0))\} = 0$.

Proof. Let A be the event in curly brackets and $\bar{A}$ its complement. Define

$$\tau = \begin{cases} \rho & \text{on } A \\ \sigma & \text{on } \bar{A} \end{cases}.$$

We have

$$v(x(0))\mathbf{E}\int_0^\sigma a^t\,dt = R_\sigma(x(0)) = R_\tau(x(0)) + \mathbf{P}(A)\mathbf{E}[a^\tau R_{\sigma-\tau}(x(\tau))\,|\,A]$$

$$\leqslant v(x(0))\mathbf{E}\int_0^\tau a^t\,dt + \mathbf{P}(A)\mathbf{E}\left[a^\tau v_{\sigma-\tau}(x(\tau))\int_\tau^\sigma a^t\,dt\,\Big|\,A\right].$$

Thus

$$\mathbf{P}(A)\mathbf{E}\left\{[v(x(0)) - v_{\sigma-\tau}(x(\tau))]\int_\tau^\sigma a^t\,dt\,\Big|\,A\right\} \leqslant 0.$$

Since $v(x(0)) > v_{\sigma-\tau}(x(\tau))$ on A it follows that $\mathbf{P}(A) = 0$. $\qquad\square$

The equivalence of the two definitions of an index policy follows from the lemma, together with the fact that $v(x(\rho)) \geqslant v_{\sigma-\rho}(x(\rho))$.

The proof of the index theorem itself is deferred to the next chapter. It is based on extending the interchange argument of Lemma 2.4, to show that there is no advantage to be gained from switching away from a bandit process before the end of a stage of a modified index policy, and by allowing the reward stream portions which are interchanged to be composite portions, including contributions from more than one bandit process.

2.5 JOBS

A bandit process which yields a positive reward at a random time, and no other rewards, positive or negative, will be termed a *job*. A simple family of such bandit processes defines a version of the well-known problem of scheduling a number of jobs on a single machine. Many papers on this and other scheduling problems have been written over the past thirty years. For a review of the field the reader is referred in particular to the books by Conway *et al.* (1967) and Coffman (1976), and the conference proceedings edited by Dempster *et al.* (1982). French (1984) gives a convenient introduction.

In addition to their importance in scheduling theory SFABPs of this type retain sufficient generality to provide further insight into why the index theorem holds, as well as providing counter-examples to any conjecture that substantial relaxation of its conditions might be possible. The remaining sections of this chapter are devoted to carrying out this programme, and also showing that a little ingenuity considerably stretches the range of problems for which the conditions of the theorem may be shown to hold.

A natural extension of the notion of a job is to allow costs and rewards to occur before completion of the job. This leads to the definition of a job as a quite general bandit process, except for the existence of a state (or set of states) C representing completion. Selection of an uncompleted job at each decision time until all the

jobs have been completed may be forced by supposing that a job in state
C behaves like a standard bandit process with parameter $-M$, where M is large,
as this convention means it can only be optimal to select a completed job for
continuation if every job has been completed. Jobs of this kind will be termed
generalized jobs to distinguish these from the less elaborate jobs of the last but
one paragraph.

Restrictions of the form 'Job A must be completed before job B may be started'
are termed *precedence* constraints. They are present in many real-life scheduling
problems. Resource allocation to jobs, including generalized jobs, under
precedence constraints is one of the themes of the next three chapters.

The extension to generalized jobs will not be made without explicit mention,
and for the moment we revert to jobs with a single reward on completion. There
remains the question of the times at which the processing, or service, of a job may
be interrupted. One possibility is that no interruptions are allowed and service is
switched from one job to another only when the first job has been completed. This
will be termed the *non-preemptive* case. At the other extreme, there may be no
restrictions on switching, so that every time is a decision time. An intermediate
possibility is that switching is allowed either on completion of a job or when its
process-time (i.e. the amount of service it has so far received) is an integer multiple
of some time-unit Δ.

For a typical job let $Ve^{-\gamma t}$ be the reward on completion at time t, σ be the
service-time (i.e. the process-time required for completion), $F(t) = \mathbf{P}(\sigma \leqslant t)$, $f(t) =$
$dF(t)/dt$ (if F is differentiable) and $\tau_k = \min[\sigma, k\Delta]$. The state of the job is either
C (denoting completion), or $k\Delta$ if it has so far received service for a time $k\Delta$
without being completed. Thus $v(C) = 0$ and, for the intermediate case just
described,

$$v(r\Delta) = \sup_{k>r} \frac{R_{\tau_k - r\Delta}(r\Delta)}{W_{\tau_k - r\Delta}(r\Delta)} = \sup_{k>r} \frac{V \int_{r\Delta}^{k\Delta} e^{-\gamma t} \, dF(t)}{\int_{r\Delta}^{k\Delta} [1 - F(t)] e^{-\gamma t} \, dt}. \tag{2.6}$$

Note that in (2.6) the stopping-time $\tau_k - r\Delta$ is conditioned by the event $\{\sigma > r\Delta\}$.
Thus it has the probability density function $f(s + r\Delta)/[1 - F(r\Delta)]$ in the range
$0 < \tau_k - r\Delta = s < (k - r)\Delta$, and

$$\mathbf{P}\{\tau_k - r\Delta = k\Delta | \sigma > r\Delta\} = [1 - F(k\Delta)]/[1 - F(r\Delta)].$$

With this hint, and the help of Exercise 2.1, the derivation of (2.6) is left as a
simple manipulative exercise.

Various limiting cases of (2.6) are of particular interest. Letting $\Delta \to \infty$
corresponds to the non-preemptive case, when the only state of interest apart

from state C is state 0, and we have

$$v(0) = \frac{V \int_0^\infty e^{-\gamma t} \, \mathrm{d}F(t)}{\int_0^\infty [1 - F(t)] e^{-\gamma t} \, \mathrm{d}t}. \tag{2.7}$$

Letting $\Delta \to 0$ corresponds to no restrictions on switching, giving

$$v(x) = \sup_{t > x} \frac{V \int_x^t e^{-\gamma s} \, \mathrm{d}F(s)}{\int_x^t [1 - F(s)] e^{-\gamma s} \, \mathrm{d}s} \qquad (x \geqslant 0), \tag{2.8}$$

and thereby an alternative derivation of (1.3). The nature and optimality of the corresponding index policy are considered in Chapter 5.

The next limit of interest is as $\gamma \searrow 0$. For n jobs $J_i \{i = 1, 2, \ldots, n\}$ with completion times t_i an index policy maximizes the expected total reward

$$\mathbf{E} \sum V_i e^{-\gamma t_i} = \sum V_i - \beta \mathbf{E} \sum V_i t_i + 0(\gamma^2). \tag{2.9}$$

In the limit as $\gamma \searrow 0$ this amounts to minimizing $\mathbf{E} \Sigma V_i t_i$, i.e. minimizing the expected total cost, when V_i is the cost per unit time of any delay in completing J_i. Σt_i is termed the *flow-time* of the policy (or schedule) and $\Sigma V_i t_i$ the *weighted flow-time* with weights V_i. Putting $\gamma = 0$ in (2.7), and since for any non-negative random variable σ with distribution function F we have

$$\mathbf{E}\sigma = \int_0^\infty [1 - F(t)] \mathrm{d}t,$$

it follows that if preemption is not allowed the expected weighted flow-time (EWFT) is minimized by scheduling jobs in order of decreasing $V_i / \mathbf{E}\sigma_i$. This is Problem 1 once again, with the trivial extension of random service-times, now solved through our general theory. The formal justification of taking the limit as $\gamma \searrow 0$ is given in §3.6.

Jobs, their properties, and the problems of scheduling them, form a recurrent theme in this book. For the moment we return to our discussion of when and why the index policy is optimal.

2.6 THE INDEX THEOREM FOR JOBS WITH NO PREEMPTION

For jobs with no preemption allowed Lemma 2.4 quickly leads, as we now proceed to show, to the optimality of the index policy. The problem of maximizing the expected total reward from the jobs which remain at the time of the first job completion depends only on the job which has been completed, and which is therefore removed from the set of jobs available for selection. The time at which this occurs does not influence the policy which should be followed, since its effect is simply to multiply all future rewards by the same factor. It follows that there must be an optimal policy in the form of a deterministic listing of the order in which the jobs are to be carried out. The lemma now shows that listings in descending order of the index are optimal, since any other listing can be improved by interchanging some pair of consecutive jobs.

It is to be noted that this argument depends on any discounting of future rewards being at an exponential rate. If s, $t_1 + s$ and $t_2 + s$ are the times of the first job completion and of two subsequent job completions, respectively, under some deterministic ordering, and $d(t)$ is the discount factor for a reward at time t, the argument depends on $d(t_1 + s)/d(t_2 + s)$ being independent of s. This is true only if $d(t) = e^{-\gamma t}$ for some γ.

An essential point in the argument is that at the time of the first job completion the situation is the same whenever it occurs, except for features which have no bearing on future policy. Let us further note that for this to be true there must be no fixed limit on the time during which rewards from the various jobs may be accumulated. In fact if no rewards are allowed to accrue after time T this is equivalent to a discount factor $d(t)$ such that $d(t) = 0$ for $t > T$, which is a particular form of non-exponential discounting. Finally note that the validity of our essential point depends on the characteristics of the other jobs not changing during the service time of the first job selected. This property is the consequence for jobs with no preemption of the general assumption that a bandit process remains frozen in the same state until such time as it is selected.

2.7 KNAPSACKS

The fact that our proof of the index theorem for jobs with no preemption does not work without exponential discounting, an infinite time horizon, and frozen unselected jobs, does not, of course, mean that all these assumptions are necessary for the theorem to hold. However, except for special cases, they are actually all necessary.

Consider first the infinite time horizon. If instead of this there is a finite time horizon T at which the reward stream is terminated we can see what happens by putting $\gamma = 0$ and supposing the n jobs have deterministic service times t_i $(i = 1, 2, \ldots, n)$, with $\Sigma t_i > T$. We thus have the problem of maximizing $\Sigma_S V_i$ with respect to S subject to $\Sigma_S t_i \leqslant T$, where S is a subset of the n jobs and V_i is the reward on completion of job i. Since the total time available is limited the general aim must clearly be to include in S those jobs for which V_i/t_i is large. Indeed if either V_i or t_i is the same for all the jobs the optimal S is obtained by listing the jobs in decreasing order of V_i/t_i and defining S as the first m jobs in the list, where m is maximal subject to $\Sigma_S t_i \leqslant T$. With arbitrary values for the V_is and t_is, however, this simple recipe does not always work. Consider, for example, the case $T = 10$, $n = 3$, $(t_1, t_2, t_3) = (7, 2, 3)$, $(V_1, V_2, V_3) = (9, 3, 4)$. Clearly the optimal $S =$

$\{1,3\}$, whereas the V_i/t_i criterion leads to $S=\{2,3\}$. Moreover, by varying T and/or n it is easy to construct cases for which S includes any of the subsets of these three jobs, so there can be no alternative index giving the optimal policy in every case.

This finite horizon, undiscounted, deterministic version of our problem is well known in the literature as the *knapsack problem*. This is because we may think in terms of selecting items of differing values and weights to pack into a knapsack, subject to a maximum weight. Far from being soluble by means of an index, this problem is known to be *NP-complete*. This means that a large class of combinatorial problems, for which there is believed to be no bound on the computer time required that is expressible as a polynomial in the amount of information needed to specify the problem, may all be reduced, with sufficient ingenuity, to the knapsack problem. See, for example, Papadimitriou and Steiglitz (1982) for an account of this important class of NP-complete combinatorial optimization problems.

In general terms any solution to the knapsack problem must (i) include items for which V_i/t_i is large, (ii) avoid undershooting the weight restriction by an amount which is substantial, but less than the weight of any of the unselected items, and (iii) strike a balance between objectives (i) and (ii) when they conflict. In other words use of V_i/t_i as an index must be tempered by the requirements of good packing so as to exploit the full weight allowance.

2.8 DIFFERENT DISCOUNT FUNCTIONS

The question of which discount functions are compatible with an index theorem may also be explored by considering jobs with deterministic service times. Suppose two such jobs with the same index value have parameters (V_1,t_1) and (V_2,t_2), and $d(\cdot)$ is the discount function. Thus if these jobs are carried out in succession starting at time t the total reward from them both should not depend on which of them is done first, so that

$$V_1 d(t+t_1)+V_2 d(t+t_1+t_2)=V_2 d(t+t_2)+V_1 d(t+t_1+t_2).$$

Hence

$$\frac{V_2}{V_1}=\frac{d(t+t_1)-d(t+t_1+t_2)}{d(t+t_2)-d(t+t_1+t_2)}, \qquad (2.10)$$

a constant independent of t, and (2.10) must hold for any $t_1,t_2\in\mathbf{R}^+$ for suitably chosen V_1 and V_2. This is true only if $d(t)=A+Be^{-\gamma t}$ for some A,B and γ. The constant A may be set equal to zero with no loss of generality, since its value has no influence on the choice of an optimal policy. Thus, putting $d(0)=1$, we have $d(t)=e^{-\gamma t}$ for some γ, so that for a general index theorem to hold the discounting must be exponential. A proof of this result using Bernoulli bandit processes with two-point prior distributions is given by Berry and Fristedt (1985) for discount functions which are restricted to their class of *regular* discount sequences. As remarked earlier, a finite time horizon amounts to a form of discounting, so our discussion of this particular case, while instructive, was strictly speaking redundant.

The assumption that unselected bandit processes remain frozen also becomes entangled with discounting. Note first that any interaction between the reward streams from different bandit processes is obviously going to have to be severely restricted if an index result is to hold. What, then, if there is no interaction and the states of different bandit processes change independently over time? One possibility would be that during an interval throughout which a bandit process is not selected the values of its future stream of rewards are multiplied by some factor depending only on the length of the interval and on the bandit process concerned. In other words, what happens if different bandit processes have different discount functions?

When the discount functions are all the same a general index theorem holds only if they are exponential. Thus the best chance of an interesting index theorem with bandit processes having different discount functions is when these are all exponential. In fact the following theorem holds. The original unpublished proof by Nash used the terminology of his 1980 paper, but is essentially the one reproduced here.

Theorem 2.6. An optimal policy for a simple family $\mathscr{F}$ of two alternative bandit processes A and B with discount factors a and b is one which selects A or B according as $v_{AB}(x) >$ or $< v_{BA}(y)$ when A is in state x and B in state y, where, writing $x = x(0)$ and $y = y(0)$, and s_i and t_i for the ith decision time after time zero for A and B respectively,

$$v_{AB}(x) = \sup_{\tau > 0} \frac{\mathbf{E} \sum_{s_i < \tau} a^{s_i} r_A(s_i)}{\mathbf{E} \int_0^\tau b^t \, dt}, \quad \text{and} \quad v_{BA}(y) = \sup_{\tau > 0} \frac{\mathbf{E} \sum_{t_i < \tau} b^{t_i} r_B(t_i)}{\mathbf{E} \int_0^\tau a^t \, dt}.$$

Proof. Suppose $c < 1$. Define A^* to be a bandit process with the same properties as A except that (i) the discount factor for A^* is c, (ii) $s_i^* = s_i \log_c b$ $(i = 1, 2, \ldots)$, (iii) $r_A^*(s_i^*) = (a/b)^{s_i} r_A(s_i)$. Similarly let B^* have (i) discount factor c, (ii) $t_i^* = t_i \log_c a$, (iii) $r_B^*(t_i^*) = (b/a)^{t_i} r_B(t_i)$. There is an obvious one-one correspondence between policies for $\mathscr{F}$ and for the SFABP $\mathscr{F}^* = \{A^*, B^*\}$. If under policy g for $\mathscr{F}$ bandit process A yields a reward $r_A(s_i)a^{s_i + t_j}$ at process times s_i for A and t_j for B, then under the corresponding policy g^* for $\mathscr{F}^*$ bandit process A^* yields the same reward at process time s_i^* for A^* and process time t_j^* for B^*, since

$$r_A^*(s_i^*) c^{s_i^* + t_j^*} = \left(\frac{a}{b} \right)^{s_i} r_A(s_i) b^{s_i} a^{t_j} = r_A(s_i) a^{s_i + t_j}.$$

Similarly B^* yields precisely the same rewards as B. Thus a policy g for $\mathscr{F}$ is optimal iff the same is true of the corresponding policy g^* for $\mathscr{F}^*$, which is to say, iff g^* is an index policy, since A^* and B^* both have the same discount factor, and therefore the index theorem applies to $\mathscr{F}^*$. Observe finally that g^* is an index policy iff g is as specified in the statement of the theorem. □

The index functions v_{AB} and v_{BA} of Theorem 2.6 differ from those obtained when the discount factors are all the same in the important respect that each of them depends on the properties of both A and B. This has the consequence that the theorem does not extend directly to families of more than two bandit processes. For such an extension it would be necessary, with an obvious extension of notation, for $v_{BC}(y) = v_{CB}(z)$ if $v_{AB}(x) = v_{BA}(y)$ and $v_{AC}(x) = v_{CA}(z)$, and this is not necessarily so, as is easily seen for example in the case of jobs with no preemption.

Nash (1980) defines a *generalized bandit problem* in terms of a modified SFABP in the Markov setting, for which the reward when bandit process B_i is selected at time t in state x_i may be written in the form

$$a^t \prod_{j \neq i} Q(B_j, x_j) R(B_i, x_i),$$

the non-negative factors $Q(B_j, x_j)$ depending on the states of the other bandit processes. This is equivalent to a semi-Markov SFABP for which process time is incorporated in the definition of the state of a bandit process and the process time of B_i is increased by $\log_a Q(B_j, x_j)$. The original (unpublished) proof of Theorem 2.6 by Nash used the idea of a generalized bandit problem, which has recently found favour with Glazebrook and Fay (1987), as a means of dealing with a set of jobs only one of which has to be completed. Two other tricks which may sometimes be used for the same purpose are described in the next section.

2.9 STOCHASTIC DISCOUNTING

Theorem 2.6 is for a problem to which the index theorem does not obviously apply, the method of proof being to show that the problem is equivalent to a second problem, on a different time scale, satisfying the conditions for the index theorem. A similar device may be used for the gold-mining problem (Problem 2). The difficulty here is that when the machine breaks down no further gold can be mined, either from the mine on which the breakdown occurred or from any other mine, so that the reward streams available from unselected mines do not necessarily remain frozen. This can be resolved, as pointed out by Kelly (1979), by exploiting the well-known equivalence between discounting a future reward, and receiving the same reward with a probability less than one, which may be regarded as stochastic discounting.

A deterministic policy for this problem is a fixed schedule of the mines to be worked on each day until the machine breaks down. Under a particular policy let R_i be the amount of gold mined on day i if there is no breakdown on day i, and P_i the probability of no breakdown on day i, conditional on no breakdown during the previous $i-1$ days. Write $T_i = \log_e P_i$. Thus the expected total reward may be written as

$$\sum_{i=1}^{\infty} R_i \exp\left[-\sum_{j \leqslant i} T_j \right].$$

This is also the total reward from a deterministic SFABP $\mathscr{F}$ for which R_i is the reward at the ith decision time (including time zero), T_i is the interval between the ith and $(i+1)$th decision times, and the discount parameter is 1. There is an obvious one–one correspondence beteen gold-mines and the bandit processes in $\mathscr{F}$.

For mine i let r_{ij} be the amount of gold mined on the jth day that it is worked if there is no breakdown on or before this day, p_{ij} be the probability of no breakdown on the jth day conditional on no earlier breakdown, and $t_{ij} = -\log_e p_{ij}$, generalizing somewhat the original specification of the problem. The initial index value for mine i is

$$v_{i0} = \max_{N \geqslant 1} \frac{\displaystyle\sum_{j \leqslant N} r_{ij} \exp\left\{ -\sum_{k \leqslant j} t_{ik} \right\}}{1 - \exp\left\{ -\displaystyle\sum_{k \leqslant N} t_{ik} \right\}} = \max_{N \geqslant 1} \frac{\displaystyle\sum_{j \leqslant N} r_{ij} \prod_{k \leqslant j} p_{ik}}{1 - \displaystyle\prod_{k \leqslant N} p_{ik}}. \tag{2.11}$$

If $r_{ij} p_{ij}(1-p_{ij})^{-1}$ is decreasing in j the maximum occurs for $N=1$ and $v_{i0} = r_{i1} p_{i1}(1-p_{i1})^{-1}$. If $r_{ij} p_{ij}(1-p_{ij})^{-1}$ is increasing in j the maximum occurs for $N = \infty$.

Since $\mathscr{F}$ is a deterministic SFABP the expression (2.11) for the index v_{i0} may be generalized immediately to the situation when the sequence $\{(r_{ij}, p_{ij}) : j = 1, 2, \ldots\}$ is stochastic. Theorem 2.3 still applies.

If the r_{ij}s are all negative it is not very helpful to think of this problem in terms of gold-mining, though as we shall see there are some genuine applications to research planning. Equation (2.11) holds with no restriction on the sign of the r_{ij}s, but a second derivation which holds when every $r_{ij} \leqslant 0$ is of some interest.

The single-machine scheduling problem with generalized jobs was formulated as a SFABP by appending to the end of each job a standard bandit process with a large negative parameter $-M$. This has the effect of ensuring that every job is completed before one of these standard bandit processes is selected. For the gold-mining problem with every $r_{ij} \leqslant 0$ a similar device may be used. Each mine corresponds to a generalized job, and breakdown of the machine to completion of the job being processed. To ensure that no further costs are incurred once a job has been completed we simply append to each job at the point of completion a standard bandit process with parameter zero. This has the desired effect of formulating the problem as an equivalent SFABP, except that the discount factor is one and the index values are all zero. This difficulty may be overcome by inserting a discount factor $a\,(<1)$ between each pair of successive decision times, and then treating the limit, as a tends to one, of the indices for the discounted problem, when multiplied by $1-a$, as indices for the original undiscounted problem. This procedure leads to the same index (2.11) as the method based on stochastic discounting. On the other hand, we may of course wish to consider a discounted version of the problem. This is not so easily achieved if we use stochastic discounting, as there seems to be no obvious way of combining the two forms of discounting. For this reason the full extent of the optimality of index policies for the discounted gold-mining problem has yet to be established.

One example of a gold-mining problem with non-positive r_{ij}s is the problem facing a chemist who wishes to synthesize a compound by one of n possible routes. Each route has a number of stages, at each of which synthesis is possible, $-r_{ij}$ being the cost of the jth stage of route i.

A second feature of pharmaceutical research is that compounds are subjected to a series of tests, known as screens, before being accepted for clinical trials. These tests are for therapeutic activity and absence of toxic side-effects, and result in the rejection of the vast majority of compounds. If a compound is going to fail it is of course desirable that the cost of finding this out should be as small as possible, with consequent implications for the order of the screens. For example, if a sequence of activity screens and a sequence of toxicity screens have been fixed the problem of merging the two sequences into one may be modelled as a gold-mining problem with $n=2$ and non-positive r_{ij}s. A more detailed discussion of this problem may be found in §3.3 of Bergman and Gittins (1985).

2.10 ONGOING BANDIT PROCESSES

A bandit process changes state only when the continuation control is applied to it, and yields a reward only at those times. It is also of interest, as pointed out by Varaiya, Walrand and Buyukkoc (1985), to consider reward processes with similar properties except that rewards also accrue when the freeze control is applied, rather than just when the continuation control is applied, which will be termed *ongoing* bandit processes. A family of ongoing bandit processes might, for example, model the problem faced by a businessman with interests in a number of projects, each of which yields a constant cash flow except when he spends some time on it, thereby changing its characteristics in some way. The ongoing bandit processes represent the projects; continuing one of the ongoing bandit processes is equivalent to working on the corresponding project.

Formally, then, an ongoing bandit process is a semi-Markov decision process with state-space Θ and a σ-algebra $\mathscr{X}$ of subsets of Θ, together with control elements 0 and 1 for every state. Application of control 0 initiates a time interval in which every point is a decision time and throughout which the state does not change, terminating at the next application of control 1. If (t_1, t_2) is such an interval with the process in state x, the reward which accrues is $\lambda(x)\int_{t_1}^{t_2} a^t \, dt$, where $\lambda(\cdot)$ is an $\mathscr{X}$-measurable function of x. When control 1 is applied a zero reward accrues. Subject to measurability conditions and Conditions A and B, the functions $r(\cdot, 1)$, $P(\cdot \mid \cdot, 1)$, and $F(\cdot \mid \cdot, \cdot, 1)$ are arbitrary.

The nice thing about an ongoing bandit process is that there is a corresponding bandit process which produces the same expected total return, apart from a contribution which is independent of the policy followed, if the same policy is applied to both processes. To see this, note that if control 1 is applied at time s in state x, leading to a transition to state y at time t, a reward stream at rate $\lambda(x)$ from s to ∞ is lost in return for a reward stream at rate $\lambda(y)$ from t to ∞. Thus the change in expected reward is the same as if control 1 is applied to a bandit process in state x at time s for which

$$r(x) = \mathbf{E}\left[\lambda(y) \int_{t-s}^{\infty} a^u \, du - \lambda(x) \int_{0}^{\infty} a^u \, du \,\middle|\, x \right] + r(x, 1). \qquad (2.12)$$

It follows that if Θ, $\mathscr{X}$, $P(\cdot \mid \cdot, 1)$ and $F(\cdot \mid \cdot, \cdot, 1)$ are the same for both processes, and $r(\cdot)$ for the bandit process is defined by (2.12), they are equivalent in the sense just stated.

This equivalence means that the index theorem for a SFABP may be extended to allow some or all of the constituent bandit processes to be ongoing bandit processes. From (2.5) and (2.12) it follows that the index for an ongoing bandit process in state $x(=x(0))$ is

$$v(x) = \sup_{\tau > 0} \frac{\mathbf{E}[a^\tau \lambda(x(\tau)) - \lambda(x)] + R_\tau(x)}{1 - \mathbf{E}a^\tau},$$

where

$$R_\tau(x) = \mathbf{E}\left[\sum_{t < \tau} r(x(t)) \mid x(0) = x \right].$$

2.11 MULTIPLE PROCESSORS

To conclude our discussion of conditions for the optimality of index policies for SFABPs, and other decision processes which may be reduced to this form, we now consider what happens when, in the language of scheduling, there is more than one processor, or machine. This means modifying the rules for a family of alternative bandit processes so that the number of bandit processes being continued at any one time is equal to the number of processors. It turns out that quite strong additional conditions are required for an index theorem to hold. Once again we proceed by looking at what happens for jobs with no preemption. Suppose too that a is close to 1. For an infinite time-horizon the problem thus becomes one of minimizing expected weighted flow-time. Let n be the number of jobs, m the number of machines, c_i the cost per unit time of delay in completing job i, and s_i the service time of job i.

As noted earlier, when there is just one machine an optimal schedule orders deterministic jobs in decreasing order of c_i/s_i. When there is more than one machine an optimal schedule clearly must preserve this feature so far as the jobs processed by each individual machine are concerned. However, this condition does not guarantee optimality, and in general there is no simple condition which does. A counter-example illustrates the difficulty which arises.

Counter-example 2.7.

$$n = 3, \ m = 2, \ (c_1, s_1) = (c_2, s_2) = (1, 1), (c_3, s_3) = (2, 2).$$

All three jobs have the same value for the index c_i/s_i, so one might suppose that any selection of two jobs for immediate service would be equally good. However, this is not so, as comparison of the two schedules set out in Figure 1 shows.

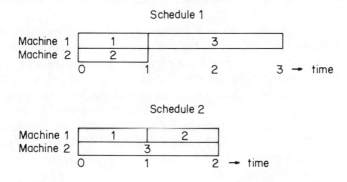

Figure 1. Two alternative schedules for three jobs on two machines.

For Schedule 1 the weighted flow-time is 8 and for Schedule 2 it is 7.

Fairly obviously Schedule 2 does better than Schedule 1 because it uses all the available processing capacity until every job is finished. This observation may be formalized as a theorem, which may be proved by simple algebraic manipulations.

Theorem 2.8. For m machines and n deterministic jobs with c_i/s_i the same for all jobs, minimization of weighted flow time is equivalent to minimization of $\Sigma\delta_j^2$, where $Sm^{-1} + \delta_j$ is the total process time for machine j, and $S = \Sigma s_i$.

This result gives a useful insight as to how to construct a schedule minimizing the weighted flow-time, though it does not provide an explicit solution. The computational complexity involved in finding a solution is NP-complete, like the knapsack problem. In fact for $m = 2$ finding a solution amounts to finding a subset of the n jobs for which the total service time is as close as possible to $S/2$. This is an alternative standard specification of the knapsack problem.

Besides fixing the value of c_i/s there are two further ways of reducing the generality of the problem, wihout reverting to a single machine, in the hope of finding an easily stated solution. These are represented by the special cases where, for all i, (i) $s_i =$ a constant, or (ii) $c_i =$ a constant. In cases (i) and (ii) the obvious extension of the single machine index theorem holds, the optimizing indices being c_i and s_i^{-1} respectively. For case (i) this is a statement of the obvious, and easily extended to allow the service times to be identically distributed indepedent variables. It is less obvious for case (ii).

Theorem 2.9. The flow-time for a set of n deterministic jobs on any fixed number m of machines is minimized by scheduling in order of increasing service time. Such policies are the only optimal policies, except that for every positive integer r the jobs in the rth position from the end of the schedule for each machine may be permuted arbitrarily between machines without changing the flow-time.

Proof. For $m = 1$ the theorem is a special case of the single-machine index theorem, and also follows from a simple interchange argument. For $m \geqslant 2$ note first that the flow-time is clearly the same for every policy in the set S of policies described in the theorem. Note secondly that any policy P not belonging to the set S must be such that for a suitably chosen pair of machines the policy which P defines for the jobs processed by those machines is itself outside the set S. It is therefore sufficient to prove the theorem for $m = 2$, since this result implies that the flow-time for P may be reduced by modifying P just for the pair of machines mentioned. Since there is only a finite number of schedules there must be one which minimizes flow-time.

Suppose then that $m = 2$. We shall show that if a given schedule Q minimizes flow-time then for every positive integer r the rth last job on each of the machines must

(a) start no later than the $(r-1)$th last job on the other machine, and
(b) be no longer than the $(r-1)$th last job on the other machine.

(a) holds since otherwise the flow-time may be reduced by transferring the r last jobs on the first machine to the end of the schedule for the second machine, and the $r-1$ last jobs on the second machine to the end of the schedule for the first machine. (b) holds because otherwise it would follow from (a) that the flow-time could be reduced by interchanging the rth last job on the first machine and the $(r-1)$th last job on the second machine.

From (a) it follows that

(c) if n is even, $\frac{1}{2}n$ jobs are scheduled on each of the two machines, and if n is odd one more job is scheduled on one machine than on the other.

Since the theorem holds for $m = 1$ we also have

(d) the jobs scheduled by Q on any one machine are in order of increasing s_i.

The $m = 2$ result is a simple consequence of (b), (c) and (d).   □

In view of the simplicity of the solutions for deterministic jobs when either the s_is or the c_is are equal, it is of interest to consider whether $c_i/\mathbf{E}s_i$ serves as an index when service times are random, the c_is are equal, and the expected service times $\mathbf{E}s_i$ are equal. Thus our conjecture is that under the stated conditions every initial selection of jobs for service is optimal. This is not so, as the following counter-example shows.

Counter-example 2.10. $n = 4$, $m = 2$, $c_1 = c_2 = c_3 = c_4 = 1$.

$$\mathbf{P}(s_i = 0) = \mathbf{P}(s_i = 2) = \tfrac{1}{2}(i = 1, 2). \ s_3 = s_4 = 1.$$

Let $W = $ (weighted) flow time. Thus if jobs 1 and 2 are processed first, $\mathbf{P}(W = 2) = \tfrac{1}{4} = \mathbf{P}(W = 10)$, $\mathbf{P}(W = 5) = \tfrac{1}{2}$, so $\mathbf{E}W = 5\tfrac{1}{2}$. If jobs 3 and 4 are processed first, $\mathbf{P}(W = 4) = \tfrac{1}{4} = \mathbf{P}(W = 8)$, $\mathbf{P}(W = 6) = \tfrac{1}{2}$, so $\mathbf{E}W = 6$. As in the previous example, the more efficient schedule does better because it enables both machines to be used until every job has been completed.

Let us now try to bring these results together. If expected weighted flow-time is to be minimized then $c(t) = \Sigma_{I(t)} c_i$ must be reduced rapidly, where $I(t)$ is the set of jobs which remain uncompleted at time t. The quantity $c(t)$ is, of course, the rate of increase of weighted flow-time at time t, or simply the *weight* at time t. The aim then is an early reduction of weight. When there is just one machine and job i is being processed the rate of weight reduction should be regarded as equal to $c_i/\mathbf{E}s_i$ for a time $\mathbf{E}s_i$. This view of the matter is confirmed by the index theorem. With m machines working on a set $J(t)$ of jobs at time t the rate of weight reduction is $\Sigma_{J(t)}(c_i/\mathbf{E}s_i)$. Again the aim is to ensure that large values of this quantity occur sooner rather than later. A perfect realization of this aim occurs if

(a) for each machine $c_{i(t)}/\mathbf{E}s_{i(t)}$ is decreasing in t, where $i(t)$ is the job on which the machine is working at time t,
(b) $(i, j) \in J(t) \Rightarrow c_i/\mathbf{E}s_i = c_j/\mathbf{E}s_j, \forall t$, and
(c) $J(t)$ consists of m jobs, so that no machine is idle, until the simultaneous completion of the last m jobs. Note that condition (c) is included in (a) and (b) if we regard an idle machine as working on a job for which $c_i = 0$.

For $m = 1$ conditions (*b*) and (*c*) are satisfied trivially and condition (*a*) is the specification of an index policy, which thus gives the desired perfect realization. For $m > 1$ only exceptionally is a perfect realization possible. Counter-examples 2.7 and 2.10 are two of these exceptional cases. More generally there is some degree of conflict between conditions (a), (b) and (c). For example, there is a conflict between (b) and (c) if Counter-example 2.7 is modified by increasing c_1 and c_2 by the same amount, or between (a) and (c) if c_1 and c_2 are reduced by the same amount in Counter-example 2.10. Simple criteria for optimally resolving such conflicts are available only when the generality of the problem is reduced in some way; for example, when the s_is are identically distributed, or the c_is are equal and the s_is deterministic.

To relate these insights to the general setting of a SFABP with m servers it is useful to think in terms of the stages in the realization of a bandit process which were used to define a modified index policy. For a given bandit process let the *current* stage at time t be the stage which finishes earliest after time t. Let $v^*(i, t)$ be the index value for bandit process i at the beginning of the stage which is current at time t. Thus $v^*(i, t)$ may be regarded as the equivalent constant reward rate for bandit process i at time t. By extension, $\Sigma_{J(t)} v^*(i, t)$ is the equivalent constant reward rate generated by the SFABP at time t, where $J(t)$ denotes the set of bandit processes in progress at time t. As discussed in §2.4 for the single server case, an optimal policy must be one for which large values of $\Sigma_{J(t)} v^*(i, t)$ occur sooner rather than later. A perfect realization of this aim occurs if

(a) $v^*(i(j, t), t)$ is decreasing in t ($j = 1, 2, \ldots, m$), where $i(j, t)$ is the bandit process being served by server j at time t, and
(b) $(i, j) \in J(t) \Rightarrow v^*(i, t) = v^*(j, t), \forall t$.

For $m = 1$ condition (a) is satisfied by an index policy. This follows from Theorem 3.4(v), a similar result to Lemma 2.4. Since condition (b) is satisfied trivially it follows that an index policy is a perfect realization of the aim just stated. This is what we would expect in view of the index theorem. For $m > 1$ there is in general some conflict between conditions (a) and (b), as already shown by our discussion of jobs with no preemption. Note that for such a job it follows from equation (2.7) that $c_i / \mathbf{E} s_i$ is the limit as a tends to one of $v^*(i, t)$, provided $t < s_i$, so that our discussion of jobs with no preemption does serve as an illustration of the general problem.

A further illustration is provided by the problem of minimizing weighted flow-time for discrete-time jobs with no restriction on preemption. Weber, in his important paper (1982) on multi-processor stochastic scheduling, has shown (see also Gittins , 1981), that the index rule is optimal for any m if the weights are the same for every job, the service times are independently and identically distributed, and the completion rate is a monotone function of process time. Here the completion rate is the hazard rate of reliability theory, i.e. the probability of completion in the next time unit conditional on the job not so far being completed. It is perhaps rather surprising that such strong conditions are required, but there seems to be no general result with significantly weaker conditions. For example, for decreasing completion rates which are not necessarily the same for every job a myopic policy may not be optimal, even with equal weights. This is shown by the following counter-example.

Counter-example 2.11. $n = 3, m = 2, c_1 = c_2 = c_3 = 1$.

$$p_{11} = p_{21} = p_{3j} = \tfrac{1}{10} \quad (j = 1, 2, \ldots), p_{1j} = p_{2j} = \tfrac{1}{100} \quad (j = 2, 3, \ldots),$$

where $p_{ij} = \mathbf{P}(s_i = j \mid s_i > j - 1)$.

In §4.4 it is shown that for jobs with decreasing completion rates the stopping time in the definition of the index is the interval between decision times. For discrete-time jobs and no preemption restrictions this means that each of the stages of a modified index policy lasts for just one time unit. Since only integer values of t are relevant this in turn means that $v^*(i, t)$ is simply the index for job i at time t. It follows from (2.6) that $v^*(i, t) = c_i p_{ii(t)}$, where $i(t)$ is the process time of job i at time t, provided job i is still uncompleted at time t.

In the present case, since $c_1 = c_2 = c_3 = 1$ the index value for an uncompleted job i at process time j is equal to the completion rate p_{ij}. Since $p_{11} = p_{21} = p_{31}$ it follows that there are myopic policies which start by serving any two of the three jobs. It is a simple matter to check that the expected weighted flow-time for the policy 12 which serves jobs 1 and 2 at time 1 is greater than for the other two myopic policies 13 and 23. This difference is to be expected since under 12 the indices for the two bandit processes continued at time 2 differ by at least 0.09, whereas under 13 and 23 these indices are equal with probability 0.892. Thus policies 13 and 23, which are in fact both optimal, do not violate condition (b) so much as does 12.

EXERCISES

2.1. Let $\mathscr{F} = \{B_i : 1 \leqslant i \leqslant n\}$ be a SFABP, g an index policy for $\mathscr{F}$, and $T_i(t)$ $(i = 1, 2, \ldots, n)$ the process time for B_i at time t when policy g is applied to $\mathscr{F}$. Show that it follows from Lemma 2.5 that the expected total reward from $\mathscr{F}$ under policy g may be written as

$$R_g(\mathscr{F}) = \mathrm{E} \int_0^\infty \max_{1 \leqslant j \leqslant n} \left\{ \inf_{0 \leqslant s \leqslant t} v(B_i, x(T_i(t))) \right\} a^t \, dt.$$

This result is due to Mandelbaum (1986).

2.2. Use integration by parts to show that the expectation of a non-negative random variable X with distribution function F may be written as

$$\int_0^\infty (1 - F(x)) \, dx.$$

2.3. Establish the identity (2.6).

2.4. Prove the assertion on page 26 that $d(t_1 + s)/d(t_2 + s)$ is independent of s only if $d(t) = e^{-\gamma t}$ for some γ.

2.5. Prove the assertion following equation (2.10) that 'this is true only if $d(t) = A + Be^{-\gamma t}$ for some A, B and γ'.'

2.6 Why does the second derivation of equation (2.11) given at the end of §2.9 not necessarily work if some of the r_{ij}s are positive?

2.7. Prove Theorem 2.8.

CHAPTER 3

Central Theory

3.1 INTRODUCTION

This chapter sets out the main index theorems and their proofs. Two important ideas underly these proofs. The first is the calibration of bandit processes on the scale defined by the standard bandit processes, as discussed in Chapter 1 in connection with Problem 4. The second is the notion of *forwards induction*.

In Lemma 2.4 we observed that, given portions of the realizations of different bandit processes, each terminating when the bandit process concerned reaches a given stopping set, the sequence which yields the highest expected total reward is the one which always selects the portion with the highest available expected reward per unit of expected discounted time. For a simple family of alternative bandit processes forwards induction amounts to extending this procedure in two ways: (i) the stopping sets, as well as the bandit processes themselves, are chosen so as to maximize the average reward rate; (ii) the bandit process portions available for selection at each stage are the reward-rate-maximizing portions of each bandit process, starting from its current state. The term forwards induction refers to the fact that this is an inductive process which proceeds forwards in time, in contrast to the backwards induction which is more typical of dynamic programming.

The chapter is largely based on an earlier preliminary account of allocation index theory (Gittins, 1980), which includes the proofs of Theorems 3.6, 3.18, 3.22, 3.25 and 3.26. More recent work is also included, in particular Whittle's (1980) index theorem (3.15) for bandit superprocesses, and the results by Glazebrook (1982) and Katehakis and Veinott (1985) on ε-optimality of ε-index policies.

For a SFABP calibration by standard bandit processes immediately establishes (Theorem 3.1) the form which an optimal index must take. After a number of results on the properties of index values, and what happens when portions of realizations of bandit processes are moved relative to each other, it is then not difficult to show that forwards induction defines an optimal policy, and that this policy is also an index policy (Theorem 3.6). The two ε results just mentioned are then given, the second of them falling neatly out of the forwards induction argument.

After some discussion of the limiting case as $\gamma \searrow 0$ we then consider index theory for bandit superprocesses, following Nash (1973) in defining these to be bandit processes with more than one continuation control, though here without including a rule for choosing between these controls as part of the definition. Again the form of an optimal index for a simple family of alternative superprocesses (SFAS) is determined by the calibration procedure, though a counter-example shows that index policies are not optimal without some restriction on the set of allowed superprocesses.

For a SFAS a well-defined policy must not only pick out one of the superprocesses at each decision-time but also must select a control from the control-set for that superprocess. Whittle (1980) showed that, provided every superprocess in $\mathscr{F}$ has the property (Condition D) that the optimal control for a SFAS comprising just the superprocess and a standard bandit process with the parameter λ is the same for all λ small enough for the optimal policy to select the superprocess, then an index policy is optimal for the SFAS $\mathscr{F}$. For superprocesses with this property the optimal index policy is a forwards induction policy. This means that if it is optimal to select a particular superprocess in a given state the optimal control does not depend on the other superprocesses available for selection. Whittle's proof of Theorem 3.6 is given in §9.2. It involves spotting the form of the optimal payoff function and then appealing to the uniqueness of the solution to the dynamic programming recurrence equation (Theorem 2.2 parts (ii) and (iii)), and extends very neatly to the case of a SFAS (Theorem 3.15). The proof of Theorem 3.15 given here is longer, but more direct, and brings out more clearly the role of Condition D. This theorem is then applied to families of alternative *stoppable* bandit processes, following Glazebrook (1979c). These are bandit processes which allow the option of retirement with state-dependent final reward.

Theorem 3.18 is an index theorem for generalized jobs (see §2.5) subject to precedence and arrival constraints. It is less general than Theorem 3.15 but has the advantages of conditions which in some cases are more easily checked, and of leading to ε-optimality for ε-index policies. The proof is based on the notion of forwards induction. Theorem 3.18 is shown to cover the cases of arbitrary precedence constraints when preemption is not allowed, and of precedence constraints forming an out-tree (Theorem 3.22). Theorem 3.22 extends to the case of stoppable jobs (Theorem 3.24), and also to some arrivals processes (Theorems 3.25 and 3.26).

The most important of these arrivals processes is a Poisson arrivals process. The chapter concludes with the nice result that for ungeneralized jobs with Poisson arrivals, and under the criterion of minimum time-averaged weighted flow-time, the optimal policy is the index policy defined by putting $\gamma = 0$ in the expression (2.8). Thus in this limiting case the optimal policy is unaltered by Poisson arrivals.

The chapter is quite hard reading and the reader may find it best to concentrate in the first instance on the results rather than their proofs.

3.2 A NECESSARY CONDITION FOR AN INDEX

Our chief interest is in indices which define optimal allocation rules irrespective of which particular family of alternative bandit processes is available for selection.

Thus, if $\mathscr{B}_a$ is the entire uncountable set of bandit processes with the discount factor a, we should in the first instance like an index for each bandit process $B \in \mathscr{B}_a$ which defines optimal policies for every SFABP formed by a finite subset of $\mathscr{B}_a$. Let us call such an index *an index for $\mathscr{B}_a$*.

Given the existence of an index μ for $\mathscr{B}_a$ which takes every real value as B varies, it follows that any other index for $\mathscr{B}_a$ is a strictly increasing function of μ. Otherwise we should have different indices defining incompatible index policies. In fact the following theorem holds.

Theorem 3.1. Any index for $\mathscr{B}_a(0 < a < 1)$, if there is one, must be a strictly increasing function of $v(B, x)$ (see §2.3).

Proof. The main idea in the proof is to use the standard bandit processes as a means of calibrating the other bandit processes. We first note that the statement of the theorem is true if $\mathscr{B}_a$ is replaced by Λ_a, the set of all standard bandit processes with discount factor a. This is because if Λ is a standard bandit process with parameter λ (and just one state) then $v(\Lambda) = \lambda$.

Now consider a SFABP with bandit processes B and Λ, where Λ is a standard bandit process with the parameter λ. To complete the proof of the theorem it is sufficient to show that for such a SFABP there is an optimal policy which continues Λ at all times if and only if $v(D) \leqslant \lambda$.

There is an obvious one–one correspondence between policies for the SFABP $\{B, \Lambda\}$ and freezing rules for the bandit process D. The expected total reward from $\{B, \Lambda\}$ under the policy which corresponds with the freezing rule f for B may be expressed as

$$R_f(\{B, \Lambda\}) = \mathbf{E} \sum_{i=0}^{\infty} a^{t_i + f_i} r(t_i) + \lambda \mathbf{E} \sum_{i=0}^{\infty} a^{t_i} \int_{f_{i-1}}^{f_i} a^t \, dt$$

$$= R_f(B) + \mathbf{E} \sum_{i=0}^{\infty} a^{t_i} \int_{f_{i-1}}^{f_i} a^t \, dt.$$

The total reward from $\{D, \Lambda\}$ under the policy which continues Λ at all times may be expressed as

$$\lambda \int_0^{\infty} a^t \, dt = \lambda \mathbf{E} \sum_{i=0}^{\infty} a^{f_i} \int_{t_i}^{t_{i+1}} a^t \, dt + \lambda \mathbf{E} \sum_{i=0}^{\infty} a^{t_i} \int_{f_{i-1}}^{f_i} a^t \, dt$$

$$= \lambda W_f(B) + \lambda \mathbf{E} \sum_{i=0}^{\infty} a^{t_i} \int_{f_{i-1}}^{f_i} a^t \, dt.$$

Clearly

$$\lambda \int_0^{\infty} a^t \, dt \left\{ {\geqslant \atop <} \right\} R_f(\{B, \Lambda\})$$

according as

$$\lambda \begin{Bmatrix} \geqslant \\ = \\ < \end{Bmatrix} \frac{R_f(B)}{W_f(B)} = v_f(B),$$

provided $f_0 \neq \infty$, which ensures that $W_f(B) \neq 0$. It follows that the policy which continues Λ at all times is optimal if and only if

$$\lambda \geqslant \sup_{\{f:f_0 \neq \infty\}} v_f(B),$$

or, alternatively, if and only if

$$\lambda \geqslant \sup_{\{f:f_0 = 0\}} v_f(B) = v'(B),$$

since the common factor a^{f_0} may be cancelled from the numerator and denominator of $v_f(B)$.

The final stage in the proof of the theorem is the following lemma.

Lemma 3.2 For any bandit process B, $v(B) = v'(B)$.

Proof. The lemma will follow by restricting to the stopping rules the class of freezing rules considered in the argument of the preceding paragraph, provided it can be shown that, given any freezing rule f, there is a stopping time τ such that $R_\tau(\{B, \Lambda\}) \geqslant R_f(\{B, \Lambda\})$. This we now proceed to do, first observing that a direct appeal to Theorem 2.2(i) does not work, since the SFABP $\{B, \Lambda\}$ does not obey Condition A. The first step is to show that given any policy f for which $\mathbf{P}\{0 < f_0 < \infty\} > 0$ there is a policy f^* such that

$$\mathbf{P}\{0 < f_0^* < \infty\} = 0 \quad \text{and} \quad R_{f^*}(\{B, \Lambda\}) \geqslant R_f(\{B, \Lambda\}). \tag{3.1}$$

The policy f is defined by the probability distributions of the random variables

$$f_0; \quad f_1 - f_0; \quad \text{and} \quad f_i - f_{i-1} \quad (i = 2, 3, \dots),$$

conditional on

$$x(t_0); \quad x(t_0), f_0, t_1, x(t_1);$$

and

$$x(t_0), f_0, t_1, x(t_1), f_{j-1} - f_{j-2}, t_j, x(t_j) \quad (j = 2, 3, \dots, i).$$

In each case the conditioning random variables describe the previous history of the bandit process D at the appropriate decision time.

Consider now a freezing rule f^*, under which control 1 is applied for the ith time at time $t_{i-1} + f_{i-1}^*$, and such that the random variables $f_i - f_{i-1}$ and $f_i^* - f_{i-1}^*$ are equal $(i = 1, 2, \dots)$. Thus, for example, $f_1 - f_0$ and $f_1^* - f_0^*$ have the same distributions conditional on $x(t_0), f_0, t_1$ and $x(t_1)$. Suppose that f_0^* is an arbitrary

Lebesgue-measurable function of $x(t_0)$. Thus

$$R_{f_*}(\{D,\Lambda\}) - \lambda \int_0^\infty a^t \, dt = \mathbf{E}\left\{ a^{f\delta} \mathbf{E}\left[\sum_{i=0}^\infty a^{f_i^* - f_0^*}\left(a^{t_i} r(t_i) - \lambda \int_{t_i}^{t_{i+1}} a^t \, dt \right) \middle| x(t_0) \right] \right\}$$

$$= \mathbf{E}\left\{ a^{f\delta} \mathbf{E}\left[\sum_{i=0}^\infty a^{f_i - f_0}\left(a^{t_i} r(t_i) - \lambda \int_{t_i}^{t_{i+1}} a^t \, dt \right) \middle| x(t_0) \right] \right\},$$

using the identity $\mathbf{E}(X) = \mathbf{E}[\mathbf{E}(X\,|\,Y)]$ for random variables X and Y defined on the same probability space. Clearly, then, $R_{f_*}(\{D,\Lambda\})$ is maximized with respect to f_0^* by setting

$$f_0^* = \{_\infty^0\} \text{ according as } \mathbf{E}\left[\sum_{r=0}^\infty a^{f_r - f_0}\left(a^{t_i} r(t_i) - \lambda \int_{t_i}^{t_{i+1}} a^t \, dt \right) \middle| x(t_0) \right] \{\gtreqless\} 0.$$

Thus f^* has the required properties (3.1) if f_0^* is chosen in this way.

From the properties (3.1) of f^* it follows that in showing that there is a stopping time τ such that $R_\tau(\{B,\Lambda\}) \geqslant R_f(\{B,\Lambda\})$ attention may be confined to policies f which never switch back to B if they choose Λ at time zero. A similar argument applies at any decision time, in terms of the expected further reward from then onwards, conditional on previous history, to show that policies which switch from A to B at any stage need not be considered. This restriction on the set of allowed policies means that the SFABP $\{B,\Lambda\}$ satisfies Condition A. The lemma, and hence also the theorem, now follow from Theorem 2.2(i).   □

3.3 SPLICING BANDIT PROCESS PORTIONS

Theorem 3.1 is a consequence of the index theorem (3.6) for a SFABP, which in turn is a special case of the index theorem (3.15) for superprocesses. In one sense, therefore, the separate proofs of Theorems 3.1 and 3.6 are superfluous. The gradual build-up to greater generality should, however, make the basic ideas stand out more sharply, and also presents an opportunity to introduce and motivate some further terminology and notation.

As remarked earlier the role of a policy for a FABP is to order the reward stream portions provided by the individual bandit processes by splicing them together to form a single reward stream. A good policy is one which places reward stream portions with a high equivalent constant reward rate early in the sequence, subject to any constraints on precedence, in paticular the precedence sequence defined by the process time for reward stream portions from the same bandit process. The decision-maker's problem is like that of the cricket captain who would like his most effective batsmen to bat early in the innings, where they are likely to have the greatest possible effect on the outcome of the match.

It was shown in the proof of Lemma 3.2 that there is a stopping time τ for B such that the policy for $\{B,\Lambda\}$ which continues B up to time τ and then continues Λ indefinitely is an optimal policy. That part of the realization of B which occurs before time τ will be termed the *truncation of B at process time τ*, and denoted by B^*. Our argument showed that there is no advantage to be gained by, so to speak, promoting in the batting order above any part of B^* an interval during which Λ is continued, by placing such an interval earlier in the time sequence. We need a terminology in which such arguments may conveniently be expressed.

When a freezing rule f is applied to a non-standard bandit process B the time at which control 1 is applied for the $(i+1)$th time exceeds the process time at that point by f_i, where $f_i \geqslant 0$. We are usually interested in bandit processes in the context of a family of alternative bandit processes. If A is a member of such a family the quantity f_i is now the total time during which control 1 has so far been applied to the other bandit processes in the family. Consequently there is no question of f_i taking a negative value. If the bandit process A forms part of a SFABP comprising A and the truncation B^* of a bandit process B at some value τ of its process time we have the situation considered in the previous paragraph, with A in place of Λ. It follows that $f_i \leqslant \tau$ for all i. We note in passing that B^* may be regarded as a bandit process for which all the rewards which accrue after process time τ have large negative values, so that it is never actually selected for continuation beyond then.

Writing $p_i = f_i - \tau$, the expected total reward from A is

$$R_f(A) = \mathbf{E} \sum_{i=0}^{\infty} a^{t_i + p_i + \tau} r(t_i) = \mathbf{E} \left[a^\tau \mathbf{E} \sum_{i=0}^{\infty} a^{t_i + p_i} r(t_i) | \tau \right],$$

where the inner expectation is conditional on the entire realization of bandit process B up to process time τ, and will be written as $R_p(A)$. Under this condition, then, the non-positive random variables $p_i (i = 0, 1, 2, \ldots)$ are past-measurable in terms of the bandit process A, and $-\tau \geqslant p_0 \geqslant p_1 \geqslant p_2 \geqslant \cdots$. By analogy with a freezing rule they will be said to define a *promotion rule p for A with respect to time τ*. In fact this is simply a freezing rule with the property that the total time for which control 0 is applied to bandit process A does not exceed τ. A promotion rule for which $p_i = 0 (i = 0, 1, 2, \ldots)$ will be termed a *null* promotion rule (or a null freezing rule).

Since the specification of a freezing rule depends on the state-space and other characteristics of the bandit process concerned it is not generally possible to apply it to another bandit process. However, standard bandit processes form an important exception to this rule, since there is just one state in the state-space and every time is a decision time. A freezing (or promotion) rule f (or p), defined for any bandit process, will thus be said to be applied to a given standard process if the sequence of random times t_i and f_i (or p_i) $(i = 0, 1, 2, \ldots)$, specifying the control applied at any time, have the same joint distributions for the two processes. This notion is implicit in the statement of the following lemma.

Lemma 3.3. For a bandit process B, the increase in the expected total reward which is caused by applying the promotion rule p instead of the null promotion rule (both defined with respect to the same time s) is less than or equal to the corresponding increase for a standard bandit process with the parameter $v(B)$.

The basic idea here is very simple. Given a simple family of two standard bandit processes Λ_1 and Λ_2 both with the parameter $v(B)$, the expected reward is the same for any policy. In particular it is the same for all policies defined by applying a promotion rule for Λ_2 with respect to time s, and selecting Λ_1 whenever Λ_2 is not selected, including the policies defined by promotion rule p and by the null promotion rule. In symbols

$$R_p(\{\Lambda_1, \Lambda_2\}) = R_0(\{\Lambda_1, \Lambda_2\}).$$

Thus if the lemma is not true

$$R_p(\{\Lambda_1, B\}) > R_0(\{\Lambda_1, B\}).$$

This inequality means that there must be some stopping time σ for B such that the equivalent constant reward rate from B up to process time σ when the promotion rule p is followed is greater than the constant reward rate $v(B)$ from Λ_1. Thus, writing f for the freezing rule for B which coincides with p up to process time σ, and for which $f_i = \infty$ for every decision time $t_i \geq \sigma$, we have $v_f(B) > v(B)$, contradicting Lemma 3.2. An explicit proof along rather different lines follows for those who remain unconvinced.

Proof. The two increases in expected total reward which are mentioned in the statement of the lemma are: for bandit process B,

$$a^s \mathbf{E} \sum_{i=0}^{\infty} (a^{p_i} - 1) a^{t_i} r(t_i);$$

and for the standard bandit process,

$$a^s v(B) \mathbf{E} \sum_{i=0}^{\infty} (a^{p_i} - 1) \int_{t_i}^{t_{i+1}} a^t \, dt.$$

Since $p_i (i = 0, 1, 2, \ldots)$ is an increasing sequence of non-positive terms, the lemma is trivially true if $p_0 = 0$. Thus it is sufficient to show that

$$v(B) \geq \sup_{\{p : p_0 < 0\}} \frac{\mathbf{E} \sum_{i=0}^{\infty} (a^{p_i} - 1) a^{t_i} r(t_i)}{\mathbf{E} \sum_{i=0}^{\infty} (a^{p_i} - 1) \int_{t_i}^{t_{i+1}} a^t \, dt}. \tag{3.2}$$

To prove this we first define, for any non-negative integers I and J such that $I \leq J$, the class of promotion rules C_{IJ}. This consists of all those promotion rules p such that $p_0 < 0$ and, for some past-measurable random integer κ, for which

$I \leqslant \kappa \in J, p_i = 0 \forall i \geqslant \kappa$, and $p_i = p_I \forall i$ such that $I \leqslant i < \kappa$. In particular $C_{JJ} = \{p : p_i = 0 \forall i \geqslant J\}$, and clearly $C_{0J} \subset C_{1J} \subset \cdots \subset C_{JJ}$. The right-hand side of (3.2) will be denoted by μ, and μ_{IJ} denotes the similar quantity with the restriction that only promotion rules belonging to C_{IJ} are allowed.

The next step is to show that, for any J,

$$\mu_{IJ} = \mu_{I-1,J} \qquad (I = 1, 2, \ldots, J). \tag{3.3}$$

Writing H_I for the previous history $(t_0, t_1, t_2, \ldots, t_I; r(t_0), r(t_1), r(t_2), \ldots, r(t_I))$ of B at process time t_I, we have

$$\mu_{IJ} = \sup_{p \in C_{IJ}} \left\{ \frac{\mathbf{E} \sum_{i=0}^{I-1} (a^{p_i} - 1) a^{t_i} r(t_i) + \mathbf{E}\left[(a^{p_I} - 1) \mathbf{E} \sum_{i=1}^{\kappa-1} a^{t_i} r(t_i) \middle| H_I \right]}{\mathbf{E} \sum_{i=0}^{I-1} (a^{p_i} - 1) \int_{t_i}^{t_{i+1}} a^t \, dt + \mathbf{E}\left[(a^{p_I} - 1)\mathbf{E} \int_{t_I}^{t_\kappa} a^t \, dt \middle| H_I \right]} \right\}. \tag{3.4}$$

By definition of a promotion rule, p_I depends on H_I and is restricted to values between 0 and p_{I-1}. Thus it follows that the quantity in curly brackets on the right-hand side of (3.4) is maximized with respect to p_I if and only if

$$p_I = \left\{ \begin{matrix} p_{I-1} \\ 0 \end{matrix} \right\} \quad \text{according as} \quad \frac{\mathbf{E} \sum_{i=1}^{\kappa-1} a^{t_i} r(t_i) | H_I}{\mathbf{E} \int_{t_I}^{t_\kappa} a^t \, dt | H_I} \left\{ \begin{matrix} \geqslant \\ \leqslant \end{matrix} \right\} \mu_{IJ}.$$

In the case of equality, p_I may be assigned any value between 0 and p_{I-1}. Equation (3.3) follows immediately.

If we now write down the expression for μ_{0J}, the factor $a^{p_i} - 1$ which occurs in every term in both numerator and denominator is, for every i, equal to either $a^{p_0} - 1$ or to zero. It may therefore be cancelled, so that

$$\mu_{0J} = \mu_{JJ} = \sup_{\{\kappa : 0 < \kappa \leqslant J\}} \frac{\mathbf{E} \sum_{i=0}^{\kappa-1} a^{t_i} r(t_i)}{\mathbf{E} \int_0^{t_\kappa} a^t \, dt}, \tag{3.5}$$

where κ is past-measurable.

From (3.5) and Lemma 3.2 it follows that

$$\mu_{JJ} \leqslant v(D) \quad (J = 1, 2, \ldots). \tag{3.6}$$

Condition A implies that μ_{JJ} tends to μ as J tends to infinity, so that, letting J tend to infinity in (3.6), $\mu \leqslant v(D)$. $\qquad \square$

3.4 EQUIVALENT CONSTANT REWARD RATES AND FORWARDS INDUCTION FOR ARBITRARY DECISION PROCESSES

A *forwards induction* policy for a decision process is one which at each decision time maximizes the equivalent constant reward rate up to an arbitrary stopping time. Thus it is a more sophisticated version of a myopic policy, which maximizes the equivalent reward rate up to the next decision time. Neither is in general optimal, essentially because of their lack of long-range vision. The forwards induction method does, however, produce optimal policies for simple families of alternative bandit processes, and other similar decision processes. As we shall see, these policies are index policies, the indices being maximized equivalent constant reward rates for the bandit processes, or other components, concerned. The term *forwards induction* refers to the iterative construction forwards in time of such a policy, in contrast to the familiar backwards induction of dynamic programming. In this section a forwards induction policy is formally defined. We begin with some notation, and a portmanteau theorem, which are required for this and other purposes.

Given any semi-Markov decision process D, together with a policy g, a bandit process may be defined by introducing the freeze control 0 with the usual properties, and requiring that at each decision time either the control 0 or the control given by g be applied. Call this the bandit process D_g (or (D, g) if the suffix notation is likely to cause confusion). Thus, if g is deterministic, stationary, and Markov, application of the continuation control 1 to D_g when D is in state x is equivalent to applying control $g(x)$ to D. The quantities $t_i(i = 0, 1, 2, \ldots)$, $x_g(t)$, $r_g(t)$ and $R_g(D)$ have the same definitions as $t_i(i = 0, 1, 2, \ldots)$, $x(t)$, $r(t)$ and $R(B)$, but with D_g in place of the bandit process B. Also $R(D) = \sup_g R_g(D)$, $R_{g\tau}(D) = R_\tau(D_g)$, $W_{g\tau}(D) = W_\tau(D_g)$, $v_{g\tau}(D) = R_{g\tau}(D)/W_{g\tau}(D)$, $v_g(D) = \sup_{\{\tau > 0\}} v_{g\tau}(D)$, and $v(D) = \sup_g v_g(D)$. Similarly $R_{gf}(D) = R_f(D_g)$, etc. As for bandit processes, $v(D, x)$, for example, or simply $v(x)$ where the context makes it clear which is the relevant decision process, denotes the value of $v(D)$ when D is initially in state x. As in the definitions of §2.3, for the Markov case all expected discounted times, for which the first letter of the notation is W, are redefined by multiplying by the discrete time correction factor $\gamma(1 - a)^{-1}$, and all equivalent constant reward rates, for which the first letter of the notation is v, are redefined by dividing by this factor.

The theorem which follows is a rough parallel to Theorem 2.2, with the focus now on $v(D)$ rather than $R(D)$, bringing together a number of results associated with the attainment of the suprema over τ and over g. The proofs could well be skipped by the reader who wants to press on.

Theorem 3.4. For a semi-Markov decision process D satisfying Condition A, and for which the control set is always finite:

(i) for any $\xi \in \Theta$, $v(\xi) = v_{g\tau}(\xi)$ for some deterministic stationary Markov policy g and stopping time $\tau(>0)$;

(ii) $v(\cdot)$ is an $\mathscr{X}$-measurable function;

(iii) the stopping set $\Theta_0(\in \mathcal{X})$ defining the τ described in (i) may be chosen to be either $\{x : v(x) < v(\xi)\}$ or $\{x : v(x) \leqslant v(\xi)\} - \{\xi\}$;
also for any g and τ as described in (i),

(iv) $P\{v_{g, \tau - t}(x(t)) \geqslant v(\xi)$ for every decision time $t < \tau \,|\, x(0) = \xi\} = 1$, and

(v) $P\{v(x(\tau)) > v(\xi) \,|\, x(0) = \xi\} = 0$, where, for any t, $x(t)$ is the state of D at time t under policy g.

Proof. Let D have a control set $\Omega(x)$ which is finite for any $x \in \Theta$. For any $\xi \in \Theta$ we define the semi-Markov decision process $D(\xi)$ as follows. The state space and associated σ-algebra are again Θ and $\mathcal{X}$. For any $x \in \Theta$ the control set is $\Omega(x) \cup \Omega(\xi)$. Application of a control belonging to $\Omega(x)$ leads to the same reward and joint probability distribution for the next state and decision time, as does application of the same control to D when it is in state ξ. Thus, in effect, $D(\xi)$ coincides with D but with the additional option of switching to state ξ at any decision time. We adopt the unrestrictive convention that the switch option is never used if $D(\xi)$ is already in state ξ.

Also a deterministic stationary Markov policy for $D(\xi)$ is defined by (a) a deterministic stationary Markov policy g for D, which gives the control applied to $D(\xi)$ when a member of $\Omega(\xi)$ is not selected, (b) a subset $\Theta_0(\in \mathcal{X})$ of Θ comprising those states of $D(\xi)$ for which the switch option is used, and (c) a function e from Θ_0 to $\Omega(\xi)$ which selects the control when $D(\xi)$ is in a state belonging to Θ_0. Theorem 2.2 tells us that there is an optimal policy for $D(\xi)$ which is of this type. Moreover, e may be restricted to be single-valued. The reason is that, for a given $\Theta_0, D(\xi)$ is equivalent to a modified semi-Markov decision process obtained by merging Θ_0 with state $\xi (\notin \Theta_0$, by our convention). If $D(\xi)$ starts in state ξ the sequence of decision times $\rho_i (i = 1, 2, \ldots)$ at which it is in a state belonging to Θ_0 under such a policy form a renewal process.

The maximum expected total reward may be expressed, writing $\rho_0 = 0$, $\tau = \rho_1$, $x(t)$ for the state at time t with the convention that $x(\rho_i) = \xi (i = 1, 2, \ldots)$, and $r_g(x)$ for the (undiscounted) reward $r(x, g(x))$ from D when control $g(x)$ is applied in state x, as

$$R(D(\xi), \xi) = \sup_{g, \Theta_0} \sum_{i=0}^{\infty} E \sum_{\rho_i \leqslant t_j < \rho_{i+1}} a^{t_j} r_g(x(t_j))$$

$$= \sup_{g, \Theta_0} \sum_{i=0}^{\infty} E a^{\rho_i} E \sum_{\rho_i \leqslant t_j < \rho_{i+1}} a^{t_j - \rho_i} r_g(x(t_j))$$

$$= \sup_{g, \Theta_0} \sum_{i=0}^{\infty} (E a^\tau)^i E \sum_{t_j < \tau} a^{t_j} r_g(x(t_j))$$

$$= \sup_{g, \Theta_0} \frac{E \sum_{t_j < \tau} a^{t_j} r_g(x(t_j))}{1 - E a^\tau}, \tag{3.7}$$

where each of the expectation operators is conditioned by the event $x(0) = \xi$. For a given g there is a one–one correspondence between sets Θ_0 and deterministic stationary Markov stopping times τ. Thus, using the definition of $v(\xi)$ and writing γ for $-\log_e a$,

$$v(\xi) = \gamma \sup_{g, \Theta_0} \frac{\mathbf{E} \sum_{t_j < \tau} a^{t_j} r_g(x(t_j))}{1 - \mathbf{E} a^\tau}$$

$$= \gamma R(D(\xi), \xi). \tag{3.8}$$

As already mentioned, the supremum in (3.7) is attained for some Θ_0 and deterministic stationary Markov g. Since ξ is arbitrary, part (i) of the theorem therefore follows from (3.8).

From Theorem 2.2 it follows that $R(D(\xi), x)$ is an $\mathcal{X}$-measurable function of x, and thus the same must be true of $R_x(D(\xi), x)$, where the subscript x indicates that an optimal policy is followed subject to the restriction that the control applied at time zero belongs to $\Omega(x)$ rather than $\Omega(\xi)$. From part (iii) of Theorem 2.2 it also follows that the set $\Theta_0(\in \mathcal{X})$ for which, together with a suitable policy g, the supremum is attained in equation (3.7) may be chosen to be either $\{x : R_x(D(\xi), x) < R(D(\xi), \xi)\}$ or $\{x : R_x(D(\xi), x) \leqslant R(D(\xi), \xi)\} - \{\xi\}$.

Now a real-valued function h defined on Θ is $\mathcal{X}$-measurable if and only if the set $\{x : h(x) \leqslant c\}$ belongs to $\mathcal{X}$ for all c. Since $\mathcal{X}$ includes all those subsets of Θ which contain just one element it is therefore sufficient, to prove parts (ii) and (iii) of the theorem, to show that, for all $x, \xi \in \Theta$,

$$R_x(D(\xi), x) < (=) R(D(\xi), \xi) \Leftrightarrow v(D, x) < (=) v(D, \xi). \tag{3.9}$$

From part (i) of Theorem 2.2 it follows that

$$R_x(D(\xi), x) = \sup_{g, \tau > 0} \left\{ \mathbf{E} \sum_{t_j < \tau} a^{t_j} r_g(x(t_j)) + R(D(\xi), \xi) \mathbf{E} a^\tau \right\}, \tag{3.10}$$

where g and τ are deterministic stationary and Markov at positive decision times, τ is the time at which the switch option is first used, and the expectation operators are conditioned by the event $x(0) = x$. From (3.8) we have, for any τ,

$$R(D(\xi), \xi) = v(\xi) \mathbf{E} \int_0^\tau a^t \, dt + R(D(\xi), \xi) \mathbf{E} a^\tau \tag{3.11}$$

so that, subtracting (3.11) from (3.10),

$$R_x(D(\xi), x) - R(D(\xi), \xi) = \sup_{g, \tau > 0} \left\{ \left[\frac{\mathbf{E} \sum_{t_j < \tau} a^{t_j} r_g(x(t_j))}{\mathbf{E} \int_0^\tau a^t \, dt} - v(D, \xi) \right] \mathbf{E} \int_0^\tau a^t \, dt \right\}.$$

$$\tag{3.12}$$

Now clearly the right-hand side of equation (3.12) is negative (zero) iff the supremum over g and $\tau(>0)$ of the expression in square brackets is negative (zero) i.e. iff $v(D, x) < (=)v(D, \xi)$. Thus statement (3.9) is true, and we have proved parts (ii) and (iii) of the theorem.

Part (iv) is simply Lemma 2.5 with $B = D_g$. The proof of (v) is similar to that of Lemma 2.5. The essential point is that if a higher equivalent constant reward rate can be achieved after time τ then the average rate up to time τ may be increased by increasing τ. Specifically, let

$$v(x(\tau)) = v_{h\sigma}(x(\tau)),$$

where h and σ depend on $x(\tau)$. Let e be the policy which coincides with g up to time τ, and then with h up to time $\tau + \sigma$. Let A be the event $v(x(\tau)) > v(\xi)$ and $\bar{A}$ its complement. Finally, let

$$\rho = \begin{cases} \tau + \sigma & \text{on } A \\ \tau & \text{on } \bar{A} \end{cases}.$$

It is a simple matter to show that $v_{e\rho}(\xi) > v_{g\tau}(\xi) = v(\xi)$, contradicting the definition of $v(\xi)$ unless $P(A) = 0$. Thus $P(A) = 0$ as required. $\square$

Theorem 3.4 enables us to give a formal definition of a forwards induction policy for a semi-Markov decision process D which satisfies Condition A. Write $t_0(= 0), t_1, t_2, \ldots$ for the sequence of decision times. At each decision time t_i find a policy g_i such that $v_{g_i}(x(t_i)) = v(x(t_i))$. Let u_i be the control applied to D in state $x(t_i)$ and at time t_i under g_i. The policy which applies the control u_i at each decision time t_i will be described as a *forwards induction* policy.

Alternatively, a policy may be defined by analogy with a modified index policy. For this purpose the initial state will be denoted by x_0 instead of $x(0)$.

The first step is to find a policy γ_1 such that $v_{\gamma_1}(x_0) = v(x_0)$. Let σ_1 be the stopping time for the bandit process D_{γ_1} defined by the stopping set $\{x : v(x) < v(x_0)\}$. Thus, by part (iii) of Theorem 3.4, $v_{\gamma_1 \sigma_1}(x_0) = v(x_0)$.

Let x_1 be the (random) state of the bandit process D_{γ_1} at time σ_1, γ_2 a policy such that $v_{\gamma_2}(x_1) = v(x_1)$, and σ_2 the stopping time for the bandit process D_{γ_2} defined by the stopping set $\{x : v(x) < v(x_1)\}$ when the initial state is x_1. The state x_i, policy γ_{i+1}, and stopping time $\sigma_{i+1}(i = 2, 3, \ldots)$ are defined inductively by replacing $x_1, \gamma_1, \sigma_1, \gamma_2$ and σ_2 by $x_i, \gamma_i, \sigma_1 + \sigma_2 + \cdots + \sigma_i, \gamma_{i+1}$ and $\sigma_1 + \sigma_2 + \cdots + \sigma_{i+1}$, respectively, in the previous sentence. The policy for D which starts by applying policy γ_1 up to time σ_1, then applies policy γ_2 from σ_1 up to $\sigma_1 + \sigma_2, \gamma_3$ from $\sigma_1 + \sigma_2$ up to $\sigma_1 + \sigma_2 + \sigma_3$, and so on, will be termed a *modified* forwards induction, or FI*, policy. The successive intervals during which policies $\gamma_1, \gamma_2, \gamma_3$ and so on, are applied will be described as the stages of such a policy. An ε-FI* policy is defined in similar fashion in a succession of stages, with the difference that the policies γ_i and stopping times σ_i are now defined so as to satisfy the inequalities

$$v_{\gamma_i \sigma_i}(D, x_{i-1}) \geqslant v(D, x_{i-1}) - \varepsilon,$$

where ε is small and positive, rather than zero.

3.5 MORE SPLICING AND PROOF OF THE INDEX THEOREM FOR A SFABP

The proofs of the key Theorems 3.6 and 3.18 depend on considering arbitrary splicings of the available reward streams, and showing that they can be dismembered and respliced so as to form reward streams in conformity with the theorems so as to yield at least as high an expected reward. We need names and notation for the reward stream portions which arise in this process.

The notation of the previous section makes it easy to extend the notion of truncation to any semi-Markov decision process. For example, $R_{g\tau}(D)$ is the expected total reward from the truncation of D at time τ when policy g is used.

That portion of the realization of a bandit process or decision process which occurs from a truncation time onwards will be termed the *residual* bandit process or decision process from that time. For example, the expected total reward from the residual semi-Markov decision process from time τ when policy g is used on D at all times is $R_g(D) - R_{g\tau}(D)$. This simple expression is, of course, attributable to the fact that the same policy g is applied both before and after τ.

Now let us consider what happens to a typical member B of a family $\mathscr{F}$ of alternative bandit processes when a policy g is applied to $\mathscr{F}$. Conditional on the realizations of all the other bandit processes in $\mathscr{F}$ a certain freezing rule will be applied to B. Without conditioning in this way, the effect on B is to apply a random freezing rule, the randomization arising from the random realizations of all the other bandit processes. This random freezing rule will be termed *the freezing rule for B defined by g*.

In similar fashion, the random freezing rule which is applied to the residual bandit process from a truncation time σ onwards when a freezing rule f is applied to a bandit process B will be termed the freezing rule *defined by f*, and denoted by f_+. The freezing rule for B which coincides with f up to time σ, and thereafter applies control 0 to B at all times will be termed the *truncation of f at time σ* and denoted by f_-. This notation is illustrated by the proof of the next lemma, which is in the same vein as parts (iv) and (v) of Theorem 3.4.

For a non-standard bandit process B, initially in state ξ, let successive decision times occur at process times $t_0(=0), t_1, t_2, \ldots$. Let σ be the earliest decision time t_i for which $v(B, x(t_i)) < v(B, \xi)$.

Lemma 3.5. If f is a freezing rule for which $\mathbf{P}\{\exists i \text{ such that } \sigma \leqslant t_i < \infty \text{ and } f_i < \infty\} > 0$, then $v_f(B, \xi) < v(B, \xi)$.

Proof. We have

$$v_f(B, \xi) = \frac{R_f(B, \xi)}{W_f(B, \xi)} = \frac{R_{f_-}(B, \xi) + \mathrm{E}R_{f_+}(B, x(\sigma))}{W_{f_-}(B, \xi) + \mathrm{E}W_{f_+}(B, x(\sigma))}. \qquad (3.13)$$

Now

$$\frac{R_{f_-}(B, \xi)}{W_{f_-}(B, \xi)} = v_{f_-}(B, \xi) \leqslant v(B, \xi),$$

and

$$\frac{R_{f_+}(B, x(\sigma))}{W_{f_+}(B, x(\sigma))} = v_{f_+}(B, x(\sigma)) \leqslant v(B, x(\sigma)) < v(B, \xi),$$

provided $W_{f_+}(B, x(\sigma)) > 0$. This condition is satisfied since $\mathbf{P}\{\exists i \text{ such that } \sigma \leqslant t_i < \infty \text{ and } f_i < \infty\} > 0$, and the lemma therefore follows from equation (3.13). □

Now suppose that a policy g is applied to a family $\mathscr{F}$ of alternative bandit processes $B_1, B_2, \ldots, B_n$ up to some stopping time τ. Let $\tau_j (j = 1, 2, \ldots, n)$ be the process time of B_j at time τ. Let f^j be the freezing rule for B_j defined by g, $t_i^j (i = 0, 1, 2, \ldots)$ be the process time of the ith application of control 1 to B_j, and $f_i^j + t_i^j$ the true time when this occurs under f^j. Let f_-^j be the freezing rule for B_j which coincides with f^j except that f_i^j is replaced by infinity for all i such that $t_i^j \geqslant \tau_j$; thus we may describe f_-^j as the truncation of the freezing rule f^j at time τ, or at process time τ_j. The expected total reward from the truncation of $\mathscr{F}$ at time τ under policy g is the sum over j of the expected total reward from the truncation of B_j at process time τ_j under f^j. In symbols,

$$R_{g\tau}(\mathscr{F}) = \sum_{j=1}^{n} R_{f_-^j}(B_j).$$

Similarly, the expected total reward from the residual FABP from time τ when policy g is applied to $\mathscr{F}$ at all times is the sum over j of the expected total reward from the residual bandit process from process time τ_j when the freezing rule f^j is applied to B_j at all times. In symbols,

$$R_g(\mathscr{F}) - R_{g\tau}(\mathscr{F}) = \sum_{j=1}^{n} [R_{f^j}(D_j) - R_{f_-^j}(D_j)] = \sum_{j=1}^{n} \mathrm{E}R_{f_+^j}(D_j, x_j(\tau_j)).$$

This discussion prepares the way for the proof of the following theorem.

Theorem 3.6. (The Index Theorem for a SFABP.) The classes of index policies, forwards induction policies, and optimal policies are identical for a simple family $\mathscr{F}$ of alternative bandit processes $B_j (j = 1, 2, \ldots, n)$.

In proving this we shall need the following lemma.

Lemma 3.7. (i) $v(\mathcal{F}) = \max_j v(B_j)$.
(ii) If $f^{\underline{j}}$ is the freezing rule for $B_j (j = 1, 2, \ldots, n)$ defined by a policy γ for $\mathcal{F}$ and truncated at time σ, and $v_{\gamma\sigma}(\mathcal{F}) = v(\mathcal{F})$, then $v_{f^{\underline{j}}}(B_j) = v(\mathcal{F})$ for all j such that $W_{f^{\underline{j}}}(B_j) > 0$.

Proof.

$$R_{\gamma\sigma}(\mathcal{F}) = \sum_{j=1}^{n} R_{f^{\underline{j}}}(B_j) = \sum_{j=1}^{n} W_{f^{\underline{j}}}(B_j) v_{f^{\underline{j}}}(B_j)$$

$$\leqslant \sum_{j=1}^{n} W_{f^{\underline{j}}}(B_j) v(B_j) \leqslant \max_i v(B_i) \sum_{j=1}^{n} W_{f^{\underline{j}}}(B_j) = \max_i v(B_i) W_{\gamma\sigma}(\mathcal{F}). \tag{3.14}$$

Hence

$$v(\mathcal{F}) = \sup_{g,\tau} v_{g\tau}(\mathcal{F}) = \sup_{g,\tau} [R_{g\tau}(\mathcal{F})/W_{g\tau}(\mathcal{F})] = R_{\gamma\sigma}(\mathcal{F})/W_{\gamma\sigma}(\mathcal{F}) \leqslant \max_i v(B_i).$$

However, if

$$\max_i v(B_i) = v(B_k) = v_{\tau_k}(B_k)$$

for some stopping time τ_k for B_k, and by part (i) of Theorem 3.4 there must be such a τ_k, and if g_k is the policy for $\mathcal{F}$ which applies control 1 to bandit process B_k at all times, then

$$v(\mathcal{F}) \geqslant v_{g_k\tau_k}(\mathcal{F}) = v_{\tau_k}(B_k) = \max_i v(B_i).$$

Consequently $v(\mathcal{F}) = \max_i v(B_i)$, as required. This means that the inequalities in (3.14) may be replaced by equalities, and the second part of the lemma is an immediate consequence. $\qquad\square$

Proof of Theorem. That index policies are forwards induction policies is an immediate consequence of part (i) of the lemma. The converse is an immediate consequence of part (ii).

Using the notation just established, let $\mathcal{G}$ denote the residual SFABP obtained by truncating $\mathcal{F}$ at time τ_k when policy g_k is used. If $\tau_k = \infty$ we make the convention that $\mathcal{G}$ comprises the $n-1$ bandit processes which are left when B_k is removed from $\mathcal{F}$. Since the theorem is trivially true when $n = 1$, we may suppose $n > 1$.

Thus $\mathcal{G}$ is complementary to the truncated bandit process B_k^* obtained by truncating B_k at process time τ_k. Complementary, that is, in the sense that whenever control 1 is applied to a bandit process in $\mathcal{F}$ it may be regarded as being applied either to B_k^* or to one of the bandit processes in $\mathcal{G}$. The complementarity of B_k^* and $\mathcal{G}$ leads to a corresponding division of the expected total reward when an arbitrary policy g is applied to $\mathcal{F}$. Under the policy g let $I_g(t) = 1$ or 0 depending on whether control 1 is applied to B_k^* or to $\mathscr{G}$ at a decision time t, and let t_i $(i = 0, 1, 2, \dots)$ denote the ith decision time. Thus

$$R_g(\mathscr{F}) = \mathbf{E} \sum_{i=0}^{\infty} a^{t_i} I_g(t_i) r_g(t_i) + \mathbf{E} \sum_{i=0}^{\infty} a^{t_i} (1 - I_g(t_i)) r_g(t_i). \qquad (3.15)$$

The first of these components is the expected reward $R_f(B_k)$ resulting from bandit process B_k under the freezing rule f which is defined by applying policy g to $\mathscr{F}$ and truncating at process time τ_k. The second component may be expressed in the form

$$\mathbf{E}\left[\mathbf{E} \sum_{i=0}^{\infty} a^{t_i} (1 - I_g(t_i)) r_g(t_i) \,|\, B_k^* \right],$$

where the inner expectation is conditional on the entire realization of B_k^*. This condition fixes the random SFABP $\mathscr{G}$ and, together with the policy g for $\mathscr{F}$, defines a policy h for $\mathscr{G}$ and a promotion rule p for the superprocess $(\mathscr{G}, h)$ with respect to τ_k. Thus the second component in equation (3.15) may also be written as $\mathbf{E}R_{hp}(\mathscr{G})$, so that

$$R_g(\mathscr{F}) = R_f(B_k) + \mathbf{E}R_{hp}(\mathscr{G}). \qquad (3.16)$$

Consider now a policy for $\mathscr{F}$ which starts by applying control 1 to B_k until process time τ_k. At this point the state of $\mathscr{F}$ coincides with the initial state of $\mathscr{G}$, provided $\tau_k < \infty$. From time τ_k onwards our policy is defined to coincide with h. The entire policy is denoted by $*g$ since it coincides with a FI* policy for the first stage of such a policy. Denoting by 0 the null promotion rule for $(\mathscr{G}, h)$ with respect to τ_k, we have

$$R_{*g}(\mathscr{F}) = R_{\tau_k}(B_k) + \mathbf{E}R_{h0}(\mathscr{G}). \qquad (3.17)$$

The first stage in our proof of the optimality of index policies is to show that

$$R_{*g}(\mathscr{F}) \geqslant R_g(\mathscr{F}). \qquad (3.18)$$

From part (i) of Lemma 3.7 we have

$$R_f(B_k) = W_f(B_k) v_f(B_k) \leqslant W_f(B_k) v_{\tau_k}(B_k) = W_f(B_k) v(B_k)$$
$$= W_f(B_k) v(\mathscr{F}), \qquad (3.19)$$

and

$$R_{\tau_k}(B_k) = W_{\tau_k}(B_k) v_{\tau_k}(B_k) = W_{\tau_k}(B_k) v(\mathscr{F}). \qquad (3.20)$$

Lemma 3.3 gives

$$R_{hp}(\mathscr{G}) - R_{h0}(\mathscr{G}) \leqslant [W_{hp}(\mathscr{G}) - W_{h0}(\mathscr{G})] v(\mathscr{G}).$$

From the definition of $\mathscr{G}$, and using part (v) of Theorem 3.4, it follows that

$v(\mathcal{G}) \leqslant v(\mathcal{F})$, and since $W_{hp}(\mathcal{G}) \geqslant W_{h0}(\mathcal{G})$ this means that

$$R_{hp}(\mathcal{G}) - R_{h0}(\mathcal{G}) \leqslant [W_{hp}(\mathcal{G}) - W_{h0}(\mathcal{G})]v(\mathcal{F}). \tag{3.21}$$

Now putting together (3.16), (3.17), (3.19), (3.20) and (3.21) we have

$$R_g(\mathcal{F}) - R_{*g}(\mathcal{F}) \leqslant [W_f(B_k) - W_{\tau_k}(B_k)]v(\mathcal{G}) + \mathbf{E}[W_{hp}(\mathcal{G}) - W_{h0}(\mathcal{G})]v(\mathcal{F}).$$

However

$$W_f(B_k) + \mathbf{E}W_{hp}(\mathcal{G}) = W_{\tau_k}(B_k) + \mathbf{E}W_{h0}(\mathcal{G}) = \int_0^\infty a^t \, dt,$$

so this completes the proof of (3.18).

The remainder of the proof of the optimality of index policies now follows quickly from Theorem 2.2. As in proving Lemma 3.2 we must deal with the possibility that $\mathcal{F}$ may include one or more standard bandit processes, and hence not obey Condition A. This complication is soon disposed of. In the first place there is clearly no point in continuing any standard bandit process except the one with the highest parameter value. An obvious generalization of the argument leading to Lemma 3.2 now shows that given any policy for $\mathcal{F}$ there is a policy which, if it ever continues the standard bandit process, does so permanently from that time onwards, and for which the expected total reward is at least as great. Hence attention may be restricted to policies having this property. This restriction has the effect of ensuring Condition A, and we may therefore use Theorem 2.2.

To show that only index policies are optimal it is sufficient to show that (3.18) holds with strict inequality unless the bandit process selected by g at time zero is one with an index value at least as great as any other. If g does not have this property then $W_{hp}(\mathcal{G}) > W_{h0}(\mathcal{G})$ and, from part (ii) of Lemma 3.7, $v(\mathcal{F}) > v(\mathcal{G})$. Thus (3.21) holds with strict inequality, and therefore so does (3.18). $\square$

3.6 NEAR OPTIMALITY OF NEARLY INDEX POLICIES, AND THE $\gamma \searrow 0$ LIMIT

The proof of the index theorem given in the previous section quickly leads, as noted by Katehakis and Veinott (1985), to a useful bound on the suboptimality of a policy which is approximately an index policy.

For an arbitrary policy applied to the SFABP $\mathcal{F}$ define the sequence of times $u_0, u_1, u_2, \ldots$ inductively as follows, where $B_{j(i)}$ is the bandit process selected for continuation at u_i.

$u_0 = 0$; u_{i+1} is the first decision time after u_i at which either there is a switch between bandit processes or $v(B_{j(i)})$ drops below its value at u_i, unless there is no such decision time, in which case $u_{i+1} = \infty$.

Let x_{ki} be the state of bandit process B_k at time u_i, $x_i = (x_{1i}, x_{2i}, \ldots, x_{ni})$, and $s_i = u_{i+1} - u_i$ $(k = 1, 2, \ldots, n; \ i = 0, 1, 2, \ldots)$. Thus s_i is a stopping time for $B_{j(i)}$

starting from state $x_{j(i),i}$, and in general depends on the states of the other $n-1$ bandit processes. If

$$v_{s_i}(B_{j(i)}, x_{j(i),i}) \geqslant \max_k v(B_k, x_{ki}) - \varepsilon \qquad (i = 0, 1, 2, \ldots),$$

h will be termed an ε-index* policy. This definition may fairly easily be shown to imply that h is an ε-FI* policy.

Corollary 3.8. If h is an ε-index* policy then

$$R_h(\mathscr{F}) \geqslant R(\mathscr{F}) - \varepsilon\gamma^{-1}.$$

Proof. Let g be an arbitrary policy. The proof of the inequality (3.18) may be mimicked with the truncation of $B_{j(0)}$ at process time s_0 in place of B_k^*, and a corresponding redefinition of the residual SFABP $\mathscr{G}$. In place of (3.20) we now have

$$R_{s_0}(B_{j(0)}) \geqslant W_{s_0}(B_{j(0)})[v(\mathscr{F}) - \varepsilon];$$

the inequality $v(\mathscr{G}) \leqslant v(\mathscr{F})$ now follows from Lemma 3.7(i) rather than Theorem 3.4(v); and instead of (3.18) the inequality that follows is

$$R_{h1g}(\mathscr{F}) - R_g(\mathscr{F}) \geqslant -\varepsilon W_{s_0}(B_{j(0)}) = -\varepsilon\gamma^{-1}(1 - \mathbf{E}e^{-\gamma s_0}), \qquad (3.22)$$

where $h1g$ denotes the policy derived by modifying policy g as described, so that the first stage of the modified policy coincides with h.

Successive modifications hig $(i = 1, 2, \ldots)$ to policy g may be defined inductively, coinciding with policy h up to time u_i. The construction of $h(i+1)g$ from hig is the same as the construction of $h1g$ from g (which we may write as $h0g$), starting now with $\mathscr{F}$ in state x_i at time u_i, at which time bandit process $B_{j(i)}$ is selected, and continued for a time s_i. Repeating the argument leading to (3.22), we now have

$$R_{h(i+1)g}(\mathscr{F}) - R_{hig}(\mathscr{F}) \geqslant -\varepsilon\gamma^{-1}\{\mathbf{E}[e^{-\gamma u_i}(1 - \mathbf{E}e^{-\gamma s_i} | u_1, x_1, u_2, x_2, \ldots, u_i, x_i)]\}$$

$$= -\varepsilon\gamma^{-1}(\mathbf{E}e^{-\gamma u_i} - \mathbf{E}e^{-\gamma u_{i+1}}), \quad (i = 0, 1, 2, \ldots). \quad (3.23)$$

Since Condition A implies that $\mathbf{E}e^{-\gamma u_i} \to 0$ as $i \to \infty$, and Condition B implies that $R_{hig}(\mathscr{F}) \to R_h(\mathscr{F})$ as $i \to \infty$, the corollary follows from (3.23) by summing over i and then taking the infimum of the left-hand side over all policies g. $\quad\square$

An alternative definition of an approximate index policy is one for which the index of the bandit process continued at each decision time is within ε of the maximal index at that time. Call such a policy an ε-*index* policy. Glazebrook (1982d) has established the following bound when $\mathscr{F}$ is Markov.

Theorem 3.9. If h is an ε-index policy for an SFABP $\mathscr{F}$ with the discount parameter γ, then

$$R_h(\mathcal{F}) \geqslant R(\mathcal{F}) - \varepsilon\gamma^{-1}(1 - e^{-\gamma})^{-1}.$$

Which of these bounds works best clearly depends on the value of γ and on which kind of ε-approximation is most readily shown to hold in a given case.

If under a given policy the decision times for a decision process are t_i, at which time a reward $r(t_i)$ occurs $(i = 0, 1, 2, \dots)$, the total discounted reward is

$$\sum_{i=0}^{\infty} r(t_i)e^{-\gamma t_i} = \sum_{i=0}^{\infty} r(t_i) - \gamma \sum_{i=0}^{\infty} t_i r(t_i) + O(\gamma^2).$$

An optimal policy is one which maximizes the expectation of this quantity, or, in the limit as $\gamma \searrow 0$, minimizes the expectation of $\Sigma t_i r(t_i)$. This is a generalization to arbitrary decision processes of the expected weighted flow-time (EWFT) criterion of §2.5. It was conjectured there that for a SFABP made up of jobs this criterion is optimized by the index policy obtained by putting $\gamma = 0$ in the expression for the index in the discounted case. The conjecture will now be shown to be true, and to extend to arbitrary SFABPs under the generalized EWFT criterion. Under the following condition the proof is easy. A continuity argument may be used to remove the condition.

Condition C. A SFABP $\{B_i : i = 1, 2, \dots, n\}$ is said to satisfy Condition C if $\exists \varepsilon > 0$ such that for any pair of distinct bandit processes B_i, B_j in states x, y, either $v(B_i, x) > v(B_j, y)$ for all $\gamma < \varepsilon$, or $v(B_i, x) < v(B_j, y)$ for all $\gamma < \varepsilon$.

Corollary 3.10. Under Condition C the generalized EWFT criterion for a SFABP is minimized by the index policy obtained by putting $\gamma = 0$ in the general expression for the index.

Proof. From the theorem it follows that the index policy is optimal. From Condition C it follows that this policy is the same for all $\gamma < \varepsilon$. The corollary follows by considering the limits as $\gamma \searrow 0$ of the payoffs (a) under the index policy, and (b) under any other policy. $\qquad\square$

Theorems 3.15, 3.18, 3.22, 3.24, 3.25 and 3.26 are all index theorems which are stated and proved later in this chapter. For each of them the obvious analogue of Corollary 3.10 holds, and may be proved in similar fashion. To save tedious repetition we shall, however, now be silent on this subject until we reach Theorem 3.25, for which EWFT is of particular interest. In contrast, there appears to be no obvious way of extending Corollary 3.8 to provide an approximation result in the most general setting covered by Theorem 3.15, though the extension is straightforward for the rather less general Theorem 3.18, and hence for Theorems 3.22, 3.25 and 3.26, all of which are really corollaries of Theorem 3.18.

It is interesting to note that no approximation result follows from Theorem 3.9 on letting $\gamma \searrow 0$. A different approach seems to be necessary (for example see Exercise 3.1).

3.7 BANDIT SUPERPROCESSES AND SIMPLE FAMILIES OF ALTERNATIVE SUPERPROCESSES

A family $\mathcal{F}$ of alternative bandit processes is formed from a set $\{B_1, B_2, \ldots, B_n\}$ of bandit processes with the same discount factor by requiring that at each decision time the freeze control 0 is applied to all but one of the constituent bandit processes, so that $\mathcal{F}$ generates a single reward stream depending on the policy followed. Suppose now that the starting point is a set of decision processes $\{D_1, D_2, \ldots, D_n\}$ all with the discount factor a. From each D_i we now form a second decision process S_i by adding to the control set $\Omega_i(x_i)$ at state x_i the freeze control 0 which, as in the case of a bandit process, leaves the state unchanged and generates zero rewards for as long as it is applied. A similar composition of the constituents $\{S_1, S_2, \ldots, S_n\}$ may be achieved by again requiring that at each decision time the freeze control is applied to all but one of the S_is. The only difference is that at each decision time a choice must now be made both of an S_i and of a control (other than the freeze control) from its control set. The constituents S_i of the composite decision process $\mathcal{F}$ defined in this way will be termed *bandit superprocesses*, or simply *superprocesses*, and $\mathcal{F}$ itself a *family of alternative superprocesses (FAS)*. As in the case of SFABP, if there are no additional constraints on the set of superprocesses available for selection at a decision time then we have a *simple* family of alternative superprocesses (*SFAS*).

An index policy for a SFAS $\mathcal{F}$ made up of superprocesses $\{S_i : i = 1, 2, \ldots, n\}$ must pick out at each decision time both a superprocess S_j and a control from the control set for the corresponding D_j. Thus consider the functions $\mu_i\{(x_i, u_i) \to \mathbf{R}; x_i \in \Theta_i, u_i \in \Omega(x_i)\}$ $(i = 1, 2, \ldots, n)$, where Θ_i is the state-space for S_i and $\Omega_i(x_i)$ is the control set for D_i in state x_i. An *index policy for $\mathcal{F}$ with respect to $\mu_1, \mu_2, \ldots, \mu_n$* is one which at any decision time when S_i is in state $x_i (i = 1, 2, \ldots, n)$ applies control u_j to superprocess S_j, for some S_j and u_j such that

$$\mu_j(x_j, u_j) = \max_{\{i, u_i \in \Omega_i(x_i)\}} \mu_i(x_i, u_i).$$

Note that this maximum exists because of our standing assumption that control sets are finite.

In the case of bandit processes we showed that the dynamic allocation index $v(B, x)$ is the only function (apart from monotone transformations of itself) defining an index policy which is optimal for an SFABP formed by an arbitrary finite subset of $\mathcal{B}_a$, the set of all bandit processes with discount factor a. In other words $v(.,.)$ is the only index for $\mathcal{B}_a$. For superprocesses we should like a function which is an index for as large as possible a set $\mathcal{S}_a$ of superprocesses with discount factor a. An obvious candidate is the function $v(.,.,.)$ defined by the equation

$$v(S, x, u) = \sup_{\{g:g(x)=u\}} v_g(D, x) \quad (S \in \mathscr{S}_a, x \in \Theta_D, u \in \Omega_D(x)), \quad (3.24)$$

where D is formed from A by removing the freeze control. In fact the following theorem holds.

Theorem 3.11. If $\mathscr{S}_a \supset \Lambda_a$ and an index exists for $\mathscr{S}_a$ then it must be a strictly increasing function of $v(S, x, u)$.

The proof involves calibrating $\mathscr{S}_a$ by means of the standard bandit processes Λ_a and runs along similar lines to the proof of Theorem 3.1. Because of Theorem 3.11 an index policy with respect to $v(.,.,.)$ will be referred to simply as an *index policy*.

As in the case of a SFABP, index policies for a simple family $\mathscr{F}$ of alternative superprocesses $\{S_j : j = 1, 2, \ldots, n\}$ are forwards induction policies. The following lemma sets this observation in context.

Lemma 3.12. (i) $v(\mathscr{F}) = \max_j v(S_j)$.
(ii) If g^j_- is the policy for $S_j (j = 1, 2, \ldots, n)$ defined by a policy γ for $\mathscr{F}$ and truncated at time σ, and $v_{\gamma\sigma}(\mathscr{F}) = v(\mathscr{F})$, then $v_{g^j_-}(S_j) = v(\mathscr{F})$ for all j such that $W_{g^j_-}(S_j) > 0$.
(iii) A policy for $\mathscr{F}$ is an index policy iff it is a forwards induction policy.

Proof. The proof of parts (i) and (ii) is virtually identical to the proof of Lemma 3.7. The place of $v_{\tau_k}(B_k)$ in the proof is now taken by $v_{g_k \tau_k}(S_k) = v(S_k)$. That index policies are forwards induction policies is an immediate consequence of (i). The converse is an immediate consequence of (ii). □

The question now is just how inclusive can we define $\mathscr{S}_a$ to be if $v(.,.,.)$ is to be an index for $\mathscr{S}_a$. If $\mathscr{S}_a = \mathscr{B}_a$ then $S \in \mathscr{S}_a \Rightarrow S$ is a bandit process and $v(S, x, u) = v(S, x, 1) = v(S, x)$. Thus $v(.,.,.)$ is an index for $\mathscr{S}_a$ by Theorem 3.6. We can actually extend $\mathscr{S}_a$ well beyond $\mathscr{B}_a$, though not to include every superprocess with discount factor a. Clearly any superprocess which leads to optimal policies that do not conform to the index $v(.,.,.)$ must be excluded from $\mathscr{S}_a$. This possibility is illustrated by the following example.

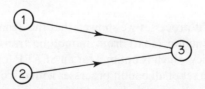

Figure 2. Precedence constraints for Example 3.13.

Example 3.13. Jobs 1, 2 and 3 are subject to the above precedence constraints, so that job 3 cannot be started until jobs 1 and 2 have been completed. Jobs 1, 2 and 3 require, respectively, 1, 2 and 1 unit(s) of service time up to the first decision time after commencement of the job, and then terminate with probabilities $\frac{1}{2}$, 1 and 1, yielding rewards 0, 1 and $M(> 5\frac{2}{3})$. After these times every job behaves like a standard bandit process with parameter 0, whether it has terminated or not. A reward at time t is discounted by the factor $(\frac{1}{2})^t$.

The three jobs form a decision process D and, with the addition of the freeze control 0, a superprocess S. The controls in a control set, other than control 0, may be identified with the jobs available for selection, and the policies of interest identified by job sequences, with the convention that a job equivalent to a 0 standard bandit process is selected only if no further positive rewards are available. Thus, if x is the initial state,

$$v(S, x, 1) = v_{123}(D, x) = \frac{\frac{1}{2}[(\frac{1}{2})^3 + M(\frac{1}{2})^4]}{1 + \frac{1}{2}(\frac{1}{2} + (\frac{1}{2})^2) + (\frac{1}{2})^4} = \frac{2 + M}{46},$$

$$v(S, x, 2) = v_{213}(D, x) = \frac{(\frac{1}{2})^2 + (\frac{1}{2})^5 M}{1 + \frac{1}{2} + (\frac{1}{2})^2 + (\frac{1}{2})^4} = \frac{8 + M}{58}.$$

Note that the condition $M > 5\frac{2}{3}$ ensures that the stopping times in the definitions of $v_{123}(D, x)$ and $v_{213}(D, x)$ (see equation (2.5)) are at process time 1 for job 1 if it fails to terminate, and otherwise are at $t = 4$, the completion time for job 3.

Thus

$$v(S, x, 1) > v(S, x, 2) \Leftrightarrow M > 21,$$

and S is compatible with the index $v(\cdot, \cdot, \cdot)$ only if any optimal policy for a SFAS $\mathscr{F}$ including S which selects S in state x also selects job 1 if $M > 21$. This condition is not satisfied, as may be seen by defining $\mathscr{F}$ to be the SFAS of which the superprocess S is the only member. In this case the optimal policy is either 123 or 213. Since the expected reward from job 3 is the same for both policies, and the only other reward is from job 2 and occurs earlier under 213 than under 123, policy 213 must be optimal. It follows that the superprocess S is incompatible with the index $v(\cdot, \cdot, \cdot)$, and therefore must be excluded from $\mathscr{S}_{1/2}$.

The key to an appropriate definition for $\mathscr{S}_a$ is the observation that if an index picks out a particular superprocess S in state x, then it must pick out the same control(s) from $\Omega_D(x)$ irrespective of which other superprocesses are available for selection. This suggests that we look for a condition which ensures that if an optimal policy selects S in state x it always selects the same control(s) from $\Omega_D(x)$. This would have the effect of reducing S to a bandit process, so that Theorem 3.6 applies. In fact it is sufficient, as shown by Whittle (1980), to ensure that this reduction occurs when the only available alternative to S is a standard bandit process.

Condition D. Consider the SFAS $\{S, \Lambda\}$, where Λ is a standard bandit process with parameter λ. Suppose that when S is in state x it is optimal to select S and apply control u. If for all $x \in \Theta_D$, $\lambda \in \mathbf{R}$, and SFASs $\{S, \Pi\}$ (Π a standard bandit process) for which it is optimal to select S in state x, this implies that it is optimal to apply u to S in state x, then S will be said to satisfy Condition D.

Note 3.14. If a superprocess S satisfies Condition D a control $u \in \Omega_D(x)$ is optimal for $\{S, \Lambda\}$ iff $v(S, x, u) = v(S, x) \geqslant \lambda$. The 'if' and 'only if' parts of this statement follow, respectively, by supposing that first Λ and then Π in the statement of Condition D have the parameter $v(s, x)$.

3.8 THE INDEX THEOREM FOR SUPERPROCESSES

Theorem 3.15. The classes of index policies, forwards induction policies, and optimal policies are identical for a simple family $\mathscr{F}$ of alternative superprocesses $S_i (i = 1, 2, \ldots, n)$ satisfying Condition D.

Proof. The identity of the first two classes of policy is given by Lemma 3.12.

Now let g_i be a deterministic stationary Markov policy for $S_i (i = 1, 2, \ldots, n)$ such that

$$v(S_i, x_i, g_i(x_i)) = v(S_i, x_i) \, (x_i \in \Theta_i).$$

Thus $v(S_i, x_i)$ is the index for the simple family $\mathscr{A}$ of alternative bandit processes $(S_i, g_i)(i = 1, 2, \ldots, n)$. The index policy g for $\mathscr{A}$, which (for definiteness) selects the bandit process with the lowest suffix i in the event of a tie for the largest index value, also defines an index policy for $\mathscr{F}$, which we also denote by g. Let P be the class of those policies for $\mathscr{F}$ which coincide with g from the first decision time after time zero onwards. The proof hinges on the following lemma.

Lemma 3.16. A policy in P is optimal in that class iff it is an index policy.

Proof. Suppose $h \in P$, and that at time zero h applies control $u (\neq 0)$ to S_k, and thus control 0 to every other superprocess. Let the initial state of S_k be y_k, and replace y_k by two states y_k^1 and y_k^2 with properties identical to y_k, except that return to y_k^1 is impossible, so that the superprocess S_k^* with the new state space and initial state y_k^1 is therefore in state y_k^2 whenever S_k returns to y_k. Let ug_k denote the policy for S_k^* which initially applies control u and thereafter applies policy g_k (regarding y_k^2 as equivalent to y_k). Thus policy h for $\mathscr{F}$ defines a policy (also denoted by h) for the SFABP $\mathscr{B}$ formed from $\mathscr{A}$ by replacing (S_k, g_k) by (S_k^*, ug_k).

For both $\mathscr{A}$ and $\mathscr{B}$ optimal policies are index policies (Theorem 3.6), and an index policy for $\mathscr{A}$ is equivalent to an index policy for $\mathscr{F}$. To show then that h is optimal in P iff it is an index policy it suffices to show that the payoff $R(\mathscr{A})$ from

$\mathcal{A}$ under an index policy is greater than or equal to the payoff $R(\mathcal{B})$ from $\mathcal{B}$ under an index policy, with equality iff $v(S_k^*, ug_k) = v(\mathcal{F})$.

There are two possible ways in which this condition may fail to be fulfilled: (i) there may be a control $v(\neq u)$ such that $v(S_k) = v((S_k^*, vg_k)) = v(\mathcal{F})$; (ii) there may be no such control, so that $v(S_k) < v(\mathcal{F})$. In case (ii) policy h does not define an index policy for $\mathcal{B}$, and consequently any index policy δ for $\mathcal{B}$ defines a policy for $\mathcal{F}$ which yields a higher payoff. There may be a zero probability that policy δ ever selects S_k^* for continuation, and consequently policy δ is indistinguishable from the obvious analogous policy for $\mathcal{A}$, thus yielding a payoff no greater than an index policy for $\mathcal{A}$, which is therefore better than policy h, as required.

Alternatively, there is a non-zero probability that under policy δ control v is first applied to S_k^* at some finite time T. Again policy δ is indistinguishable from the analogous policy for $\mathcal{A}$ up to time T. To show that for this variant of case (ii) policy h yields a lower payoff than an index policy for $\mathcal{A}$, it is therefore sufficient to show that the reward yielded by δ from time T onwards, conditional on the state of $\mathcal{F}$ at T, is less than the expected reward from this point under an index policy for $\mathcal{A}$. This will follow immediately if we can show that $R(\mathcal{A}) > R(\mathcal{B})$ in case (i), to which we now turn.

To make the necessary comparison between the SFABPs $\mathcal{A}$ and $\mathcal{B}$ in case (i) it is useful to have a notation which clearly exhibits their common component. To this end write $\mathcal{A} = \{A, \mathcal{C}\}$ and $\mathcal{B} = \{B, \mathcal{C}\}$, where $A = (S_k, g_k)$, $B = (S_k^*, ug_k)$, and $\mathcal{C}$ denotes the superprocess formed by the alternative bandit processes $(S_j, g_j)(j = 1, 2, \ldots, n; j \neq k)$. Regarding $\mathcal{C}$ as a SFABP: let $R(t)$ be the expected total reward from $\mathcal{C}$ up to time t, and $x(t)$ the state at time t, under a forwards induction policy g; let $t_i(i = 1, 2, \ldots)$ denote the end of the ith stage of such a policy, with $t_0 = 0$; and write $\mu_i = v(\mathcal{C}, x(t_i))$ $(i = 0, 1, \ldots)$. Let $R_A(t)$ and $R_B(t)$ be the payoffs up to time t from A and B, respectively, when control 1 is applied at all times, so that t is also the process time. Let $\alpha(t)$ and $\beta(t)$ denote the states of A and B, respectively, at process time t, and let

$$\tau_A(\mu) = \inf\{t : v(A, \alpha(t)) < \mu, t \text{ is a decision time for } A\},$$

$$\tau_B(\mu) = \inf\{t : v(B, \beta(t)) < \mu, t \text{ is a decision time for } B\}.$$

Consider now the index policy g_A for $\mathcal{A}$ which selects bandit process A for continuation whenever $v(A) \geqslant v((S_j, g_j))$ for all $j \neq k$, and chooses between the bandit processes in $\mathcal{C}$ according to g (which is an index policy by Theorem 3.6). The total return from bandit process A under policy g_A is

$$R_A(\tau_A(\mu_0)) + \sum_{i=1}^{\infty} e^{-\gamma t_i}[R_A(\tau_A(\mu_i)) - R_A(\tau_A(\mu_{i-1}))], \qquad (3.25)$$

the factors $e^{-\gamma t_i}$ arising from the periods during which one of the bandit processes in $\mathcal{C}$ is continued, thereby delaying the sequence of rewards from A, and causing them to be further discounted. Similarly, the total return from $\mathscr{C}$ is

$$e^{-\gamma \tau_A(\mu_0)}R(t_1)+\sum_{i=1}^{\infty}e^{-\gamma \tau_A(\mu_i)}[R(t_{i+1})-R(t_i)]. \tag{3.26}$$

Note that either of the increasing sequences t_i and $\tau_A(\mu_i)$ may attain the value infinity for a finite value of i. Writing $I_1=\min\{i:t_i=\infty\}, I_2=\min\{i:\tau_A(\mu_i)=\infty\}$, we make the obvious conventions:

$$e^{-\gamma t_i}=0 \quad \text{and} \quad R(t_{i+1})=R(t_i) \qquad (i\geqslant I_1),$$

$$e^{-\gamma t_{\tau_A}(\mu_i)}=0 \quad \text{and} \quad R(\tau_A(\mu_{i+1}))=R(\tau_A(\mu_i)) \qquad (i\geqslant I_2).$$

With these conventions the expressions (3.25) and (3.26) are valid in all cases.

The total return from the SFABP $\mathscr{A}$ is obtained by adding the expressions (3.25) and (3.26), giving

$$\sum_{i=0}^{\infty}\{R_A(\tau_A(\mu_i))(e^{-\gamma t_i}-e^{-\gamma t_{i+1}})+e^{-\gamma \tau_A(\mu_i)}[R(t_{i+1})-R(t_i)]\}. \tag{3.27}$$

The total return from $\mathscr{B}$ is given by the expression derived from (3.27) by replacing A by B throughout, which we refer to as the *B-analogue* of (3.27).

The sequence of rewards from bandit process A, and the process times at which they occur, are independent of the similar quantities for the bandit process formed by applying policy g to $\mathscr{C}$. Moreover μ_i is a function of $x(t_i)$, so that applying the expectation operator conditional on t_i and $x(t_i)$ to the ith term in the summation in (3.27), and denoting this operator by $\mathbf{E}_i$, gives

$$\mathbf{E}_iR_A(\tau_A(\mu_i))\mathbf{E}_i(e^{-\gamma t_i}-e^{-\gamma t_{i+1}})+\mathbf{E}_ie^{-\gamma \tau_A(\mu_i)}\mathbf{E}_i[R(t_{i+1})-R(t_i)]. \tag{3.28}$$

From the definitions of t_i, t_{i+1} and μ_i it also follows that

$$\mathbf{E}_i[R(t_{i+1})-R(t_i)]=e^{-\gamma t_i}\mu_i\mathbf{E}_i\int_0^{t_{i+1}-t_i}e^{-\gamma s}\,ds$$

$$=\gamma^{-1}\mu_i\mathbf{E}_i(e^{-\gamma t_i}-e^{-\gamma t_{i+1}}),$$

so that (3.28) may be rewritten in the form

$$\mathbf{E}_i[R_A(\tau_A(\mu_i))+e^{-\gamma \tau_A(\mu_i)}\gamma^{-1}\mu_i]\mathbf{E}_i(e^{-\gamma t_i}-e^{-\gamma t_{i+1}}). \tag{3.29}$$

Applying $\mathbf{E}_i$ to the corresponding term in the expression for the total return from $\mathscr{B}$ gives us the B-analogue of (3.29).

Now, by definition of A and g_k, and using Condition D, it follows that, for all $\lambda < v(M_k)$, $\mathbf{E}[R_A(\tau_A(\lambda))+e^{-\gamma \tau_A(\lambda)}\gamma^{-1}\lambda]$ is the maximum expected total reward from the SFAS $\{S_k,\Lambda\}$. Moreover, using in addition the definition of B and Note 3.14,

$$\mathbf{E}[R_A(\tau_A(\lambda))+e^{-\gamma \tau_A(\lambda)}\gamma^{-1}\lambda]\geqslant \mathbf{E}[R_B(\tau_B(\lambda))+e^{-\gamma \tau_B(\lambda)}\gamma^{-1}\lambda] \tag{3.30}$$

for $\lambda < v(S_k)$, with equality iff u is optimal for $\{S_k, \Lambda\}$ in state y_i, i.e. iff $v(\mathcal{F}) = v(S_k) = v((S_k^*, ug_k))$. Since we are assuming that $v((S_k^*, ug_k)) < v(\mathcal{F})$ it follows that (3.30) must hold with strict inequality.

We therefore have

$$\mathbf{E}_i[R_A(\tau_A(\mu_i)) + e^{-\gamma\tau_A(\mu_i)}\gamma^{-1}\mu_i] > \mathbf{E}_i[R_B(\tau_B(\mu_i)) + e^{-\gamma\tau_B(\mu_i)}\gamma^{-1}\mu_i] \quad (3.31)$$

for every i for which μ_i is defined. Now taking the expectation of (3.27), subtracting the expectation of its B-analogue, and using (3.28), (3.29), their B-analogues, and (3.31), it follows that $R(\mathcal{A}) > R(\mathcal{B})$.

Similarly, if $v(\mathcal{F}) = v(S_k) = v((S_k^*, ug_k))$ then $R(\mathcal{A}) = R(\mathcal{B})$. $\qquad\square$

The theorem now quickly follows from Theorem 2.2, parts (ii) and (iii). Since g is optimal in P the function $R_g(\mathcal{F}, \cdot)$ satisfies the functional equation for the decision process $\mathcal{F}$, and it follows from Theorem 2.2(ii) that g is optimal. The rest of the theorem follows by putting together Lemma 3.16 and Theorem 2.2(iii). $\qquad\square$

3.9 STOPPABLE BANDIT PROCESSES

This term was coined by Glazebrook (1979,c) to describe the process obtained by adding to the control set of a bandit process a *stop* control, control 2, which causes the process to behave as a standard bandit process with a parameter $\mu(x)$ depending on the state x. Strictly speaking this is a superprocess rather than a bandit process. If control 2 is selected the state of the stoppable bandit process, and the state of any SFAS to which it belongs, do not change, so we can, as usual with a standard bandit process, ignore the possibility of switching from the use of control 2. Thus to use control 2 has the same effect as terminating the decision process with a final reward of $\gamma^{-1}\mu(x)$, and this is why it is called the stop control.

A SFAS made up of stoppable bandit processes might be a suitable model for an industrial research project in which there are several possible approaches, each represented by one of the stoppable bandit processes. Using the stop control corresponds to stopping research and exploiting commercially the know-how gained by one of the approaches. The buyer's problem discussed by Bergman (1981) provides another example of alternative stoppable bandit processes. This problem is defined in §9.5.

A stoppable bandit process will be said to have an *improving* stopping option if $\mu(x(t))$ is almost surely non-decreasing in process time t. The process time remains the total time for which the continuation control 1 has been applied.

Lemma 3.17. Condition D holds for a stoppable bandit process S with an improving stopping option.

Proof. Suppose that S belongs to a family $\mathcal{F}$ of two alternative superprocesses, the other of which is a standard bandit process Λ with parameter λ. We must show that for those values of λ for which an optimal policy for $\mathscr{F}$ selects S when it is in state x, the control which such a policy applies to S is independent of λ. This can be done by showing that the control applied is optimal iff it is optimal for the decision process D formed by removing the freeze control 0 from the control set for S.

If $\mu(x) \geqslant \lambda$ then, since $\mu(x(t))$ is increasing in t, an optimal policy for $\mathscr{F}$, starting with S in state x, need never select Λ. Thus $\mathscr{F}$ reduces to S and the required relationship between optimal policies holds trivially. If $\mu(x) < \lambda$ it cannot be optimal for $\mathscr{F}$ to stop S as it is better to select Λ. It therefore remains only to show that if $\mu(x) < \lambda$, and for $\mathscr{F}$ it is optimal to apply control 1 to S, then it is also optimal to do so for D.

Let g be an optimal stationary policy for $\mathscr{F}$ in state x which starts with control 1, and τ the time at which g first continues Λ (g may never select Λ, in which case $\tau = \infty$). Since g is optimal it must be at least as good as the policy which always continues Λ. It must therefore achieve at least as high an expected reward up to time τ. In symbols

$$R_\tau((\mathscr{F}, g)) \geqslant \lambda \gamma^{-1} \mathbf{E}(1 - e^{-\gamma\tau}).$$

Thus

$$R_\tau((D, g)) = R_\tau((\mathscr{F}, g)) > \mu(x) \gamma^{-1} \mathbf{E}(1 - e^{-\gamma\tau}),$$

i.e. we can achieve a higher expected reward up to time τ by applying g to $\mathscr{F}$ (or equivalently to D) than by using control 2. Since $\mu(x(t))$ increases with t, the policy which coincides with g up to time τ and then switches permanently to control 2 must therefore achieve a higher payoff than one which uses control 2 at the outset. Thus control 1 is also optimal for D in state x if $\mu(x) > \lambda$.

From Theorem 3.15 and Lemma 3.17 it follows that index policies are optimal for simple families of stoppable bandit processes, for each of which $\mu(x(t))$ increases with t.

3.10 THE INDEX THEOREM FOR A FABP WITH PRECEDENCE CONSTRAINTS

From the discussion of §3.7 and Theorem 3.15 it follows that Condition D is both necessary and sufficient for a superprocess to be compatible with the optimality of a general index policy. However, the condition is not always easy to check, and it is useful to have alternative, and relatively easily checked, conditions which are sufficient for it to hold. This section and the next one explore conditions of this kind for families of generalized jobs (i.e. bandit processes with a completion state, see §2.5) subject to precedence and arrival constraints. A potentially important side-benefit is a method of proof of an index theorem for a FABP which closely parallels the proof of Theorem 3.6, and hence provides an extension to ε-FI* policies for a FABP of the approximation given by Corollary 3.8.

Consider a family $\mathscr{F}$ of generalized jobs $B_j (j = 1, 2, \ldots, n)$ subject just to precedence constraints. A little preliminary discussion is required before the relevant conditions can be stated.

Let g be a policy for $\mathscr{F}$, and τ a stopping time for the bandit process $\mathscr{F}_g$. Thus τ is the first time at which $\mathscr{F}$ reaches some stopping subset Θ_0 of its state space when g is the policy used. Let $\tau_i (i = 1, 2, \ldots, n; \tau_1 + \tau_2 + \cdots + \tau_n = \tau)$ be the process time of bandit processes B_i when $\mathscr{F}$ first reaches Θ_0 under g. If g is the policy used, τ_i is g-measurable in the sense that whether or not $\tau_i \geqslant t$ depends, with probability one and for all $t \in \mathbf{R}^+$, only on the history of $\mathscr{F}_g$ before the process time of B_i exceeds t. It is to be noted that τ_i is well-defined whether or not g is the policy used. For a given set of realizations of the n alternative bandit processes the value taken by τ_i is well-defined whether or not g is the policy used. It is simply the process time of B_i up to the first time when $\mathscr{F}$ would have reached Θ_0 under g. However, if some other policy h is used τ_i may or may not be measurable in the above sense. We express all this by saying that τ_i is by definition g-measurable but may or may not be h-measurable. Speaking rather loosely, τ_i is not h-measurable if under policy h there is a possibility that for some t the process time of B_i may exceed t before we know whether or not $\tau_i \geqslant t$. The truncated FABP defined by truncating $\mathscr{F}_g$ at time τ is said to be h-measurable if τ_i is h-measurable for all i.

Now let γ be a FI* policy for $\mathscr{F}$, and the stopping time $\sigma (> 0)$ be the end of the first stage of γ. Let $\mathscr{K}$ be the truncated FABP and $\mathscr{G}$ the residual FABP which are defined by truncating $\mathscr{F}_\gamma$ at time σ. The roles of these objects in the proof of the next theorem parallel those of bandit process B_k and the residual SFABP $\mathscr{G}$ in the proof of Theorem 3.6. It is to be noted, in the light of our discussion of g- and h-measurable stopping times, that σ and realizations of $\mathscr{K}$ and of $\mathscr{G}$ are well-defined whether or not policy γ is followed up to time σ. Any policy g applied to $\mathscr{F}$ and realization of $\mathscr{K}$ fix the residual FABP $\mathscr{G}$, and define a policy e for $\mathscr{G}$ and a promotion rule for the bandit process $\mathscr{G}_e$ with respect to σ.

Extending the notation, $R_{hg\tau}(\mathscr{F})$ is the total over $j = 1, 2, \ldots, n$ of the expected rewards from the occasions before process time τ_j at which control 1 is applied to bandit process B_j, where τ and τ_j are defined with respect to the policy g, as in the previous paragraph, and h is the policy followed. We have

$$R_{hg\tau}(\mathscr{F}) = \mathbf{E} \sum_{i=0}^{\infty} a^{t_i} I_{hg\tau}(t_i) r_h(t_i),$$

where $t_i \ (i = 0, 1, 2, \ldots)$ is the ith decision time under policy h, and $I_{hg\tau}(t_i) = 1$ or 0 depending on whether control 1 is applied at time t_i to a bandit process B_j whose process time is less than, or greater than or equal to, τ_j. Also

$$W_{hg\tau}(\mathscr{F}) = \mathbf{E} \sum_{i=0}^{\infty} I_{hg\tau}(t_i) \int_{t_i}^{t_{i+1}} a^t \, dt, \text{ and } v_{hg\tau}(\mathscr{F}) = R_{hg\tau}(\mathscr{F})/W_{hg\tau}(\mathscr{F}).$$

Thus $v_{gg\tau}(\mathscr{F}) = v_{g\tau}(\mathscr{F})$, so that

$$\sup_h v_{h\gamma\sigma}(\mathscr{F}) \geqslant v_{\gamma\sigma}(\mathscr{F}) = v(\mathscr{F}).$$

Condition E. The FABP $\mathscr{F}$ is said to satisfy Condition E if, for any initial state and FI* policy γ,

$$\sup_h v_{h\gamma\sigma}(\mathscr{F}) = v(\mathscr{F}).$$

This is the first of our sufficient conditions for Condition D to hold. This is a corollary of the following theorem. Note that the FABP may be divided into subfamilies $\mathscr{F}_j$ $(j = 1, 2, \ldots, m)$ $(1 \leqslant m \leqslant n)$ with the property that there are no precedence constraints between bandit processes belonging to different subfamilies. Thus $\mathscr{F}$ is also a simple family of alternative superprocesses, the superprocesses being the $\mathscr{F}_j$s.

Theorem 3.18. (The Index Theorem for a FABP with Precedence Constraints.) The classes of index policies, forwards induction policies, and optimal policies are identical for a family $\mathscr{F}$ of alternative generalized jobs with no arrivals (regarded as a SFAS), and which satisfies Condition E.

Proof. The identity of the first two classes of policy follows from Lemma 3.12.

The rest of the proof is on similar lines to the proof of Theorem 3.6. The truncated family of alternative bandit processes $\mathscr{K}$ takes the place of the truncation at process time τ_k of the bandit process B_k. The symbol $\mathscr{G}$ now denotes a residual family of alternative bandit processes which is not necessarily simple. This plays a role similar to that of the simple family $\mathscr{G}$ of alternative bandit processes in the proof of Theorem 3.6.

Consider, then, an arbitrary policy g for $\mathscr{F}$. Together with a realization of $\mathscr{K}$ this fixes the random FABP $\mathscr{G}$, and defines a policy h for $\mathscr{G}$ and a promotion rule p for the bandit process $\mathscr{G}_h$ with respect to σ. Thus, separating the contributions to the expected total reward yielded by $\mathscr{K}$ and $\mathscr{G}$ respectively,

$$R_g(\mathscr{F}) = \mathbf{E} \sum_{i=0}^{\infty} a^{t_i} I_{g\gamma\sigma}(t_i) r_g(t_i) + \mathbf{E} \sum_{i=0}^{\infty} a^{t_i} (1 - I_{g\gamma\sigma}(t_i)) r_g(t_i)$$

$$= R_{g\gamma\sigma}(\mathscr{F}) + \mathbf{E} R_{hp}(\mathscr{G}).$$

Consider also the policy $\gamma\sigma h$ for $\mathscr{F}$, which coincides with the FI* policy γ up to time σ, and then applies policy h to the residual FABP $\mathscr{G}$. We must first check that this defines a feasible policy. Since γ is feasible no precedence constraints within $\mathscr{K}$ are violated, or precedence constraints between $\mathscr{K}$ and $\mathscr{G}$. Under policy $\gamma\sigma h$ the process times of each bandit process in $\mathscr{F}$ when control 1 is first applied to any bandit process in $\mathscr{G}$ are no less, for any realization of $\mathscr{K}$, than they are under policy g. Since g is feasible it follows that no precedence constraints within $\mathscr{G}$ are violated. Thus $\gamma\sigma h$ violates no precedence constraints, and is therefore feasible. We have

$$R_{\gamma\sigma h}(\mathscr{F}) = R_{\gamma\sigma}(\mathscr{F}) + \mathbf{E}R_{h0}(\mathscr{G}).$$

Here the contributions to the expected total reward from $\mathscr{H}$ and $\mathscr{G}$ have again been separated, and 0 denotes the null promotion rule with respect to σ. We shall show that

$$R_{\gamma\sigma h}(\mathscr{F}) \geqslant R_g(\mathscr{F}). \tag{3.32}$$

It follows from Condition E that

$$R_{g\gamma\sigma}(\mathscr{F}) = W_{g\gamma\sigma}(\mathscr{F})v_{g\gamma\sigma}(\mathscr{F}) \leqslant W_{g\gamma\sigma}(\mathscr{F})v(\mathscr{F}),$$

and by definition

$$R_{\gamma\sigma}(\mathscr{F}) = W_{\gamma\sigma}(\mathscr{F})v_{\gamma\sigma}(\mathscr{F}) = W_{\gamma\sigma}(\mathscr{F})v(\mathscr{F}).$$

The rest of the proof of (3.32) is very similar to that of (3.18) with $R_{g\gamma\sigma}(\mathscr{F})$ in place of $R_f(B_k)$, $R_{\gamma\sigma}(\mathscr{F})$ in place of $R_{\tau_k}(B_k)$, and $\gamma\sigma h$ in place of $*g$, and using part (i) of Lemma 3.12 instead of Lemma 3.7. The inequality $v(\mathscr{G}) \leqslant v(\mathscr{F})$ now becomes strict.

The remainder of the proof is also virtually identical to the final two paragraphs of the proof of Theorem 3.6. □

Corollary 3.19. If h is an ε-FI* policy then

$$R_h(\mathscr{F}) \geqslant R(\mathscr{F}) - \varepsilon\beta^{-1}.$$

The proof closely follows that for Corollary 3.8. It is based on an ε-modification of the inequality (3.32) rather than (3.18).

Corollary 3.20. Given a FABP $\mathscr{F}$ with no arrivals and which satisfies Condition D, the superprocess S formed by adding the freeze control to the control set satisfies Condition D.

Proof. It is sufficient to show that a policy for the FABP $\{S, \Lambda\}$, formed by adding a λ standard bandit process Λ to S with no additional precedence constraint, is optimal iff it is an index policy. This will follow from the theorem if we can show that $\{S, \Lambda\}$ itself satisfies Condition D for all λ.

There are two cases to consider. Either (i) $v(\{S, \Lambda\}) = \lambda > v(\mathscr{F})$, or (ii) $v(\{S, \Lambda\}) = v(\mathscr{F}) \geqslant \lambda$. In case (i) it follows from Lemma 3.12 that only Λ is selected during the first stage of a modified forwards induction policy for $\{S, \Lambda\}$. Condition D then follows since the reward rate achieved by a standard bandit process is the same over any time period. In case (ii) $\{S, \Lambda\}$ inherits Condition D from $\mathscr{F}$.

A second sufficient condition for Condition D, in fact a sufficient condition for Condition E, follows from the observation that if $v_{h\gamma\sigma}(\mathscr{F}) > v_{\gamma\sigma}(\mathscr{F})(= v(\mathscr{F}))$ for some feasible policy h, it must be the case that the truncated FABP $\mathscr{K}$ is not h-measurable. This may be shown by supposing the contrary and showing that this leads to a contradiction.

Thus suppose that, for some h, $\mathscr{K}$ is h-measurable. There must then be some policy g for $\mathscr{F}$, and freezing rule f, truncated at process time σ, for $\mathscr{F}_g$, for which the following joint distributions are equal: firstly, the joint distribution of the decision times at which control 1 is applied under f to $\mathscr{F}_g$, and of the bandit processes continued and the rewards obtained at those times; secondly, the joint distribution of the decision times t_i at which $I_{h\gamma\sigma}(t_i) = 1$ when policy h is applied to $\mathscr{F}$, and of the bandit processes continued and the rewards obtained at those times. Since γ is feasible g violates no precedence constraints between $\mathscr{K}$ and $\mathscr{G}$, and since h is feasible g violates no precedence constraints within $\mathscr{K}$. Since f is truncated at process time σ, g may be defined arbitrarily from time σ onwards, and consequently may be chosen to violate no precedence constraints within $\mathscr{G}$, and hence to be feasible. We have $R_{h\gamma\sigma}(\mathscr{F}) = R_f(\mathscr{F}_g)$, $W_{h\gamma\sigma}(\mathscr{F}) = W_f(\mathscr{F}_g)$, and $v_{h\gamma\sigma}(\mathscr{F}) = v_f(\mathscr{F}_g)$. By part (i) of Theorem 3.4 there must be a stopping time τ for $\mathscr{F}_g$ such that $v_\tau(\mathscr{F}_g) \geqslant v_f(\mathscr{F}_g)$, and if $v_{h\gamma\sigma}(\mathscr{F}) > v(\mathscr{F})$ this would imply that $v_\tau(\mathscr{F}_g) > v(\mathscr{F})$, contrary to the definition of $v(\mathscr{F})$.

Thus a sufficient condition for Condition E to hold is that, for any state which $\mathscr{F}$ may reach, the truncated FABP $\mathscr{K}$ defined by a FI* policy starting from that state be h-measurable for all h. If this is true then $\mathscr{F}$ is said to satisfy *Condition F*.

Corollary 3.21. Condition F, and therefore Theorem 3.18, holds if all the bandit processes in $\mathscr{F}$ are jobs with a single reward on completion, and preemption is not allowed.

Proof. For this case the state of $\mathscr{F}$ at each decision time is defined by the set of completed jobs. It follows that any deterministic stationary Markov policy g for $\mathscr{F}$ is defined by listing the jobs in the non-random sequence in which they are to be carried out. Moreover, any stopping time for the bandit process $\mathscr{F}_g$ must coincide with the end of a non-random job in this sequence, since a stopping time for a bandit process is defined as the time when it first reaches some subset of the state-space. Consequently the first stage of a FI* policy for $\mathscr{F}$ may be defined by listing in order those jobs which are completed during the first stage of such a policy. This non-random subset of the jobs in $\mathscr{F}$ therefore constitutes the truncated FABP $\mathscr{K}$. This is clearly h-measurable for any policy h, and the corollary is proved.   □

Partial results along the lines of Corollary 3.21 have been obtained without the benefit of Theorem 3.18 by Sidney (1975), Kadane and Simon (1977), and Glazebrook and Gittins (1981). The problems of determining an optimal schedule when preemption is not allowed, and of estimating the penalty in confining attention to non-preemptive schedules when preemption is allowed, have been considered by Glazebrook in a series of papers which are listed in §9.3. Note that when there are precedence constraints the calculation of indices is typically fairly complicated even with no preemptions (e.g. see Exercise 3.4).

If we can show that preemption, although permitted, does not occur under an optimal policy, Theorem 3.18 will still hold. This is true, for example, if the jobs are ordinary jobs, and for each job the expression

$$[1 - F(t)]^{-1} e^{\gamma t} \int_t^\infty f(s) e^{-\gamma s} \, ds$$

is increasing in t (see Theorem 4.8). Alternatively we can look for conditions on the set of precedence constraints which ensure that Condition E holds, despite the possibility of preemption, as we shall see in the following section.

Note first that, whereas Condition D is both necessary and sufficient for compatibility with a general index policy, Condition E is only a sufficient condition. Our earlier discussion of Example 3.13 may be amplified to provide an example of a sub-family of alternative bandit processes which satisfies Condition D, and is therefore compatible with a general index policy, although it does not satisfy Condition E.

Consider the sub-family $\mathscr{F}$ formed by adding to the superprocess S formed by the three jobs of Example 3.13 a standard bandit process Φ with parameter ϕ, where

$$\frac{8 + M}{58} < \phi < \frac{2 + M}{46}$$

(which means, of course, that $M > 21$), and the SFAS $\{\mathscr{F}, \Lambda\}$, where Λ is a λ standard bandit process. The greatest equivalent constant reward rate achievable from S by starting with job 2 is $v(S, x, 2) = (8 + M)/58$. Since $v(S, x, 2) < \phi$ the reward rate obtained by selecting Φ instead is greater, and it follows that for no value of λ does an optimal policy for $\{\mathscr{F}, \Lambda\}$ start with job 2. Similarly, since $v(S, x, 1) > \phi$, if λ is such that an optimal policy for $\{\mathscr{F}, \Lambda\}$ starts by selecting any of the bandit processes in $\mathscr{F}$ (i.e. if $\lambda \leqslant v(S, x, 1)$) it must select job 1. Thus $\mathscr{F}$ satisfies Condition D so far as the initial state x is concerned, and it is easy to check that the same is true for each of the other six possible states.

On the other hand the FABP D (i.e. S without the stop control) does not satisfy Condition E, and this property is transmitted to $\{\mathscr{F}, \Lambda\}$. That D does not satisfy Condition E follows immediately from Theorem 3.18 and the fact that the index policy is not optimal for D. It may also be checked directly.

3.11 PRECEDENCE CONSTRAINTS FORMING AN OUT-TREE

If preemption is permitted Theorem 3.18 does not always hold when the jobs are ordinary jobs, as is shown by Example 3.13. An essential feature is that there are initially two ways, so to speak, of making progress towards the large reward which accrues on completion of job 3: by working on job 1; or on job 2. The probability of completing job 3, and the time when this occurs, do not depend on the initial choice between jobs 1 and 2. Thus in the determination of an optimal policy this choice depends only on the relative profitability of time spent on job 1 or on job 2. On the other hand, in constructing the first stage of a FI* policy it is also useful to know as soon as possible whether either of jobs 1 or 2 is going to remain uncompleted, and thus stop us from completing job 3. The reason is that as soon as this is known we will declare the first stage to have terminated, and the sooner this happens the less our inability to complete job 3 will bring down the expected reward per unit of discounted time during the first stage. Only one time unit need be spent on job 1 to find out whether it will be completed, as compared with two time units for job 2. Thus there is a reason for choosing job 1, in constructing an FI* policy, which is irrelevant when constructing an optimal policy. This difference leads to the counter-example.

The two ways of progressing towards completion of job 3 correspond to two arcs leading to node 3 in the directed graph representing the jobs and precedence constraints of the FABP *D*. It is natural to consider what happens if this feature is avoided by restricting our attention to directed graphs with no circuits and for which the in-degree of each node is at most one. Figure 3 shows a directed graph of this type, which we shall refer to as an *out-tree* (the term *arborescent* is perhaps more standard, though less descriptive). A FABP with precedence constraints which may be represented by an out-tree forms a simple family of sub-families of alternative bandit processes, the number of sub-families being equal to the number of jobs available for service at time zero. In the case of the graph shown in Figure 3 there are three sub-families, as indicated. Clearly each of these sub-families of jobs is a superprocess.

Figure 3. Precedence constraints forming an out-tree.

As we shall see, for precedence constraints of this form Condition E is satisfied, and consequently Theorem 3.18 holds. A little further terminology prepares the way.

Provided that the FABP $\mathscr{F}$ has been divided into as many sub-families as possible there is one job in each sub-family which must be completed before any other job in the sub-family may be started. This job will be termed the *initial* job in the sub-family, and a sub-family with the property just described will be termed a *minimal* sub-family (or a minimal family if it coincides with $\mathscr{F}$). If the minimal sub-families of $\mathscr{F}$ are $\{\mathscr{F}_j : j = 1, 2, \ldots, m\}$ the initial job in $\mathscr{F}_j$ is denoted by J_j.

Once the initial job in a sub-family has been completed there will, unless this is the only job in the sub-family, be one or more other jobs in the sub-family which may be started; for example, two jobs in the case of the sub-family $\mathscr{F}_3$ whose precedence constraints are shown in Figure 3. The sub-family then sub-divides into further minimal sub-families. Each of the newly available jobs is the initial job for one of these subordinate sub-families. Indeed, if we extend the notion of a minimal sub-family to include not just those which are present initially, but also all those which may arise as jobs are completed, each of the n jobs in $\mathscr{F}$ is the initial job for a minimal sub-family. $\mathscr{F}_j$ will denote the minimal sub-family whose initial job is $J_j (j = 1, 2, \ldots, n)$. When $\mathscr{F}$ is a SFABP, $n = m$, but otherwise $n > m$.

If the initial job J_j of a sub-family $\mathscr{F}_j$ has not been completed, the state of $\mathscr{F}_j$ is given by the state x_j of J_j. Write $v^\dagger(J_j, x_j) = v(\mathscr{F}_j, x_j)$. Let the successive decision times for J_j occur at process times $t_0^j (= 0), t_1^j, t_2^j, \ldots$. Let τ_j be the process time at which J_j is completed. Let

$$\rho_j = \min\{\tau_j, \min_i [t_i^j : v^\dagger(J_j, x_j(t_i^j)) < v(\mathscr{F})]\}.$$

Let M denote that subset of the first n integers defined by the equation

$$M = \{j : J_j \text{ has not been completed and is available for service}\}.$$

The state of the overall FABP $\mathscr{F}$ at any particular stage is given by M, together with the set of job states $\{x_j : j \in M\}$. $M(t)$ refers to the state at time t.

A policy for $\mathscr{F}$ will be described as a *minimal sub-family index policy* if at any decision time the bandit process (or job) J_k selected for continuation is such that

$$k \in M \quad \text{and} \quad v^\dagger(J_k, x_k) = \max_{j \in M} v^\dagger(J_j, x_j).$$

Our next result may now be stated very simply.

Theorem 3.22. The classes of minimal sub-family index policies, forwards induction policies, and optimal policies are identical for a family $\mathscr{F}$ of alternative jobs with no arrivals and precedence constraints in the form of an out-tree.

Two proofs of this theorem will be given. The first proof is via Theorem 3.18, and gives a second example of the use of Condition F. It also generalizes easily to include the possibility of arrivals, as shown in the following section. The second proof, due to Glazebrook (1976a), is more direct, using only Theorem 3.6.

Proof 1. To establish Condition F Lemma 3.23 will be required, which is a generalized version of Lemma 3.5. Let g be an arbitrary policy for $\mathscr{F}$, e an arbitrary freezing rule for the bandit process $\mathscr{F}_g$, and f^j the freezing rule for J_j $(j = 1, 2, \ldots, n)$ defined by g and e.

Lemma 3.23. If $\mathbf{P}\{\exists j \text{ and } i \text{ such that } \rho_j \leqslant t_i^j < \tau_j \text{ and } f_i^j < \infty\} > 0$, then $v_{ge}(\mathscr{F}) < v(\mathscr{F})$.

The proof is a straightforward extension of the proof of Lemma 3.5. The equivalent constant rate of return from any truncation of $\mathscr{F}$ obtained by truncating every J_j before the corresponding process time ρ_j is at most $v(\mathscr{F})$. The expected return per unit time from $\mathscr{F}_j$ after process time ρ_j for J_j is less than $v(\mathscr{F})$ for every j. The lemma follows immediately.   □

Now let σ mark the end of the first stage of an FI* policy for $\mathscr{F}$, and let $\mathscr{G}$ be the associated residual FABP. Let σ_j be the process time and $x_j(\sigma_j)$ the state of J_j at time σ, for $j \in M(\sigma)$. By definition $v(\mathscr{G}) < v(\mathscr{F})$, and it follows from part (i) of Lemma 3.12 that

$$v^\dagger(J_j, x_j(\sigma_j)) = v(\mathscr{F}_j, x_j(\sigma_j)) < v(\mathscr{F}) \ (j \in M(\sigma)).$$

From this inequality, together with Lemma 3.23 and part (ii) of Lemma 3.12, it follows that $\sigma_j = \rho_j$ for $j \in M(\sigma)$. From Lemma 3.23 it also follows that $v^\dagger(J_k, x_k(t)) \geqslant v(\mathscr{F})$ for every decision time t up to the completion of J_k, if job J_k precedes any $J_j \in M(\sigma)$. These observations define $M(\sigma)$, and σ_j for $j \in M(\sigma)$, and thereby the truncated FABP $\mathscr{K}$, in a manner which is h-measurable for any policy h, and Condition E is therefore satisfied. Thus $\mathscr{F}$ satisfies the conditions for Theorem 3.18.

We next note that there is an ambiguity in the definition of a simple family $\mathscr{F}$ of sub-families of alternative bandit processes, arising from the possibility of dividing a sub-family into smaller sub-families. This leads to a corresponding ambiguity in the definition of an index policy, though it follows from Theorem 3.18 that the various possible definitions must be equivalent, since Condition E holds. The general rule is that the definition which maximizes the number of sub-families is to be preferred, since the simplifying concept of an index policy then reduces the problem to the greatest extent. Applying this principle in the present instance defines once again the minimal sub-family index policies. These are therefore index policies, and the theorem follows immediately from Theorem 3.18.   □

Note that there is no need for an index policy to be defined in terms of the same set of sub-families for every possible state of $\mathscr{F}$. Note also that Theorem 3.6 is the special case of Theorem 3.22 for which the set of precedence constraints is empty.

Applying Theorem 3.22 both to the overall family $\mathscr{F}$ and to the sub-families $\mathscr{F}_j$ $(j = 1, 2, \ldots, n)$ it follows that a minimal sub-family index policy, and therefore an optimal policy, for a FABP with precedence constraints only, and those in the form of an out-tree, may be constructed as follows.

(i) Identify all those minimal sub-families which are subject only to the restriction that none of the other members of such a minimal sub-family may be continued until the initial job of the sub-family has been completed. Call these *terminal* minimal sub-families. Every set of precedence constraints forming an out-tree must terminate with minimal sub-families of this kind. An example is obtained by removing the initial job from the sub-family $\mathscr{F}_1$ whose precedence constraints are shown in Figure 3.

(ii) Replace each terminal minimal sub-family by the bandit process obtained by associating with it the policy which is defined by applying an index policy as soon as the initial job has been completed. A bandit process defined in this way may be regarded as a generalized job which reaches its completion state as soon as each of its constituent jobs has been completed.

(iii) If there are still precedence constraints between the jobs in the new FABP obtained by replacing terminal minimal sub-families by jobs in this way, repeat stages (i) and (ii).

(iv) Iterate until $\mathscr{F}$ reduces to a SFABP.

(v) Form a bandit process B by associating this SFABP with an index policy.

The reward process generated by B is the same as that generated by $\mathscr{F}$ under a minimal sub-family index policy. This minimal sub-family index policy is given by the nested index policies defined by step (ii).

An alternative proof of Theorem 3.22 which parallels the construction just given is instructive. This proof uses only Theorem 3.6.

Proof 2. We first show that the nested index policies defined by the construction are optimal. The proof of this is by induction on the number of nodes d in the out-tree of precedence constraints.

Suppose first that $d = 1$. Thus the only precedence constraint is that there is a set of jobs, $J_1, J_2, \ldots, J_r$, say, which may not be started until one of the other jobs, job J say, has been completed. Once job J has been completed the system reduces to a SFABP and Theorem 3.6 applies. It is therefore optimal to assign priorities between $J_1, J_2, \ldots, J_r$ using the index rule. For definiteness, suppose the job with the lowest subscript is preferred in the event of a tie between index values. This assignment of priorities is stage (ii) of the construction and leaves us with a new, equivalent, set of jobs with no precedence constraints. This forms a SFABP and an index policy is therefore optimal. Thus nested index policies are optimal for $d = 1$.

Now suppose that nested index policies are optimal for $d \leqslant k$. Consider the case $d = k + 1$. As soon as any job which precedes another job is completed the system reduces to the case $d = k$. It may therefore be reduced to an equivalent SFAB by applying nested index policies from the time of the first such job completion onwards. For this SFABP an index policy is again optimal, so that nested index policies are optimal for $d = k + 1$, completing the induction.

That no policy other than a nested index policy is optimal follows from the fact that only index policies are optimal for a SFABP and from the nature of the proof just given. That the nested index policy is the usual index policy for $\mathscr{F}$, regarded as a SFAS, follows from the fact that it is an index policy, and from Theorem 3.11. That the classes of index policies and forwards induction policies are the same follows from Lemma 3.12. $\qquad\square$

This second proof of Theorem 3.22 has the advantage that it may be extended to the case of *stoppable jobs* subject to precedence constraints. A job is a bandit process with a completion state C, attainment of which means that the job behaves like a standard bandit process with the parameter $-M$. A stoppable bandit process is a bandit process with an additional control 2, application of which in state x causes behaviour like that of a $\mu(x)$ standard bandit process. A stoppable job is a bandit process which brings these features together. State C is replaced by a completion set m_C. Once the job reaches a state x in m_C it never leaves that state. Control 2 remains available for states in m_C. Thus the $-M$ standard bandit process available on completion of a job, and introduced only to allow each job arbitrarily large process times, becomes irrelevant.

Precedence constraints may be defined between stoppable jobs. Job J precedes job K if the freeze control must be applied to job K until job J reaches its completion set. We have the following theorem.

Theorem 3.24. The classes of minimal sub-family index policies, forwards induction policies, and optimal policies are identical for a family of alternative stoppable jobs with no arrivals and precedence constraints in the form of an out-tree, provided that each of the stoppable jobs has an improving stopping option.

The proof is virtually identical to the second proof of Theorem 3.22 except that Theorem 3.15 and Lemma 3.17 are used in place of Theorem 3.6. The construction of nested index policies works as before. This time we have to check that the stoppable jobs formed by stage (ii) of the construction have an improving stopping option. They do if we define the stopping option for a composite job to be the best of the stopping options available from the original jobs of which it is composed.

3.12 FABPS WITH ARRIVALS

We shall now extend Theorem 3.22 to provide an index theorem for a FABP with arrivals. The basic idea is that under certain circumstances bandit processes which arrive after time zero may be regarded as subject to precedence constraints in the form of an out-tree. To establish this equivalence a generalization of the notion of an out-tree of precedence constraints will be needed.

A bandit process (or job) B will be described as generating a *random out-tree of successors* or, equivalently, to have *bandit process-* (*or job-*)*linked arrivals* under the following circumstances. The state-space for B has a countable sub-set Θ_C representing completion. When B reaches a state $x \in \Theta_C$ a finite batch $\mathscr{S}(B, x)$ of further jobs joins the set of jobs available for selection. $\mathscr{S}(B, x)$ includes a standard bandit process with parameter $-M(M \gg 1)$. Each member of $\mathscr{S}(B, x)$ has similar properties (though not necessarily the same state-space, law of motion, and reward process) and produces a second set of successor jobs on reaching the completion set, and so on. Precedence relations between the jobs in $\mathcal{S}(B, x)$ (or between the jobs of any similar set) may occur, but not between those jobs and any other job, except the precedence relations defined by the random out-tree.

Attention will be restricted to out-trees whose growth is consistent with Condition A, apart from standard bandit processes; this ensures that the set of jobs available for selection at any decision time is finite, as required for Theorem 3.18. The number in the set may, however, tend to infinity with time.

Since x is in general random so is $\mathcal{S}(B, x)$, and the effect of the definition may be described by saying that the set of jobs immediately preceded by job B is random. Alternatively, $\mathcal{S}(B, x)$ may be regarded as the random set of jobs which arrives when job B is completed. Note that, although Θ_C has been described as a completion set, the possibility that the physical process represented by job B may continue after other jobs have arrived may be accounted for by including in $\mathcal{S}(B, x)$ a job which represents this continuation.

The proof of Theorem 3.18 goes through when, in addition to non-random precedence constraints, each job generates its own random out-tree of successor jobs. The truncated FABP $\mathcal{K}$ corresponding to the first stage of a modified forwards induction policy is once again well-defined irrespective of the policy followed. In general $\mathcal{K}$ now includes portions of the realizations of some of the successor jobs in the out-tree. Thus it does not remain unaltered when out-trees of successor jobs are added to a given FABP, and neither, therefore, is it clear whether Condition E continues to hold when out-trees are added. However, provided Condition E (or the stronger Condition F) holds, the proofs of Theorem 3.18 and Corollary 3.20 go through with no change except that bandit processes must now be labelled, $1, 2, \ldots$ rather than $1, 2, \ldots, n$.

Note that, on the other hand, the proof of Corollary 3.21 does not work if random out-trees of successor jobs are added. The reason is that the prohibition of preemption no longer reduces the specification of a deterministic stationary Markov policy to the listing of a non-random job sequence. At present it is an open question whether the generalized version of the corollary holds.

Except for the notational changes involved in removing the bound n on the total number of jobs, the proof of Lemma 3.23 and Proof 1 of Theorem 3.22 also go through unchanged provided the batches $\mathcal{S}(B, x)$ of jobs which arrive together are subject to no within-batch precedence constraints. Proof 2 of Theorem 3.22 continues to work if there is a finite upper bound on the number of occasions when arrivals occur, and presumably may be extended to remove this upper bound by allowing it to tend to infinity. A proof along these lines would have the advantage of extending to random out-trees of stoppable jobs, each job having an improving stopping option, by using Theorem 3.24. Proof 1 leads to the following result.

Theorem 3.25. The classes of minimal sub-family index policies, forwards induction policies, and optimal policies are identical for a FABP all of whose precedence constraints are defined by an out-tree and the presence of a bandit-process-linked arrival process.

As remarked earlier, despite initial appearances the notion of an out-tree of successor jobs, and hence application of Theorem 3.22, is not restricted to situations where the arrival of a job depends on the completion of the physical process represented by its predecessor. The overall arrival process is, however, restricted to being describable in terms of separate arrival processes associated with each of the jobs available for processing, and which are, so to speak, switched on only when the corresponding job is being processed. This rules out most arrival processes which are independent of the job being processed, and for which the inter-arrival times are independent, since in general these assumptions mean that when processing switches from job A to job B the distribution of the further time until the next arrival, assuming there are no more processing switches, depends on the time since the previous arrival, and is therefore not determined solely by the state of job B.

An important exception to this exclusion occurs when the arrivals of sucessive batches of jobs form either a Poisson process or, in the discrete-time case for which every decision time is integer-valued, a Bernoulli process. Suppose too that the compositions of the different batches are identically distributed and independent of previous history. Because of the well-known memoryless property of the Poisson and Bernoulli processes it follows that in either case the arrival pattern is the same, whether the Poisson or Bernoulli arrival process is regarded as taking place in process time for the various jobs or in real time, provided it does not, in the former case, depend on the particular job. We shall therefore refer simply to Poisson or Bernoulli arrivals, and the next result follows immediately.

Theorem 3.26. (*The Index Theorem for a FABP with Poisson or Bernoulli arrivals.*) For a FABP with Poisson arrivals (Bernoulli arrivals in the discrete-time case) and any precedence constraints in the form of a (possibly random) out-tree, the classes of minimal sub-family index policies, forwards induction policies and optimal policies are identical.

Recall that this result has been derived via the notion of a random out-tree of successor jobs, and the observation that arrivals during, rather than on completion of, the processing of a job may be handled by regarding the remaining portion of the job as one of the job's successors. It follows that any arrival during the processing of a given job J must also be regarded as a successor of job J, and the possibility of such arrivals affects the value of the index for the minimal sub-family of which job J is the initial job.

To determine index values when there are arrivals is computationally quite a task, and by no means always a tractable one. Section 5.7, for example, discusses in some detail a model for the allocation of resources between pharmaceutical research projects, and the results which may be derived from it. There is, however, one important case for which everything falls out very neatly, and this is the subject of the final section of this chapter.

3.13 MINIMUM EWFT FOR THE M/G/1 QUEUE

The setting is essentially that of §2.5 in the $\gamma = 0$ limit, and with the addition of Poisson arrivals. The process times at which switching between jobs is allowed are isolated, and not necessarily equally spaced. M/G/1 is standard queueing theory notation (Kendall, 1951) and refers to randoM (i.e. Poisson) arrivals, General independent service times, and 1 server. Suppose there are n classes $i(= 1, 2, \ldots, n)$ of jobs arriving at Poisson rates λ_i, and that class i jobs have independent service times all with the distribution F_i, each of them costing c_i per unit time between its arrival time and the completion of its service time.

The weighted flow time which accrues in a time T clearly $\to \infty$ as $T \to \infty$ under any policy, so we take as our aim the minimization of weighted flow time per unit time in the long run. Provided the expected duration of a busy period is finite, this aim is realized by a policy which minimizes EWFT in a busy period. The criterion of minimum EWFT in a busy period comes within the scope of our theory if we suppose that for every job the job-linked arrival rate for class i jobs is λ_i, and that no arrivals occur when the server is idle. This modifies the problem so that the only busy period is the one generated by the jobs waiting for service at time zero. The expected duration of a busy period is finite (e.g. see Prabhu, 1965) iff the *traffic intensity* $\Sigma \lambda_i \mu_i < 1$, where μ_i is the mean of the distribution F_i.

To obtain a $\gamma = 0$ index theorem for minimizing EWFT which is easy to prove we need a condition to hold which is similar to Condition C, under which Corollary 3.10 holds.

Condition C†. This condition refers to a FABP $\mathscr{F}$ with precedence constraints forming an out-tree and bandit process-linked arrivals. Every bandit process belongs to one of n classes $i (= 1, 2, \ldots, n)$ within each of which bandit processes have identical properties. $\mathscr{F}$ is said to satisfy Condition C† if $\exists \varepsilon > 0$ such that for any pair of bandit processes belonging to classes i and j, and in states x and y, either $v_i^\dagger(x) > v_j^\dagger(y)$ for all $\gamma < \varepsilon$, or $v_i^\dagger(x) < v_j^\dagger(y)$ for all $\gamma < \varepsilon$, unless $i = j$ and $x = y$. Here $v_i^\dagger(x)$ means the minimal sub-family index for a class i bandit process in state x.

Corollary 3.27. Under Condition C† the generalized EWFT criterion for a FABP $\mathscr{F}$ with precedence constraints forming an out-tree and bandit-process-linked arrivals is minimized by the minimal sub-family index policy with $\gamma = 0$ in the expression for the index. This corollary follows from Theorem 3.25 by the same argument as Corollary 3.10 follows from Theorem 3.6. Note that two bandit processes from the same class and in the same state have the same index value for all values of γ, so the possible occurrence of such a coincidence does not upset the argument.

Reverting to ordinary jobs and the original EWFT criterion, the state of an unfinished job may be identified with its process time, which we shall refer to as its *age*. Let $v_i(x)$ be the index, ignoring any successor jobs or arrivals, for an unfinished class j job of age x, when x is also a possible switching time.

Theorem 3.28. The EWFT in a busy period of a M/G/1 queue with job classes $i(=1, 2, \ldots, n)$ is minimized by the index policy defined by the $\gamma = 0$ indices $v_i(x)$ for the individual jobs, ignoring arrivals.

Proof. Our proof assumes Condition $C^\dagger$. From Corollary 3.27 it follows that the policy defined by the $\gamma = 0$ minimal sub-family indices $v_i^\dagger(x)$ is optimal.

We therefore need to show that $v^\dagger$ is an increasing function of v as i and x take different values. The index $v_i^\dagger(x)$ is the supremum over all policies and stopping times of the average rate of reward up to a stopping time obtainable from the sub-family initiated by a class i job of age x, where the cost per unit time c_j associated with a class $j(=1, 2, \ldots, n)$ job is treated as a reward (see the discussion following equation (2.9)). This supremum is achieved by a policy and stopping time which process each job in the sub-family until its minimal sub-family index first drops below the value $v_i^\dagger(x)$. This was shown for $\gamma > 0$ in specifying the truncated FABP $\mathscr{F}$ in Proof 1 of Theorem 3.22, and with a little care that proof may be extended to the present case with $\gamma = 0$ and jobs requiring a finite expected processing time. This observation means that attention may be restricted to policies and stopping rules which truncate all unfinished jobs of the same class at the same stage. Note also that, since there is no discounting, the average rate of reward achieved by a policy up to a stopping time depends only on the stages at which unfinished jobs are truncated, and is independent of the sequence in which the jobs are processed before the stopping time.

Let T be the vector $(T_1, T_2, \ldots, T_n)$, $R_i(x, y, T)$ the expected total reward from the minimal sub-family initiated by a class i job of age x, up to a point at which the truncation ages for unfinished jobs are y for the initial job and T_j for any other job in the sub-family belonging to class j $(=1, 2, \ldots, n)$, and $W_i(x, y, T)$ the corresponding expected total time. From the discussion of the previous paragraph it follows that

$$v_i^\dagger(x) = \sup_{\{y > x, T > 0\}} \frac{R_i(x, y, T)}{W_i(x, y, T)}. \tag{3.33}$$

In fact we could restrict attention to the case $y = T_i$, but the additional generality turns out to be useful. Write

$$R_i(x, T_i, T) = R_i(x, T), \quad W_i(x, T_i, T) = W_i(x, T).$$

Since the scheduling sequence makes no difference to $R_i(x, y, T)$ or $W_i(x, y, T)$, suppose the initial class i job is first processed until it either reaches age y or is finished, whichever occurs first. Let the age of the initial job at the end of this period of service be S, whether or not it is finished, and let $N_j(S)$ be the number of class j $(= 1, 2, \ldots, n)$ jobs which arrive during this period. Thus

$$R_i(x, y, T) = c_i P(S \leqslant y) + \sum_{j=1}^{n} R_j(0, T) E[N_j(S)]$$

$$= c_i P(S \leqslant y) + \sum_{j=1}^{n} R_j(0, T) E\{E[N_j(S)|S]\}$$

$$= c_i P(S \leqslant y) + \sum_{j=1}^{n} R_j(0, T) E\{\lambda_j(S - x)\}$$

$$= \frac{c_i[F_i(y) - F_i(x)]}{1 - F_i(x)} + \sum_{j=1}^{n} \lambda_j R_j(0, T) \frac{\int_x^y [1 - F_i(s)] \, ds}{1 - F_i(x)}.$$

$$W_i(x, y, T) = E(S - x) + \sum_{j=1}^{n} W_j(0, T) E[N_j(S)]$$

$$= \left[1 + \sum_{j=1}^{n} \lambda_j W_j(0, T) \right] \frac{\int_x^y [1 - F_i(s)] \, ds}{1 - F_i(x)}.$$

Substituting these expressions in (3.33) gives

$$v_i^\dagger(x) = \sup_{\{y > x, T > 0\}} \left\{ \left[\frac{c_i[F_i(y) - F_i(x)]}{\int_x^y [1 - F_i(s)] \, ds} + \sum_{j=1}^{n} \lambda_j R_j(0, T) \right] \right.$$

$$\left. \times \left[1 + \sum_{j=1}^{n} \lambda_j W_j(0, T) \right]^{-1} \right\}. \tag{3.34}$$

For $y = 0$,

$$v_i(x) = \sup_{\{y > x\}} \frac{c_i[F_i(y) - F_i(x)]}{\int_x^y [1 - F_i(s)] \, ds},$$

(see (equation (2.8)). Thus it follows from (3.34) that

$$v_i^\dagger(x) = \sup_{T>0} \frac{v_i(x) + \sum\limits_{j=1}^{n} \lambda_j R_j(0, T)}{1 + \sum\limits_{j=1}^{n} \lambda_j W_j(0, T)},$$

and $v^\dagger$ is therefore an increasing function of v as required.　　　□

To have established Theorem 3.28 only under Condition $C^\dagger$ is rather annoying. However, since $v_i^\dagger(x)$ is continuous in γ this condition holds if the number of decision points for each job is finite, and it seems likely that it can be removed altogether via an appropriate limit argument. In fact the first proof of the theorem (Olivier, 1972) was direct, rather than via the discounted version of the problem, and hence did not require Condition $C^\dagger$. The proof given here was obtained by Nash (1973). As Exercises 3.4 and 3.5 indicate, Theorem 3.25 may be used to achieve considerable generalization of this result.

Note that the sets $\{y > x\}$ and $\{T > 0\}$ occurring in the above proof are restricted to those times at which switching is allowed. The result extends to allow every point in time to be a decision point, as discussed in §5.6.

EXERCISES

3.1. Use an interchange argument to show that for Problem 1 an ε-index policy comes within $\varepsilon(\Sigma s_i)^2$ of achieving the minimum cost.

3.2. Let $\mathscr{F}$ be a simple family of alternative sub-families $\mathscr{F}_j$ ($j = 1, 2, \ldots, m$) of bandit processes. Show that $\mathscr{F}$ satisfies Condition E if the same is true of each of the $\mathscr{F}_j$. [Thus, for example, some of the $\mathscr{F}_j$ may consist of jobs with no preemption allowed (Corollary 3.21), and others of jobs with job-linked arrivals (Theorem 3.25).]

3.3. Use the law of large numbers to show that for the $M/G/1$ queue the weighted flow time per unit time over a long period of time is minimized by a policy which minimizes EWFT in a busy period, provided the expected duration of a busy period is finite.

3.4. A bandit process B is made up of two jobs (see §2.5). Job 1 must be completed before job 2 may be started, and neither of the jobs may be interrupted before completion. Find an expression for $v(B)$.

　　Switching now to the criterion of minimum expected weighted flow-time, let c_i be the cost per unit time of the flow-time of job i and τ_i its processing time ($i = 1, 2$). Use Corollary 3.10 to show that

$$v(B) = \max\{c_1/E\tau_1, (c_1 + c_2)/E(\tau_1 + \tau_2)\}. \tag{3.35}$$

Now extend Theorem 3.28 to show that the EWFT in a busy period of a M/G/1 queue, for which jobs may arrive in pairs linked in this way, is also minimized by an index policy for which the index of a linked pair of jobs is given by (3.35).

For further consideration of non-preemptive optimal scheduling see Kadane and Simon (1977), Glazebrook and Gittins (1981), and the other papers by Glazebrook listed in §9.3.

3.5. Extend the proof of Theorem 3.28, using Theorem 3.25, to allow each class of Poisson arrivals to be made up of indistinguishable sets of jobs subject to an out-tree of precedence constraints, rather than of individual jobs.

CHAPTER 4

General Properties of the Indices

4.1 INTRODUCTION

Knowing that the solution to an allocation problem is given by an index is a considerable step forward. The expressions (2.5) and (3.24), however, are not particularly explicit, and the task of determining numerical values remains. Two general methods are available: calibration by means of standard bandit processes using dynamic programming, as described for the multi-armed bandit in Chapter 1; and direct evaluation of (2.5) or (3.24) which, as we shall see in §6.4, works well for the multi-armed bandit, though it still involves a lot of computing. It is of interest, therefore, to investigate the general properties of allocation indices, and the circumstances in which expressions (2.5) and (3.24) may be simplified. In Chapters 1 and 2 we have already noted a number of cases which simplify because the optimal policy is myopic. This corresponds to giving the stopping time in (2.5) the smallest possible positive value. Another simplification may be obtained when this stopping time is infinite. These ideas are formalized in the final two sections of this chapter, and illustrated in the context of jobs, mainly the ordinary jobs of §2.5 with a single reward on completion. Other examples appear in Chapter 6.

4.2 DEPENDENCE ON DISCOUNT PARAMETER

The first result is that if discounting is made stronger by increasing the discount parameter γ all indices are, if anything, decreased. In general terms the reason for this is clear. An equivalent constant reward rate which is maximized up to an arbitrary stopping time may be higher than the reward rate up to the next decision-time. If it is higher this is because higher reward rates may be available later on. The effect of these later high reward rates is, however, discounted, and the stronger the discounting the less this effect becomes.

Theorem 4.1. $v(B, x, \gamma) \downarrow$ in $\gamma(>0)$ for all bandit processes B and states x.

Proof. Suppose $0 < \gamma < \beta$. Let: τ be a stopping time for B starting from state x; σ be a negative exponential random variable with parameter $\beta - \gamma$, and independent of B; $\tau^* = \min\{\tau, \sigma\}$; $I(t), I^*(t)$ and $J(t)$ be indicators for the events $\tau \geq t, \tau^* \geq t$ and $\sigma \geq t$ respectively.

82

Thus $I^*(t) = I(t)J(t)$ and, since σ is independent of B,

$$\mathbf{P}\{\tau^* \geqslant t\} = \mathbf{E}(I^*(t)) = \mathbf{E}(I(t))\mathbf{E}(J(t)) = \mathbf{P}\{\tau \geqslant t\}e^{(\gamma - \beta)t},$$

and this remains true if every event and expectation is conditioned by some event depending only on B. We therefore have

$$R_{\tau^*}(x, \gamma) = \mathbf{E} \sum_{i=0}^{\infty} I^*(t_i)e^{-\gamma t_i} r(t_i) = \sum_{i=0}^{\infty} \mathbf{E}[e^{-\gamma t_i} r(t_i)\mathbf{E}(I^*(t_i)|t_i, r(t_i))]$$

$$= \sum_{i=0}^{\infty} \mathbf{E}[e^{-\gamma t_i} r(t_i)\mathbf{E}(I(t_i)|t_i, r(t_i))e^{(\gamma - \beta)t_i}] = R_{\tau}(x, \beta),$$

and

$$W_{\tau^*}(x, \gamma) = \mathbf{E} \int_0^{\infty} I^*(t)e^{-\gamma t}\, dt = \int_0^{\infty} \mathbf{E}I^*(t)e^{-\gamma t}\, dt$$

$$= \int_0^{\infty} \mathbf{E}I(t)e^{-\beta t}\, dt = W_{\tau}(x, \beta).$$

Thus

$$v(x, \beta) = \sup_{\tau > 0} \frac{R_{\tau}(x, \beta)}{W_{\tau}(x, \beta)} = \sup_{\tau > 0} \frac{R_{\tau^*}(x, \gamma)}{W_{\tau^*}(x, \gamma)} \leqslant \sup_{f_0 = 0} \frac{R_f(x, \gamma)}{W_f(x, \gamma)} = v(x, \gamma). \qquad \square$$

Note that, since it is not defined by a stopping sub-set of the state-space of B, τ^* is not a stopping time. It does, however, define a freezing rule, and the last line of the proof goes through with the help of Lemma 3.2. The possibility of strict inequality arises from the fact that not all freezing rules for B may be defined in this way.

4.3 MONOTONE INDICES

In this section we examine the simplifications in the general expression (2.5) for the index for a bandit process which result when various inequalities hold between the values at different decision-times either of the index itself, or of the sequence of rewards. The strongest of these conditions is that of monotonicity in time, and we refer to them collectively as *monotonicity* conditions. For a general bandit process

$$v(x) = \sup_{\tau \geqslant t_1} \frac{R_{\tau}(x, \gamma)}{W_{\tau}(x, \gamma)} = \sup_{\tau \geqslant t_1} \frac{R_{t_1}(x, \gamma) + \mathbf{E}[e^{-\gamma t_1} R_{\tau - t_1}(x(t_1))]}{W_{t_1}(x, \gamma) + \mathbf{E}[e^{-\gamma t_1} W_{\tau - t_1}(x(t_1))]}, \qquad (4.1)$$

where $R_{t_1}(x, \gamma) = r(x)$ and $W_{t_1} = \gamma^{-1}\mathbf{E}[1 - e^{-\gamma t_1}]$. In the Markov case, when $t_i = i$ $(i = 1, 2, \ldots)$, this becomes

$$v(x) = \sup_{\tau \geqslant 1} \frac{r(x) + a\mathbf{E}R_{\tau - 1}(x(1))}{\gamma^{-1}(1 - a) + a\mathbf{E}W_{\tau - 1}(x(1))}. \qquad (4.2)$$

As the notation is simpler, the results of this section will first be given for the Markov case.

Since

$$v(x(1)) = \sup_{\sigma > 0} \frac{R_\sigma(x(1))}{W_\sigma(x(1))} \geqslant \frac{R_{\tau-1}(x(1))}{W_{\tau-1}(x(1))},$$ (4.3)

the following propositions are easy consequences of (4.2).

Proposition 4.2. Iff $\mathbf{P}\{v(x(1)) \leqslant v(x)\} = 1$ the supremum in (4.2) is attained when $\tau = 1$, and $v(x) = \gamma r(x)/(1-a)$.

Proposition 4.3. If $\mathbf{P}\{v(x(1)) > v(x)\} = 1$ the supremum in (4.2) is attained when $\tau > 1$, and $v(x) > \gamma r(x)/(1-a)$.

The two propositions follow from (4.2) together with the following trivial lemma.

Lemma 4.4. If a_0, b_0, a_1, b_1, are random and the inequalities $b_0 > 0$, $b_1 > 0$, $a_0/b_0 < (\text{or} >) a_1/b_1$ hold almost surely, then $\mathbf{E}(a_0 + \lambda a_1)/\mathbf{E}(b_0 + \lambda b_1) \uparrow (\text{or} \downarrow)$ in $\lambda(>0)$. Provided $\mathbf{P}\{b_i > 0\} > 0$ $(i=0,1)$ the lemma still holds if the strict inequalities are replaced by non-strict inequalities, except that the monotonicity of the conclusion is no longer strict.

Alternatively, we may simply note that the two propositions are corollaries of parts (iv) and (v) of Theorem 3.4.

Note that for the two conditions which Proposition 4.2 shows to be equivalent to hold it is sufficient that $r(x(t)) \leqslant r(x) (t = 1, 2, \ldots)$ almost surely. To see this write

$$v(x) = \frac{\gamma}{1-a} \sup_{\tau \geqslant 1} \frac{\mathbf{E} \sum_{t=0}^{\infty} a^t I\{\tau > t\} r(x(t))}{\mathbf{E} \sum_{t=0}^{\infty} a^t I\{\tau > t\}}.$$

$$= \frac{\gamma}{1-a} \sup_{\tau \geqslant 1} \frac{r(x) + \sum_{t=1}^{\infty} a^t \mathbf{P}\{\tau > t\} \mathbf{E}[r(x(t))|\tau > t]}{1 + \sum_{t=1}^{\infty} a^t \mathbf{P}\{\tau > t\}}.$$ (4.4)

That $v(x) = \gamma r(x)/(1-a)$ now follows from Lemma 4.4 if we put $a_0 = r(x)$, $b_0 = b_1 = 1$, $\mathbf{P}\{a_1 = \mathbf{E}[r(x(t))|\tau > t]\} = a^t \mathbf{P}\{\tau > t\}/\sum_{s=1}^{\infty} a^s \mathbf{P}\{\tau > s\}$, and $\lambda = \sum_{t=1}^{\infty} a^t \mathbf{P}\{\tau > t\}$. Thus the following proposition holds.

Proposition 4.5. If $\mathbf{P}\{r(x(t)) \leqslant r(x), t = 1, 2, \ldots\} = 1$, the supremum in (4.2) is attained when $\tau = 1$, and $v(x) = \gamma r(x)/(1-a)$.

Proposition 4.3 may be extended, giving

Proposition 4.6. If $\mathbf{P}\{v(x(t)) > v(x); \ t = 1, 2, \ldots\} = 1$ the supremum in (4.2) is attained when $\tau = \infty$, and $v(x) > \gamma r(x)/(1-a)$.

This is an immediate consequence of part (v) of Theorem 3.4, and includes the case when $v(x(t))$ is almost surely increasing. Note that for $v(x(t))$ to be almost surely increasing in t it is sufficient for the same to be true of $r(x(t))$. This follows from (4.2), (4.3), (4.4) and Lemma 4.4.

Propositions 4.2, 4.3, 4.5 and 4.6, and the sufficient conditions for Proposition 4.6 which have just been noted, all have their counterparts for a general semi-Markov bandit process. To achieve the translation into this more general setting the decision times $1, 2, 3, \ldots$ must be replaced by $t_1, t_2, t_3, \ldots$; for example in Proposition 4.2 $v(x(1))$ becomes $v(x(t_1))$ and $\tau = 1$ becomes $\tau = t_1$. The appropriate change in $r(x(t))$ ($t = 0, 1, 2, \ldots$) becomes apparent if we recall that $\gamma r(x(t))/(1-a)$ is the equivalent constant reward rate up to the next decision-time, starting from state $x(t)$. The corresponding quantity in the semi-Markov context is $\gamma r(x(t_i))/\mathbf{E}(1 - a^{t_{i+1} - t_i})$, which we may write as $v_{t_{i+1} - t_i}(x(t_i))$, or simply as v_i. In the four propositions we now have v_0 in place of $\gamma r(x)/(1-a)$; the condition for Proposition 4.5 is now

$$\mathbf{P}\{v_i \leqslant v_0; \ i = 1, 2, \ldots\} = 1;$$

the sufficient condition for Proposition 4.6 is now

$$\mathbf{P}\{v_{i+1} > v_i; \ i = 0, 1, \ldots\} = 1.$$

The proofs are similar to those for the Markov case. The most important difference is that the analogue of the identity (4.4) is a little more complicated, since the sequence of decision times is now random.

4.4 MONOTONE JOBS

As an illustration of what happens under monotonicity conditions suppose that our bandit processes are jobs. First note that the condition for Proposition 4.6 cannot hold for a job which yields a positive reward on completion. A more useful result, also an immediate consequence of Theorem 3.4, part (v), is

Proposition 4.7. If $\mathbf{P}\{v(x(t)) > v(x) > 0 \,|\, x(t) \neq C\} = 1$ ($t = 1, 2, \ldots$) the supremum in (4.1) is attained when $\tau = \min\{t: x(t) = C\}$, and $v(x) > \gamma r(x)/(1-a)$ ($x \neq C$).

This proposition extends to generalized jobs.

For an uncompleted ordinary job with no restriction on switching times the index at process time x is

$$v(x) = \sup_{t > x} \frac{V \int_x^t e^{-\gamma s} \, dF(s)}{\int_x^t e^{-\gamma s} [1 - F(s)] \, ds}$$

$$= \sup_{t > x} \frac{V \int_x^t f(s) e^{-\gamma s} \, ds}{\int_x^t [1 - F(s)] e^{-\gamma s} \, ds} \qquad \text{if } F \text{ is differentiable.} \qquad (4.5)$$

Since there is now no interval between decision-times the quantity corresponding to the reward rate up to the next decision-time is the instantaneous reward rate $Vf(x)/[1 - F(x)] = V\rho(x)$. $\rho(x)$ may be termed the *completion rate* for the job. Note that

$$\mathbf{P}\{\text{Completion before } x + \delta x \,|\, \text{no completion before } x\} = \rho(x)\delta x + o(\delta x).$$

In other contexts $\rho(x)$ would be termed the *hazard rate*, or the *mortality rate*.

Let us now see what we can deduce by applying our results on monotonicity. With every time a decision time, appropriate statements of the five propositions may be obtained from those for the semi-Markov case by supposing $t_{i+1} - t_i$ to be non-random and independent of i and allowing it to tend to zero.

Propositions 4.2 and 4.6 now tell us that $v(x) = \text{or} > V\rho(x)$, according as $dv(x)/dx < \text{or} > 0$, and corresponding to the sup in (4.5) being as $t \searrow x$ or for $t > x$, respectively. Proposition 4.5 gives $v(x) = V\rho(x)$ if $\rho(x) \geqslant \rho(t)(t > x)$. To apply Proposition 4.7, note first that $v(t)$ is increasing for $t > x$ iff the same is true of

$$Q(t, \gamma) = \int_t^\infty f(s) e^{-\gamma s} \, ds \Big/ \int_t^\infty [1 - F(s)] e^{-\gamma s} \, ds,$$

in which case $v(x) = VQ(x, \gamma)$. Note also that

$$[Q(t, 0)]^{-1} = \int_t^\infty \frac{1 - F(s)}{1 - F(t)} \, ds = \mathbf{E}(\sigma - t \,|\, \sigma > t), \qquad (4.6)$$

where σ is the total service-time required by the job. Thus, given a set of jobs J_i for each of which $Q_i(t, \gamma)$ is increasing in t, the optimal policy is to process them in order of decreasing $V_i Q_i(0, \gamma)$. If preemptions are not allowed this (see equation (2.7)) is the form of the optimal policy irrespective of the behaviour of $Q_i(t, \gamma)$. We have shown, then, that the effect of having $Q_i(t, \gamma)$ increasing in t is to reduce the problem to the non-preemptive case.

As discussed in §2.5 the index policy for $\gamma = 0$ minimizes the expectation of the weighted flow-time $\Sigma V_i t_i$. From (4.6) it follows that, if the expected remaining processing time $\mathbf{E}(\sigma_i - t/\sigma_i > t) \downarrow$ in t for all i, $\mathbf{E}\Sigma V_i t_i$ is minimized by a non-preemptive schedule in order of decreasing $V_i/E\sigma_i$. This result was first proved by Rothkopf (1966).

Propositions 4.2, 4.3, 4.5, 4.6 and 4.7 are all stated for single bandit processes. They are, however, corollaries of parts (iv) and (v) of Theorem 3.4, which applies to a general semi-Markov decision process. Extensions of the five propositions to exploit this greater generality may be obtained in fairly obvious fashion. This is true, in particular, for generalized jobs subject to precedence constraints in the form of an out-tree (see Theorem 3.22). For ordinary jobs the following result, due to Glazebrook (1980c), may be established more directly.

Theorem 4.8. Given a set of ordinary jobs J_i $(i = 1, 2, \ldots, n)$, subject to arbitrary precedence constraints, and for each of which

$$S_i(t, \gamma) = e^{\gamma t} \frac{\displaystyle\int_t^\infty f_i(s)e^{-\gamma s}\,ds}{1 - F_i(t)}$$

is increasing in t, there is an optimal policy which is non-preemptive.

Proof. Suppose first that switching is restricted to process-times which are integer multiples of Δ or completion-times. Call the corresponding policies Δ-*policies*. We first establish the following lemma.

Lemma 4.9. There is an optimal Δ-policy which is deterministic, stationary, Markov (DSM) and non-preemptive (NP).

Proof of Lemma. From parts (i) and (ii) of Theorem 2.2 it follows that it is sufficient to show that there is an NP policy which is optimal in the class P of those policies which are DSM throughout, and NP from the second decision time onwards.

Let $\pi \in P$. Without loss of generality suppose J_1 is the job initially selected by π. If J_1 has not been completed by time Δ, suppose π then schedules a permutation α_1 of some subset of the remaining $n-1$ jobs, followed by J_1, followed by a permutation α_2 of the remaining jobs. If J_1 terminates by time Δ, suppose π then schedules a permutation β of the other $n-1$ jobs. Let $R(\alpha_1)$ be the expected total reward from the jobs in α_1, discounted from the start of the first of these jobs. Let $D(\alpha_1) = E e^{-\gamma T_1}$, where T_1 is the total service-time required by the jobs in α_1. Similar notation applies to α_2 and β. Let

$$D(\Delta) = \int_0^\Delta f_1(t)e^{-\gamma t}\,dt, \quad D(\bar{\Delta}) = \int_\Delta^\infty f_1(t)e^{-\gamma t}\,dt.$$

The expected total discounted reward under π is

$$R_\pi = D(\Delta)[V_1 + R(\beta)] + \{e^{-\gamma\Delta}[1 - F_1(\Delta)]R(\alpha_1) + D(\alpha_1)D(\bar{\Delta})[V_1 + R(\alpha_2)]\}.$$

Writing α and β for the non-preemptive schedules given by α_1, J_1, α_2, in that order, and by J_1, β, respectively, the corresponding expected total discounted rewards are

$$R_\alpha = R(\alpha_1) + D(\alpha_1)[D(\Delta) + D(\bar\Delta)][V_1 + R(\alpha_2)]$$

and

$$R_\beta = [D(\Delta) + D(\bar\Delta)][V_1 + R(\beta)].$$

Thus

$$\frac{D(\bar\Delta)}{D(\Delta) + D(\bar\Delta)} R_\alpha + \frac{D(\Delta)}{D(\Delta) + D(\bar\Delta)} R_\beta - R_\pi$$

$$= \left\{ \frac{D(\bar\Delta)}{D(\Delta) + D(\bar\Delta)} - e^{-\gamma\Delta}[1 - F_1(\Delta)] \right\} R(\alpha_1)$$

$$= e^{-\gamma\Delta}[1 - F_1(\Delta)] \left\{ \frac{S_1(\Delta, \gamma)}{S_1(0, \gamma)} - 1 \right\} > 0.$$

It follows that either $R_\alpha > R_\pi$ or $R_\beta > R_\pi$, as required.   □

The number of DSMNP policies is $n!$, a finite number. Each of these is a Δ-policy for every Δ. Letting $\Delta \to 0$ it follows that the best of these deterministic permutations is an over-all optimal policy.   □

Note that

$$S_i(t, \gamma) = 1 - \gamma \frac{\int_t^\infty f_i(s)(s - t)\,ds}{1 - F_i(t)} + O(\gamma^2) = 1 - \gamma E(\sigma_i - t | \sigma_i > t) + O(\gamma^2),$$

where σ_i is the service-time for J_i. Thus for small values of γ the condition that $S_i(t, \gamma)$ is increasing in t is essentially the same as having an expected remaining service time which is decreasing. The limit as $\gamma \searrow 0$ corresponds to minimizing $E\Sigma V_t t_i$, and the following corollary may easily be deduced.

Corollary 4.10. If $S_i(t, \gamma)$ is increasing in t $(i = 1, 2, \ldots, n; 0 < \gamma < \delta)$, for some $\delta > 0$, then the expected weighted flow-time $E\Sigma V_i t_i$ is minimized by an NP policy.

EXERCISE

4.1. Write down and justify analogues of Propositions 4.2, 4.3, 4.5 and 4.7 for generalized jobs subject to an out-tree of precedence constraints.

CHAPTER 5

Jobs with Continuously-Varying Effort Allocations

5.1 INTRODUCTION

This chapter is divided into three distinct, and rather differently motivated, sections which are in turn divided into subsections.

The first section points out that jobs of the type defined in §2.5 are reasonable models for the projects which arise in new-product chemical research. This observation forms the basis of the RESPRO computer package which is also described in this section. RESPRO is designed as an aid to research managers faced with the problem of allocating effort between competing projects. In the interests of realism the possible relationships between the rate at which work is done on a project and the progress that is made are extended. As a result, optimal policies are no longer expressible in terms of allocation indices.

The second section of the chapter reverts to jobs as defined in §2.5, with no restriction on the times at which allocations may be changed. This leads to optimal policies for which effort is allocated to more than one project at a time. Versions of the index theorems 3.6, 3.25 and 3.26 are established by limit arguments.

Theorem 3.28, which continues to hold when there is no restriction on allocation times, shows that the indices which define optimal allocations under the expected weighted flow-time criterion continue to do so under the time-averaged weighted flow-time criterion if there are Poisson arrivals. With discounted costs, however, the problem of determining optimal allocations with Poisson arrivals is by no means trivial, even with the help of allocation indices. The final section of the chapter examines this problem in some detail for the case with two classes of jobs with processing-time distributions of the form considered in §5.2. This leads to some extensions of classical queueing theory which are perhaps of some technical interest.

5.2 COMPETING RESEARCH PROJECTS

5.2.1 Introduction

In these days of high technology, research is a quantitatively significant and strategically important area of investment both in national terms and in the

89

context of a particular company. This has led to increased attention to the question of how this investment may most effectively be managed. Research on research has itself become a growth industry—filling the pages, for example, of the journals *IEEE Transactions on Engineering Management* and *R&D Management*.

Various schemes have been proposed, and to some extent implemented, for establishing priorities between projects which compete for the same resources. These have been reviewed by, for example, Dean and Goldhar (1980), and Bergman and Gittins (1985).

Checklists of the relevant criteria often include some which are not obviously quantifiable, such as expertise in the proposed research area, and the compatibility of any new product with a company's existing product range. Usually some attempt is made to rank projects in terms of an index of profitability which balances the costs against the likely income from a technically successful project and the probability that a project actually does attain its technical objectives.

A simple family of alternative jobs as defined in §2.5, with rewards only on completion and no costs, is an obvious candidate as a model for the problem of allocating a resource available at a constant total rate between a number of competing projects. The single resource might be measurable in terms of senior scientist years or as an annual budget. Spending some of the resource on a project is equivalent to allocating processing time to the corresponding job. In effect, time is measured in units of the resource. For this to be a reasonable model the processor must be allowed to divide its effort between more than one job, as it is unrealistic to suppose that at any given time the whole of a laboratory's research effort is concentrated on the most promising project. As we shall see in §5.5, this is in any case a natural property to require of a family of alternative jobs between which switching may take place at any time.

For an uncompleted job the state, and therefore the probability distribution of the further processing time required for completion, is defined by the total processing time that it has already received. If such a job is to give a reasonable representation of a research project, the research project should have the same feature. There should not, that is to say, be any clear milestones, distinct from the time or money so far spent on the project, registering progress towards successful completion. This requirement rules out most engineering research projects leading, for example, to a new type of aero-engine or computer, since these typically involve the design, construction, and testing of a number of components; the completion of this process for any single component is an obvious milestone. The allocation of resources between projects of this type could be modelled by a simple family of the generalized jobs of §2.5. For projects for which the appropriate sequence of operations is in doubt a family of alternative super-processes would be needed (see §3.6).

New-product chemical research, and in particular pharmaceutical research, on the other hand is relatively free from obvious milestones if attention is restricted to the pre-clinical phase, characterized by the screening of large numbers of compounds before one or more candidate drugs are found. Apart from the discovery of a candidate drug, which effectively represents successful completion of the exploratory research phase, the only real milestone is the discovery of what is known as a *lead compound*, a compound, that is to say, which shows sufficient relevant activity to make it worthwhile concentrating subsequent efforts on testing compounds which are chemically similar. Whether the discovery of a lead component is a sufficiently clear-cut milestone to warrant explicit inclusion in a model of the allocation process is debatable. It probably depends on the project. Apart from this possible complication, alternative jobs offer a reasonable representation of competing pharmaceutical research projects at the exploratory stage. This is the most obvious application for the models discussed in this chapter.

The next sub-section explains why exploratory pharmaceutical research projects are likely to have completion rates which first increase and then decrease. This is followed by a description of the RESPRO package, which is a management aid to resource allocation between such projects.

5.2.2 Completion time distributions with bitonic completion rates

During the exploratory phase of a research project it is common for thousands of compounds to be tested in the search for a candidate drug. A crude stochastic model of this process may be derived by assuming each compound to have the same unknown probability p, independently of the other compounds, of proving suitable. If compounds are tested at a constant rate the discovery of candidate drugs thus occurs approximately as a Poisson process with unknown parameter θ. For a given θ the time X required to find a candidate drug therefore has the negative exponential

$$\text{density, } f(x|\theta) = \theta e^{-\theta x},$$

$$\text{distribution function, } F(x|\theta) = 1 - e^{-\theta x},$$

$$\text{and completion rate, } \rho(x|\theta) = f(x|\theta)/[1 - F(x|\theta)] = \theta.$$

Note that in other contexts the completion rate is termed the hazard rate or mortality rate.

Suppose now that θ has a prior distribution Π. Thus, integrating out the θ-dependence, X has the

$$\text{density, } f(x) = \int \theta e^{-\theta x} d\Pi(\theta),$$

$$\text{distribution function, } F(x) = 1 - \int e^{-\theta x} d\Pi(\theta),$$

and completion rate, $\rho(x) = f(x)/[1 - F(x)] = \int \theta e^{-\theta x} d\Pi(\theta) \Big/ \int e^{-\theta x} d\Pi(\theta)$.

It follows that

$$
\begin{aligned}
\frac{d\rho(x)}{dx} &= \frac{\left(\int \theta e^{-\theta x} d\Pi(\theta)\right)^2 - \int \theta^2 e^{-\theta x} d\Pi(\theta)}{\left(\int e^{-\theta x} d\Pi(\theta)\right)^2} \\
&= -\frac{\int \left[\theta - \int \phi e^{-\phi x} d\Pi(\phi)\right]^2 e^{-\theta x} d\Pi(\theta)}{\left(\int e^{-\theta x} d\Pi(\theta)\right)^2} \leqslant 0.
\end{aligned}
\tag{5.1}
$$

The decreasing completion rate reflects the fact that as more compounds are tested without finding a candidate drug it becomes more likely that compounds suitable as candidate drugs are scarce in the population under test.

As exploratory research progresses, the scientists concerned are naturally learning more about the types of compound which are likely to show the desired type of activity. Thus, rather than supposing every compound tested to have the same success probability, a more reasonable assumption is that this probability increases as more compounds are tested. The rate at which compounds are tested may also increase as confidence increases that the right ones are being selected, and as the chemists improve their methodology for synthesizing these compounds. These learning effects are all likely to be most pronounced at the beginning of a project, and lead to an underlying X-distribution for which the completion rate is increasing, rather than the negative exponential distribution and constant completion rate which are appropriate in their absence. A plausible assumption is that the underlying X-distribution is a gamma distribution with parameters r and θ, for which the density is

$$
f(x|r, \theta) = [\Gamma(r)]^{-1} \theta^r x^{r-1} e^{-\theta x},
$$

and that $r > 1$. The completion rate $\rho(x|r, \theta)$ may be shown to be increasing in x, and to tend to θ as $x \to \infty$.

If r and θ have the joint prior distribution $\Pi(r, \theta)$, the unconditional density for X may be obtained as before by integrating out the (r, θ)-dependence, giving

$$
f(x) = \int f(x|r, \theta) d\Pi(r, \theta).
$$

The argument leading to (5.1) may be adapted to show that, since $\rho(x|r, \theta)$ tends to a limit as $x \to \infty$, $\rho(x)$ is decreasing for all sufficiently large x. Again, the basic reason is that testing compounds without discovering a candidate drug makes it more likely that candidate drugs are scarce. On the other hand it is not difficult to show that if, for a given x, $\partial\rho(x|r, \theta)/\partial x$ is sufficiently large over a sufficiently probable set of (r, θ) values, then $d\rho(x)/dx$ must also be positive.

In brief, we can expect that in the early stages of a project the beneficial effects of learning will be dominant and $d\rho(x)/dx$ positive, and later on, if no candidate drug has been found, bitter experience will be the dominant effect, and $d\rho(x)/dx$ negative. X-distributions for which $\rho(x)$ increases to a maximum and then decreases will be described as having a *bitonic* completion rate. For the reasons given, completion times having distributions with bitonic completion rates feature prominently in our discussion.

5.2.3 The RESPRO package

RESPRO (research procedure) is a computer program designed to assist managers of exploratory new-product chemical research in the allocation of scientists and other resources between competing projects. The basic principles of the program and how it should be used are set out by Gittins and Roberts (1981), though some improvements have been made since then. The projects are modelled as jobs along the lines discussed in the previous two sections, and with a number of refinements designed to achieve greater realism. The manager provides estimates of the various parameters characterizing each of the projects, and inputs a succession of resource allocations between them. For each of these allocation plans the program computes the total expected return, and a *marginal profitability index* (*MPI*) for each project. The MPI for a project depends on the allocation plan and on time, and is the marginal rate at which the expected return from the project increases as more effort is allocated to it at a given point in time. The set of MPIs for each project indicates the directions in which effort should be re-allocated between them so as to increase the total expected return. The most acceptable allocation plan is not usually the one which emerges from the program as maximizing total expected return. This is because the model does not, nor would it be sensible to try to make it, take account of all the relevant considerations, many of which are not readily translated into monetary terms. Plausible plans may, however, be compared on the basis of estimated profitability by means of the program, leaving the manager to decide between them.

The MPI for a project is related to, but not equal to, the allocation index for jobs,

$$v(x) = \sup_{t > x} \frac{V \displaystyle\int_x^t f(s)e^{-\gamma s}\,ds}{\displaystyle\int_x^t [1 - F(s)]e^{-\gamma s}\,ds}, \tag{5.2}$$

given as equation (1.3), and again as (2.8). The index $v(x)$ in fact no longer defines an optimal allocation for projects under the assumptions of RESPRO. We now introduce these assumptions and the logical basis of the MPIs in the context of a generalization of the derivation of equation (1.3).

Suppose the resource units to be scaled so that at any time the total effort rate available is 1. At time t let $u(t)$ be the effort rate allocated to the given project if it has not terminated, the balance $1 - u(t)$ being allocated to a standard alternative investment yielding c per unit of effort, so that c represents the unit cost of effort allocated to the project. Thus, in terms of jobs, the processor need no longer be totally committed to a single job at any given time.

Let $e(u)$ be the effective rate of progress on the project when the effort rate is u, and $x(t) = \int_0^t e(u(s)) \, ds$ be the effective work done on the project up to time t. The idea here is that the rate of progress of a project may not be simply proportional to the size of the research team allocated to it. There are fairly obvious reasons why a very small or a very large research team might have a lower productivity per scientist than one of moderate size. Let $F(x)$ and $f(x)$ be the distribution and density functions for the effective work required to complete the project.

Let $1 - e^{-\phi t}$ be the probability that before time t a competitor makes a discovery which makes it not worthwhile continuing with the project. If a competitor does make such a discovery the project is said to be *preempted*; ϕ is the *preemption rate* for the project.

Let $Ve^{-\gamma t}$ be the expected present value of the cash flows generated by successful completion of the project at time t, and $e^{-\beta t}$ the discount factor for effort allocated to the standard alternative investment at time t. Since the drugs which different projects are designed to produce are likely to become obsolescent at different rates it is quite possible that $\gamma \neq \beta$.

Thus the expected payoff P and cost C resulting from a planned allocation rate $u(t)$ $(0 < t < \infty)$, are

$$P = V \int_0^\infty e(u(t)) f(x(t)) e^{-(\gamma + \phi)t} \, dt,$$

$$C = c \int_0^\infty u(t) [1 - F(x(t))] e^{-(\beta + \phi)t} \, dt. \tag{5.3}$$

Suppose now that the plan $u(t)$ is altered by a quantum (possibly negative) of additional effort Δ at time s (i.e. an increase to $u(t) + \delta$ in an interval $(s, s + \varepsilon)$, where both δ and ε are small). Assuming $u(t)$ to be continuous at s, and the functions $e(u)$ and $f(x)$ to have derivatives $e'(u)$ and $f'(x)$, the corresponding increments dP and dC in P and C may be written as follows.

$$\frac{dP}{V} = \Delta e'(u(s)) \left[f(x(s)) e^{-(\gamma + \phi)s} + \int_s^\infty e(u(t)) f'(x(t)) e^{-(\gamma + \phi)t} \, dt \right] + o(\Delta)$$

$$= \Delta e'(u(s))(\gamma + \phi) \int_s^\infty f(x(t)) e^{-(\gamma + \phi)t} \, dt + o(\Delta) \text{ (integrating by parts).}$$

$$\frac{dC}{c} = \Delta[1 - F(x(s))]e^{-(\beta+\phi)s} - \Delta e'(u(s)) \int_s^\infty u(t)f(x(t))e^{-(\beta+\phi)t}\,dt + o(\Delta).$$

Thus

$$dP - dC = \Delta\left\{ e'(u(s)) \int_s^\infty [V(\gamma+\phi)e^{-\gamma t} + cu(t)e^{-\beta t}]f(x(t))e^{-\phi t}\,dt \right.$$

$$\left. - c[1 - F(x))]e^{-(\beta+\phi)s} \right\} + o(\Delta).$$

The MPI for the project at time s is defined to be

$$\lim_{\Delta \to 0} \left(\frac{dP - dC}{\Delta}\right)\left(\frac{e^{(\beta+\phi)s}}{[1 - F(x(s))]}\right).$$

It is therefore the derivative with respect to effort allocated at time s of the expected net value of the project, discounted to time s, conditional on the project having been neither completed nor preempted before time s, and taking account of the cost of the resources allocated.

RESPRO is, of course, designed to allocate an effort budget between a number of projects, rather than between a single project and a standard alternative investment. The two concepts are reconciled by representing every project except the given project as, on average, a standard alternative investment. The program does this by evaluating β and c as weighted averages of the γ's and the MPIs respectively, for all the other projects. To carry out an explicit analysis for allocation between several projects along the lines of the previous paragraph is computationally intractable.

From equations (5.3) it follows that, if $e(u) = u$, $\beta = \gamma$, and $\phi = 0$, then $P - C > 0$ iff

$$c < \frac{V \int_0^\infty u(t)f(x(t))e^{-\gamma t}\,dt}{\int_0^\infty u(t)[1 - F(x(t))]e^{-\gamma t}\,dt}. \tag{5.4}$$

Thus the project should be started, rather than allocating all resources to the standard alternative investment, only if an allocation plan $u(t)$ can be found for which the inequality (5.4) holds. This is a similar condition to (1.2), and may be shown to be equivalent to it along the lines of the proof of Lemma 3.2, leading again to equation (1.3) for the index $v(x)$.

RESPRO models projects with the bitonic completion rates discussed in the previous section. Separate completion-time distributions of this type may be fitted to two successive stages of a project, corresponding to first identifying, and then modifying, a lead compound. The allocations in a plan may be automatically scaled, so as to give a fixed total expected discounted effort for the plan as a whole. This total must be set rather lower than the total discounted effort available, as in the future one naturally expects to be devoting an increasing proportion of research effort to projects other than those under current consideration. An alternative procedure is to model an arrival process of new proposals for projects. This approach is explored in §5.4.

5.3 CONTINUOUS-TIME JOBS

5.3.1 Introduction

Our standing assumption that families of alternative bandit processes are semi-Markov decision processes means that decision times and rewards may occur at any point in time, rather than just at integer-valued times as for a Markov decision process. However, there are still only a countable number of decision times, which are also reward times, except for standard bandit processes. The difference from the Markov case is that these times are themselves random. For jobs (see §2.5) rewards occur only on completion, though intermediate times before completion may also be designated as decision times. This section explores the consequences of increasing the density of these intermediate decision times until, in the limit, every time is a decision time. Jobs with this property will be termed *continuous-time* jobs. Limit arguments are used to provide appropriate extensions of Theorems 3.6, 3.25 and 3.26, and the final section of the chapter examines in some detail the form of the optimal policy when jobs with bitonic completion rates arrive in a Poisson process.

For an alternative route via variational arguments to index theorems for continuous-time jobs see Nash and Gittins (1977). The question of what happens when rewards as well as decisions occur continuously over time is deferred to §§6.5, 6.6 and 9.4.

The additional feature of a possibly non-linear function defining the effective rate of progress, and different preemption and obsolescence rates for each project, which were introduced in the previous section, are now left out, since otherwise index policies are not optimal. Extensions to the theorems of Chapter 3 are needed as these theorems hold under Condition A, which limits the frequency of decision times. The first stage is to define an index policy based on the index (5.2) for a simple family of alternative jobs.

5.3.2 Policies for continuous-time jobs

One difficulty becomes apparent if we suppose there are just two jobs, 1 and 2, with decreasing completion rates $\rho_1(x_1)$ and $\rho_2(x_2)$, so that

$$v_i(x_i) = V_i \rho_i(x_i) \qquad (i = 1, 2).$$

Suppose $V_1 \rho_1(0) = V_2 \rho_2(0)$ and, initially, that switching between jobs is allowed only when $x_i = k_i \Delta \ (k_i = 0, 1, 2, \dots)$, for some time-unit Δ, or when one of the jobs terminates. This restriction on switching modifies the form of the index. We have

$$v_i^\Delta(k\Delta) = \frac{V_i \displaystyle\int_{r\Delta}^{(r+1)\Delta} -f_i(s)e^{-\gamma s}\,ds}{\displaystyle\int_{r\Delta}^{(r+1)\Delta} [1 - F_i(s)]e^{-\gamma s}\,ds},$$

and

$$v_i(k\Delta) = V_i\rho_i(k\Delta) > v_i^\Delta(k\Delta)$$

$$> v_i((k+1)\Delta) = V_i\rho_i((k+1)\Delta) \qquad (i = 1, 2; \ k = 0, 1, 2, \ldots).$$

Thus the modified indices also decrease as the jobs age, and an optimal Δ-policy switches between jobs so as to keep the values of $V_i\rho_i(x_i(t))$ $(i = 1, 2)$ approximately equal for all t. If we now increase the number of decision-times by decreasing Δ this approximate equality becomes closer and closer, and it seems a fair bet that in the limit, when switching can take place at any time, we should have exact equality at all times. To achieve this we must allow policies which give some service to both jobs in any time-interval, however short. So what is required is an extension of the set of allowed policies to include this sort of shared service, and a theorem to show that the corresponding index policy is optimal.

We shall prove a theorem for a SFABP $\mathscr{F}$ of n jobs under regularity conditions which could almost certainly be weakened, though it is sufficiently general for practical purposes. The distribution function of the service-time of each job is assumed to be piecewise twice continuously differentiable, i.e. twice continuously differentiable except at points each of which lies in an open interval in which it is the only point at which the service-time distribution is not twice continuously differentiable. Denote by E_1 the set of these exceptional points.

Piecewise twice continuous differentiability of the service-time distribution F implies piecewise monotonicity of the index v, i.e. monotonicity in any interval which does not include any of a set E_2 of exceptional points, each of which partitions some open interval into two open sub-intervals, on one of which v is increasing and on the other decreasing. Let (x_{jm}, y_{jm}) be the mth interval on which $v_j(x)$ is decreasing, and n_{jm} be the number (assumed finite) of points $x_{hk} \in E_2$ and such that $v_j(x_{jm}) > v_h(x_{hk}) > v_j(y_{jm})$.

Service for the n jobs is assumed to be available at a constant over-all rate, which, without loss of generality, we assume to be 1. Thus if $u_i(t)$ is the rate of service allocated to job i at time t then $\Sigma_i u_i(t) = 1$. The function $u_i(\cdot)$ is Lebesgue-measurable and depends on the current state of the system, which is the vector of states for the n jobs. The state $x_i(t)$ of job i at time t is its *age*

$$\int_0^t u_i(s)\,ds$$

if it has not been completed, and C if it has been completed. If x^k is the state of the system at the kth job-completion, and this occurs at time t, the allocation vector $u(s) = (u_1(s), u_2(s), \ldots, u_n(s))$ is x^k-measurable for $s > t$ until the $(k+1)$th job-completion. x^k-measurability of $u(s)$ means that although every time is a decision-time, in the sense that $u(s)$ may be changed at any s, all the information relevant to determining $u(s)$ between successive job-completions is available at the first of those job-completions.

Piecewise twice continuous differentiability of the service-time distribution F allows us to define an index policy for $\mathscr{F}$ as follows. The definition is expressed recursively in a kind of pidgin programming language.

(i) Set $t = 0$, $x_i(t) = 0$ $(i = 1, 2, \ldots, n)$.

(ii) Determine the subset $\mathscr{S}_1$ of jobs j such that

$$v_j(x_j(t)) = \max_i v_i(x_i(t)),$$

and the subset $\mathscr{S}_2$ of $\mathscr{S}_1$ for which ageing beyond age $x_j(t)$ produces initially an increase in the index.

(iii) IF: $\mathscr{S}_1$ includes just one job k, GO TO (iv).
$\mathscr{S}_1$ includes more than one job and $\mathscr{S}_2 \neq \emptyset$, select a job $k \in \mathscr{S}_2$ and GO TO (iv).
$\mathscr{S}_1$ includes more than one job and $\mathscr{S}_2 = \emptyset$, GO TO (v).

(iv) Set $u_k(s) = 1$, $u_i(s) = 0$ $\quad (i \neq k, t \leqslant s < v)$, where

$$v = \inf\{s > t: v_k(x_k(s)) \leqslant v_i(x_i(t)) \text{ for some } i \neq k\}.$$

(Note that v is random, as $v_k(C) \leqslant v_i(x_i(t))$ for all i.) Set $t = v$, GO TO (ii).

(v) Choose $u(s)$ so that

$$V_i \rho_i(x_i(s)) = V_j \rho_j(x_j(s)) \qquad (i, j \in \mathscr{S}_1, t \leqslant s < w),$$

$\left[\text{i.e. so that} \right.$

$$\left. V_i \frac{d\rho_i(x_i(s))}{dx_i} u_i(s) = V_j \frac{d\rho_j(x_j(s))}{dx_j} u_j(s) \qquad (i, j \in \mathscr{S}_1, t \leqslant s \leqslant w) \right],$$

$$\Sigma_{\mathscr{S}_1} u_i(s) = 1, \text{ and } u_i(s) = 0 \ (i \notin \mathscr{S}_1),$$

where

$v = \sup\{w: v_i(x_i(s))$ is decreasing if $t \leqslant s < w$, and the interval $(x_i(t), x_i(w))$ includes no E_1 points for job i $(i \in \mathscr{S}_1)$;

$$v_i(x_i(s)) > v_j(x_j(s)) \qquad (i \in \mathscr{S}_1, j \notin \mathscr{S}_1, t \leqslant s < w)\}.$$

(Thus for jobs in $\mathscr{S}_1$, the indices in the interval $[t, v]$ are equal, decreasing, and greater than the indices of jobs outside $\mathscr{S}_1$.) Set $t = v$, GO TO (ii).

Note that this definition includes no explicit rule for stopping. The recursion stops when $v = \infty$ in (iv) or (v).

5.3.3 The continuous-time index theorem for a SFABP of jobs

Theorem 5.1. Index policies are optimal for a SFABP consisting of jobs.

Proof. From Theorem 3.6 we know that index policies are optimal in the class of Δ-policies, for which switching between jobs is restricted to decision-times occurring at intervals Δ. To prove the theorem it would suffice to show that (i) the payoff from one of these index Δ-policies tends as $\Delta \searrow 0$ to the payoff from a continuous-time index policy, and (ii) the maximal payoff from a Δ-policy tends to the continuous-time maximal payoff as $\Delta \searrow 0$. The proof runs roughly along these lines, though we shall need some further refinement of the definition of a Δ-policy, as well as a further class of restrictive policies which will be termed *geometric* Δ-policies. These are technical rather than particularly illuminating considerations, and on a first reading it is probably best to skim fairly rapidly through the rest of this section and the next.

Δ-policies and geometric Δ-policies are discrete policies with no sharing. The two classes of policy are defined by the sets of decision-times to which switching from a job is restricted. For both types of policy these include the completion-time of the job, the set S_1 of starting points of intervals on which the index $v(x)$ is increasing, the set S_2 of points x_2 such that x_2 is the smallest x for which

$$
v(x_1) = \frac{V \displaystyle\int_{x_1}^{x} e^{-\gamma s}\, dF(s)}{\displaystyle\int_{x_1}^{x} e^{-\gamma s}[1 - F(s)]\, ds}
$$

for some $x_1 \in S_1$, and the set S_3 of endpoints of intervals on which $v(x)$ is decreasing.

Call an interval of the form $[x_1, x_2)$, where $x_1 \in S_1$ and $x_2 = \min\{y_2 \in S_2 : y_2 > x_1\}$ ($= \infty$ if this subset of S_3 is empty), a *Type-1* interval, and an interval of the form $[x_2, x_3)$, where $x_2 \in S_1$ and $x_3 = \min\{y_3 \in S_3 : y_3 > x_2\}$ ($= \infty$ if this subset of S_2 is empty), a *Type-2* interval. Thus, if we include in S_2 the point 0 if $v(x)$ is initially decreasing, every point of the positive real line belongs either to a Type-1 interval or to a Type-2 interval for any given job, and the two types of interval alternate.

A Δ-policy is one for which switching is restricted to the sets of points already listed, together with a further set of points determined by some fixed rule independent of the particular policy, successive members of which differ by at most Δ and form the end-points of some open interval. A geometric Δ-policy is similar, except that the additional points occur only in Type-2 intervals, and their spacing is different in different Type-2 intervals. Specifically, each Type-2 interval is a sub-interval of one of the intervals of decreasing $v(x)$; a Type-2 interval for job j which is a sub-interval of (x_{jm}, y_{jm}) is divided by the additional decision points into open sub-intervals of length at most $\Delta n_{jm}^{-1} 2^{-m}$.

Suppose the SFABP $\mathscr{F}$ consists of n jobs. Let $R(\mathscr{F})$, $R^{\Delta}(\mathscr{F})$, and $R^{G\Delta}(\mathscr{F})$ denote the suprema of the payoffs obtainable from an arbitrary policy, from a Δ-policy, and from a geometric Δ-policy, respectively.

Lemma 5.2. $R^{\Delta}(\mathscr{F}) \nearrow R(\mathscr{F})$ as $\Delta \searrow 0$.

Proof. The basic idea is to mimic the effect of applying an arbitrary policy A by constructing a corresponding Δ-policy $\Delta(A)$.

Let $T_j(t)$, $T_j^{\Delta}(t)$, be the age of job j at time t under A, $\Delta(A)$, respectively. Let $s(T, j) = \sup \{ s : T_j(s) \leqslant T \}$. The construction of $\Delta(A)$ proceeds iteratively through the sequence of Δ-policy decision-times which are reached under $\Delta(A)$. Let t_i $(i = 0, 1, 2, \ldots)$ be the $(i+1)$th decision-time. Let k be the smallest (to be definite) h such that

$$s(T_h^{\Delta}(t_i), h) \leqslant s(T_j^{\Delta}(t_i), j) \qquad (j \neq h).$$

Put $u_k^{\Delta}(t) = 1$, $u_j^{\Delta}(t) = 0$ $(j \neq k, t_i \leqslant t < t_{i+1})$. (Since sharing is not allowed under a Δ-policy we must have $u_j^{\Delta}(t) = 1$ for some j.)

Let $s(t) = \min_j s(T_j^{\Delta}(t), j)$. Thus at time t under policy $\Delta(A)$, those parts of the realizations of each job which occur before time $s(t)$ under policy A have already taken place. From the definition of $\Delta(A)$ it follows that

$$0 \leqslant T_j^{\Delta}(t) - T_j(s(t)) \leqslant \Delta \qquad (t \geqslant 0; j = 1, 2, \ldots, n),$$

so that, summing over j,

$$0 \leqslant t - s(t) \leqslant n\Delta. \tag{5.5}$$

The right-hand inequality in (5.5) implies that if s_j, s_j^{Δ} are the completion times of job j under A, $\Delta(A)$, respectively, then $s_j^{\Delta} - s_j \leqslant n\Delta$. Thus

$$V_j \exp(-\gamma s_j^{\Delta}) \geqslant e^{-\gamma n\Delta} V_j \exp(-\gamma s_j) \qquad (j = 1, 2, \ldots, n).$$

Summing over j and taking the expectation over all realizations now gives

$$R_{\Delta(A)}(\mathscr{F}) \geqslant e^{-\gamma n\Delta} R_A(\mathscr{F}),$$

where $R_P(\mathscr{F})$ denotes the payoff under policy P. Thus

$$R(\mathscr{F}) \geqslant R^{\Delta}(\mathscr{F}) = \sup_{\{\Delta\text{-policies } P\}} R_P(\mathscr{F})$$

$$\geqslant \sup_A R_{\Delta(A)}(\mathscr{F}) \geqslant e^{-\gamma n\Delta} \sup_A R_A(\mathscr{F}) = e^{-\gamma n\Delta} R(\mathscr{F}),$$

and the lemma follows. $\qquad\qquad\qquad\qquad\qquad\qquad\qquad\qquad\qquad\qquad\qquad\square$

Let I denote an index policy, and $IG\Delta$ an index policy within the class of geometric Δ-policies.

Lemma 5.3. $R^{G\Delta}(\mathscr{F}) \to R_I(\mathscr{F})$ as $\Delta \searrow 0$.

Proof. From Theorem 3.6 we know that $R_{IG\Delta}(\mathscr{F}) = R^{G\Delta}(\mathscr{F})$, so it suffices to prove that $R_{IG\Delta}(\mathscr{F}) \to R_I(\mathscr{F})$.

For any given job the relationships between the unrestricted index $v(x)$ and the geometric Δ-index $v^{G\Delta}(x)$ include the following. (i) If $x_1 \in S_1$ then $v(x_1) = v^{G\Delta}(x_1)$. (ii) If x_2 and x_3 are successive decision-times belonging to a Type-2 interval, then $v(x_2) \geqslant v^{G\Delta}(x_2) \geqslant v(x_3)$. Both relationships are simple consequences of the expression (5.2) for $v(x)$; the first depends on the fact that every point in S_2 is a decision-time for the class of geometric Δ-policies.

Realizations of $\mathscr{F}$ under an index policy I, like the process-time of individual jobs, may also be divided into time-intervals of two types. In the first type of interval just one job is worked on throughout, and the interval coincides with a Type-1 interval for that job, unless it is completed during the interval. Call such an interval a *Type-1 $\mathscr{F}$-interval*. The second type of interval is made up of portions of Type-2 intervals for one or more jobs, and may include some sharing. Throughout such an interval the maximal index is decreasing. It terminates at the point when the maximal index is for the first time the index of a job for which the index increases with small increases in process-time. Call such an interval a *Type-2 $\mathscr{F}$-interval*. The entire realization of $\mathscr{F}$ is made up of a succession of intervals of one of these two types; a Type-2 interval is always followed by a Type-1 interval, though not vice versa.

Consider now what happens under an index geometric Δ-policy $IG\Delta$. Provided the same rule for resolving ties is used as under policy I, this may be described as follows. The same sequence of Type-1 and Type-2 $\mathscr{F}$-intervals occurs as under I. Each Type-1 interval consists of the same portion of the processing time of the same job as under I. The make-up of corresponding Type-2 intervals is also similar, but not precisely the same because of the approximate nature of the correspondence between $v(x)$ and $v^{G\Delta}(x)$ (relationship (ii)), and because there is no sharing under $IG\Delta$. At the end of a Type-2 $\mathscr{F}$-interval under $IG\Delta$ the process times of all those jobs which are in, or at the end of, a Type-2 (job-) interval may differ from their values at the end of the corresponding Type-2 $\mathscr{F}$-interval under I by up to the maximal interval between decision-times for the Type-2 interval for each of the respective jobs (i.e. by up to $\Delta n_{jm}^{-1} 2^{-m}$ for a job j whose process time belongs to the mth interval of decreasing index-values for that job). Similar discrepancies in process-time may occur throughout each Type-2 $\mathscr{F}$-interval. All this follows from relationships (i) and (ii).

The important point in this comparison of realizations under I and under $IG\Delta$ is that the reward from completing a job j differs in the two cases by an amount which does not tend to zero with Δ only if it occurs in different Type-2

$\mathscr{F}$-intervals, which may happen because of the discrepancy in process times at the end of such an interval. Suppose that completion of job j occurs at times which are separated, for the two index policies, by a Type-1 interval for job h starting at process time y_{hk}. Let

$$t = \inf\{x : v_j(x) < v_h(y_{hk})\},$$

and suppose $x_{jm} < t \leqslant y_{jm}$. It follows from relationship (ii) that

$$t - \Delta n_{jm}^{-1} 2^{-m} \leqslant \sigma_j \leqslant t + \Delta n_{jm}^{-1} 2^{-m}, \tag{5.6}$$

where σ_j is the service time for job j. Since there are at most n_{jm} distinct values of t in the range (x_{jm}, y_{jm}), each corresponding to different values of h and k, and leading to intervals of the form (5.6), it follows that the total Lebesgue-measure $\mathscr{L}(S_j(\Delta))$ of the set $S_j(\Delta)$ of σ_j-values leading to completion-times for job j separated by Type-1 intervals is at most $2\Delta\Sigma_{m=1}^{\infty} 2^{-m} = 2\Delta$, and thus $\to 0$ as $\Delta \to 0$. Since the probability density f_j for σ_j exists almost everywhere on this set, and is bounded above by $f_j(x_{j1})$, and no discontinuity of F_j can belong to an interval on which $v_j(x)$ is decreasing, it follows that

$$\mathbf{P}\{\sigma_j \in S_j(\Delta)\} \to 0 \text{ as } \Delta \to 0 \qquad (j = 1, 2, \dots, n).$$

Since the discounted reward from job j is bounded above by V_j, it follows that the contribution to total payoff from job j under $IG\Delta$ tends to its value under I as $\Delta \to 0$. Since this is true for all j, $R_{IG\Delta}(\mathscr{F}) \to R(\mathscr{F})$ and the lemma follows. $\square$

To complete the proof of the theorem, note that Δ-policies are geometric Δ-policies, so that

$$R^\Delta(\mathscr{F}) \leqslant R^{G\Delta}(\mathscr{F}) \leqslant R(\mathscr{F}).$$

From the two lemmas it therefore follows, on letting $\Delta \to 0$, that $R_I(\mathscr{F}) = R(\mathscr{F})$, as required. $\square$

Corollary 5.4. The index policies derived by setting the discount parameter equal to zero are optimal for jobs under the criterion of minimum expected weighted flow-time.

The proof is virtually identical to the proof of the theorem, using Corollary 3.10 in place of Theorem 3.6. Versions of the approximating Lemmas 5.2 and 5.3 continue to hold, with reference now to expected weighted flow-time instead of expected total reward, and with similar proofs to their counterparts for discounted rewards.

5.3.4 More general continuous-time index theorems for jobs

To extend Theorem 5.1 to jobs subject to precedence constraints in the form of an out-tree, including the possibility of bandit process- (i.e. job-)linked arrivals of further jobs is fairly straightforward.

Theorem 5.5. Index policies are optimal for continuous-time jobs subject to an out-tree of precedence constraints, and with job-linked arrivals.

This is a continuous-time version of Theorem 3.25, restricted to the case of ungeneralized jobs.

The definition of a continuous-time index policy is similar to that described in §5.5, where the set of jobs form a SFABP. The relevant index for a given job is now the index for the minimal sub-family superprocess for which that job is the initial job. The set $\mathcal{S}_0$ of jobs on which processing is allowed at time t is now restricted by the precedence constraints and excludes jobs which have not arrived by then. $\mathcal{S}_1$ is now the subset of $\mathcal{S}_0$ with maximal index values at time t.

To prove the theorem two lemmas are required: one extending Lemma 5.2 to show that a sequence of Δ-policies for decreasing values of Δ may be chosen so as to have payoffs which converge to the optimum payoff from a continuous-time policy as Δ tends to zero; and a second lemma showing that index policies in the class of some suitably modified version of a geometric Δ-policy yield payoffs which tend to the payoff from a continuous-time index policy as Δ tends to zero, along the lines of Lemma 5.3. It is not difficult to provide the required extension to Lemma 5.2, a version of which appears as Lemma 5.7. The extension to Lemma 5.3 is more troublesome because the relationships between the unrestricted index and the (modified) geometric Δ-index are rather more complicated and less easy to establish than those used in the proof of Lemma 5.3. To avoid some tedious detail a complete proof will be given only for the important special case of Poisson arrivals and no precedence constraints under a criterion of minimal expected time-averaged weighted flow-time. This case is simpler because we know from the proof of Theorem 3.28 that the index for a job J, defined by the minimal sub-family superprocess for which J is the initial job, is an increasing function of the ordinary bandit process index for J.

In §3.12 we noted that an independent stream of Poisson arrivals at the rate λ is equivalent to having bandit-process-linked arrivals at the Poisson rate λ for every bandit process. This is because at any time just one bandit process is being continued, and so the corresponding stream of bandit-process-linked arrivals is switched on. At any time then exactly one arrival stream is switched on, and with Poisson streams all at the rate λ it makes no difference whether the single stream is deemed to be independent or to be made up of fragments of bandit-process-linked streams. For continuous-time jobs service at a given time may be shared between more than one job. A job will be said to have a *job-linked Poisson arrival process at the rate* λ if arrivals occur at the Poisson rate λu when the job is receiving a fraction u of the total processing capacity. With this definition the equivalence of independent Poisson arrivals, and bandit-process-(job)-linked Poisson arrivals at the same rate for every bandit process, fairly obviously continues to hold.

Theorem 5.6. Index policies are optimal for continuous-time jobs with Poisson arrivals of jobs belonging to a countable number of classes of similar jobs under a criterion of minimal expected time-averaged weighted flow-time.

Proof. It is sufficient to show that an index policy produces the minimum expected weighted flow-time in a busy period, provided the expected duration of a busy period is finite. Our notation parallels that used in the proof of Theorem 5.1 with C, referring to the expected total weighted flow-time (or cost) in a busy period, in place of R, the expected total discounted reward. For example, $C^\Delta(\mathscr{F})$ is the minimal expected total weighted flow-time in a busy period under a Δ-policy. The proof proceeds via two preliminary lemmas.

Lemma 5.7. $C^\Delta(\mathscr{F}) \to C(\mathscr{F})$ as $\Delta \to 0$.

Proof of Lemma. Since the number N of jobs processed in a busy period is random the argument now involves taking expectations conditional on $N \leqslant n$. These conditional expectations are identified by the notation $|n$. The proof runs along similar lines to that of Lemma 5.2 up to the analogue of the inequality (5.5), which reads

$$C_{\Delta(A)|n}(\mathscr{F}) \leqslant C_{A|n}(\mathscr{F}) + n\Delta c_{\max}, \tag{5.7}$$

where $c_{\max}$ denotes the maximum cost per unit delay in completion, taken over all classes of job in $\mathscr{F}$.

Since the expected duration of a busy period is finite it follows that $C_{P|n}(\mathscr{F})$ tends to $C_P(\mathscr{F})$ as n tends to infinity uniformly over all policies P. Given $\varepsilon < 0$, choose n such that

$$|C_P(\mathscr{F}) - C_{P|n}(\mathscr{F})| < \varepsilon \text{ for every policy } P.$$

It follows from (5.7) that

$$C_{\Delta(A)}(\mathscr{F}) \leqslant C_A(\mathscr{F}) + n\Delta c_{\max} + 2\varepsilon \text{ for all policies } A \text{ and } \Delta > 0.$$

Thus, since $C^\Delta(\mathscr{F}) \leqslant \inf_A C_{\Delta(A)}(\mathscr{F})$, taking the infimum over all policies A gives

$$C^\Delta(\mathscr{F}) \leqslant C(\mathscr{F}) + n\Delta c_{\max} + 2\varepsilon \quad (\Delta > 0).$$

Hence,

$$\limsup_{\Delta \to 0} C^\Delta(\mathscr{F}) \leqslant C(\mathscr{F}) + 2\varepsilon.$$

The lemma follows since $C^\Delta(\mathscr{F}) \geqslant C(\mathscr{F})$ and ε is arbitrary. $\qquad\square$

Note that virtually the same proof establishes the extension of Lemma 5.2 to the discounted case with an out-free of precedence constraints and job-linked arrivals, as required in the proof of Theorem 5.5.

The next lemma is the analogue of Lemma 5.3. A slightly different version of a geometric Δ-policy is now required. Theorem 3.28 shows that under the weighted flow-time criterion the indices v and $v^\dagger$ are equivalent. We therefore proceed, as before, with reference only to ordinary bandit process indices. Again (x_{jm}, y_{jm}) is the mth interval on which $v_j(x)$ is decreasing, and we are interested in points $x_{hk} \in E_{2h}$ and such that $v_j(x_{jm}) > v_h(x_{hk}) > v_j(y_{jm})$. Let n_{jhm} be the number of these points. (Note that the suffices j and h now refer to the jth and hth job classes instead of to jobs j and h.)

For a class j job the set of allowed switching points under a geometric Δ-policy is $S_1 \cup S_2 \cup S_3 \cup S_4$. The definitions of S_1, S_2 and S_3 are as given in the proof of Theorem 5.1. The set of points S_4 is such that if $v_j(x_{jm}) > v_h(x_{hk}) < v_j(y_{jm})$ then $\exists x$ and $y \in S_4$ with the properties that

(i) $x_{jm} \geqslant x \geqslant y \geqslant y_{jm}$,

(ii) $x - y \leqslant \Delta n_{jhm}^{-1} 2^{-h-m}$,

(iii) $v_j(x) \geqslant v_h(x_{hk}) \geqslant v_j(y)$.

Lemma 5.8. $C^{G\Delta}(\mathcal{F}) \to C_I(\mathcal{F})$ as $\Delta \to 0$.

Proof. This closely follows the proof of Lemma 5.3. The analogue of (5.6) is

$$t - \Delta n_{jhm}^{-1} 2^{-h-m} \leqslant \sigma_j \leqslant t + \Delta n_{jhm}^{-1} 2^{-h-m}. \tag{5.8}$$

Now defining $S_j(\Delta)$ to be the set of σ_j-values within intervals of the form (5.8), it once again follows that $\mathcal{L}(S_j(\Delta)) \to 0$ as $\Delta \to 0$, and $S_j(\Delta_1) \subset S_j(\Delta_2)$ if $\Delta_1 < \Delta_2$, and hence that

$$\mathbf{P}\{\sigma_j \in S_j(\Delta)\} \to 0 \text{ as } \Delta \to 0.$$

Since the expected flow time for a class j job arriving during the first busy period is bounded above, for any policy (in fact bounded above by the expected duration of the busy period), it follows that the expected flow time for the rth class j job to arrive during the first busy period under $IG\Delta$ tends to its value under I as $\Delta \to 0$. Since this is true for all j and r, and both the expected duration of a busy period and the expected number of jobs in a busy period are finite, it follows by the dominated convergence theorem that $C_{IG\Delta}(\mathcal{F}) \to C(\mathcal{F})$. A dominating function is obtained by assigning to each job the expected weighted flow time during the first busy period which results from assuming that if it arrives before the end of that busy period its service is completed at the end of the busy period. Since index policies are optimal among geometric Δ-policies the lemma follows. $\qquad \square$

The theorem now follows from the two lemmas just as in the case of Theorem 5.1. Δ-policies are geometric policies, hence

$$C^\Delta(\mathcal{F}) \geqslant C^{G\Delta}(\mathcal{F}) \geqslant C(\mathcal{F}).$$

Now let $\Delta \to 0$. $\qquad \square$

5.4 OPTIMAL POLICIES FOR QUEUES OF JOBS

5.4.1 Introduction

We now turn to an examination in some detail of the nature of the optimal policy for continuous-time jobs with bitonic completion rates Poisson arrivals of further jobs, and discounted rewards on completion.

For an uncompleted job of age x the expected return from the allocation of unit effort for a short interval Δ is $V\rho(x)\Delta + o(\Delta)$. The expression (5.2) for the index $v(x)$ of such a job in the absence of precedence constraints and arrivals shows that if there are values of $s > x$ at which

$$f(s)/[1 - F(s)] = \rho(s) > \rho(x),$$

these may make the maximized time-averaged reward rate $v(x)$ greater than the immediate reward rate $V\rho(x)$. The extent of this upward averaging is attenuated by the discount factor $e^{-\gamma(s-x)}$, reflecting the fact that if the job is completed at age s after being allocated unit effort between ages x and s the return is reduced by this factor compared with completion at age x.

This observation gives a clue to the effect of Poisson arrivals on the relative magnitudes of the optimizing allocation indices for a set of jobs. Theorem 5.5 tells us that the appropriate indices are those for the minimal sub-family super-processes for which each job is the initial job. For the sake of brevity these indices will be termed simply *arrivals* indices, in contrast with the *no-arrivals* indices given by (5.2). From Theorem 3.11 it follows that the arrivals index for a job is also a maximized time-averaged reward rate, with the maximization now extended to the minimal sub-family of jobs generated by the job-linked arrival process.

With Poisson arrivals the job-linked arrival process is the same for every job, so we might expect to find the ordering of the arrivals indices for different jobs to be largely determined by time-averaged reward rates for the jobs themselves up to some stopping-time, ignoring the rewards derived from arriving jobs, but taking account of interruptions in processing a given job when the processor is assigned to a newly arrived job. These interruptions accentuate the discounting of the reward rate at age $s > x$, compared with the reward rate $V\rho(x)$ at age x, which was noted in the last but one paragraph. Thus if jobs A and B have the same no-arrivals index values, and job A has a higher current reward rate than job B, we may expect to find that job A also has a higher arrivals index than job B.

These considerations point in the right direction, without telling the whole story. This is set out in some detail for a restricted class of Poisson arrival processes in the remainder of this section. Note, however, that the accentuated discounting of future reward rates when arrivals cause interruptions occurs only for $\gamma > 0$. In the limit as $\gamma \to 0$ it does not occur. This limiting case corresponds to the time-averaged weighted flow-time criterion for which, as shown in the previous section, arrivals indices are ordered in precisely the same way as no-arrivals indices.

5.4.2 The case of two classes of jobs

For simplicity consider the FABP $\mathscr{F}$ formed by just two classes, 1 and 2, of identical jobs arriving in independent Poisson streams at rates λ_1 and λ_2, with a positive discount parameter γ. Jobs belonging to the respective classes will be termed 1-jobs and 2-jobs, and the suffices 1 and 2 distinguish the rewards and completion-time distributions for jobs from the two classes. The completion rates $\rho_i(x)$ are assumed to be continuous, bitonic (i.e. to have a unique local maximum, at age M_i for i-jobs, and to tend to zero as x tends to infinity ($i = 1, 2$). The arrivals and no-arrivals indices for an i-job of age x are denoted by $\alpha_i(x)$ and $\beta_i(x)$ respectively ($i = 1, 2$). To be definite suppose $\beta_1(0) \geqslant \beta_2(0)$.

There seems to be no convenient explicit expression for the arrivals indices $\alpha_i(x)$ ($i = 1, 2$) which define the optimal policy. The policy itself, on the other hand, may be determined explicitly. Some additional notation and a series of nine propositions are required for this purpose.

Define: $\mathscr{J}(i, x) =$ sub-family of alternative jobs defined by an i-job of age x together with job-linked Poisson arrivals.

$R_{gA}(i, x) = \mathbf{E}$ [total discounted reward from $\mathscr{J}(i, x)$ under policy g and using the stopping set A].

$s_{gA}(i, x) =$ time taken by $\mathscr{J}(i, x)$ to reach the set A under policy g.

$W_{gA}(i, x) = \gamma^{-1} \mathbf{E}[1 - \exp\{-\gamma s_{gA}(i, x)\}]$.

$A(y) =$ stopping set for $\mathscr{J}(i, x)$ made up of those points for which the initial i-job is completed, or is uncompleted and of age $y (> x)$, and every job which has arrived has an arrivals index value less than or equal to $\alpha_i(y)$.

$P(y) =$ class of policies for which no job has received more processing when the set $A(y)$ is reached than it would have done under the index policy, which we denote by f.

$s_{A(y)}(i, x) = s_{gA(y)}(i, x)$, $W_{A(y)}(i, x) = W_{gA(y)}(i, x)$ ($g \in P(y)$), noting that $s_{gA(y)}(i, x)$ is the same for all $g \in P(y)$.

$\alpha_i(x, y) = [W_{A(y)}(i, x)]^{-1} R_{fA(y)}(i, x)$, where f denotes the index policy.

$x + T(i, x) =$ age at completion of an i-job which is uncompleted at age x. Thus $T(i, x) =$ further processing time required by an i-job of age x.

In the statements of the propositions which follow, $i = 1$ or 2, $x \geqslant 0$, and $y \geqslant 0$.

Proposition 5.9. $\alpha_i(x) \geqslant \beta_i(x)$.

Proof. Both $\alpha_i(x)$ and $\beta_i(x)$ are suprema of time-averaged reward rates. In the case of $\alpha_i(x)$ the supremum is taken over a larger set.   □

Proposition 5.10. (i) $\alpha_i(x) \leqslant \max\{\beta_i(x), \beta_1(0)\}$. (ii) If $\beta_i(x) \geqslant \beta_1(0)$ then $\alpha_i(x) = \beta_i(x)$.

Proof. (i) $\alpha_i(x)$ is the time-averaged reward rate maximized over the minimal sub-family $\mathscr{F}_1$ of jobs formed by an i-job of age x, together with the Poisson arrival processes of 1-jobs and 2-jobs. The maximization is over the allocation of unit effort-rate to the jobs in the minimal sub-family up to some stopping time, subject to the constraint that effort may not be allocated to any job before it arrives. Consider now a second sub-family $\mathscr{F}_2$ comprising the same set of jobs, but now with no constraint on the jobs to which effort may be allocated. Thus $\mathscr{F}_2$ is a simple family of alternative jobs, comprising initially an i-job of age x and infinite numbers of 1-jobs and of 2-jobs all of age zero. From Lemma 3.7(i) (which may easily be shown to hold for jobs with no restriction on switching-times) it follows that

$$v(\mathscr{F}_2) = \max\{\beta_i(x), \beta_1(0), \beta_2(0)\}.$$

Clearly

$$v(\mathscr{F}_1) \leqslant v(\mathscr{F}_2),$$

since the removal of constraints can only increase the maximized reward rate, and part (i) of the proposition follows, since $\alpha_i(x) = v(\mathscr{F}_1)$ and $\beta_1(0) \geqslant \beta_2(0)$. Part (ii) is an immediate consequence of part (i), together with Proposition 5.9.   □

Proposition 5.11. If

$$z = \sup\{y \geqslant x : \alpha_i(t) \geqslant \alpha_i(x) \text{ for } t \in [x, y]\} > x,$$

then

$$\alpha_i(x) = \alpha_i(x, z).$$

Conversely, the reward rate $\alpha_i(x)$ is not achieved if any of the jobs in $\mathscr{J}(i, x)$ is stopped when its arrivals index is greater than $\alpha_i(x)$.

If $z = x$, then $\alpha_i(x) = \lim_{y \to x} \alpha_i(x, y)$.

Proof. For any superprocess the allocation index is defined as a time-averaged expected reward rate maximized over policies, and over stopping times for the bandit process which is defined by applying a given policy to the superprocess. A stopping time is defined by designating a subset of the state-space as a stopping set. That for $z > x$ the stopping set is $A(z)$ follows from the version of Lemma 3.23 appropriate to continuous-time jobs, which may be established by a straightforward adaptation of the proof. That the required policy is an index policy is a consequence of Theorem 5.5.

It follows that

$$\alpha_i(x) = \sup_{y > x} \alpha_i(x, y). \tag{5.9}$$

For the case $z = x$ a further simple adaptation of the proof of Lemma 3.23 shows that

$$\alpha_i(x) > \alpha_i(x, y) \text{ for all } y > x. \tag{5.10}$$

From (5.9) and (5.10) it follows that

$$\alpha_i(x) = \lim_{y \to x} \alpha_i(x, y). \qquad \square$$

A random variable X with distribution function F_X is said to be *stochastically greater* than a random variable Y with distribution function F_Y if $F_X(s) < F_Y(s)$ ($s \in \mathbf{R}$). If the inequality holds for values of s in some interval I we shall say that X is stochastically greater than Y *in I*.

Proposition 5.12. If the completion (or hazard) rates ρ_X and ρ_Y for the random variables X and Y are such that $\rho_X(s) < \rho_Y(s)$ ($0 < s < t$), then X is stochastically greater than Y in $(0, t)$.

Proof. The proposition follows immediately from the easily verified identity

$$F_X(s) = 1 - \exp\left\{ -\int_0^s \rho_X(u) \, du \right\}. \qquad \square$$

Proposition 5.13. If $T(i, x)$ is stochastically greater than $T(i, y)$ in the interval $(0, \Delta)$ then $\alpha_i(x, x + \Delta) < \alpha_i(y, y + \Delta)$.

Proof. Any form of joint distribution for the residual processing times of the two respective initial i-jobs may be assumed without affecting the expected reward rates, provided the marginal distributions remain unaltered and no dependence between processing times in the same sub-family is introduced. Suppose, then, that realizations of $T(i, x)$ and $T(i, y)$ are matched so that

$$\mathbf{P}(T(i, x) > T(i, y) \mid T(i, y) < \Delta) = 1,$$

which is clearly compatible with the assumed stochastic inequality. Suppose, also, that the arrival times and service times for the jobs in the arrival processes are the same for the two different initial i-jobs.

Given any policy P for allocating effort between the jobs in the minimal sub-family $\mathscr{J}(i, x)$, the same policy may be applied to $\mathscr{J}(i, y)$, the initial i-job receiving the same allocation in both cases. If $T(i, y) < \Delta$, in which case $T(i, x) > T(i, y)$, it follows that processing continues on the initial i-job of $\mathscr{J}(i, y)$ after it has been completed.

The rewards which occur before the age of the initial i-job increases by more than Δ are therefore precisely the same for $\mathscr{J}(i, x)$ and $\mathscr{J}(i, y)$ under policy P, except that the i-job is completed earlier in the case of $\mathscr{J}(i, y)$. Thus if the i-job is completed before the end of the Δ ageing interval for $\mathscr{J}(i, x)$ it is also completed within the Δ interval for $\mathcal{J}(i, y)$, and since it is completed earlier the reward V_i is discounted less. The expected total discounted reward is therefore greater for $\mathcal{J}(i, y)$. The proposition follows by taking the supremum over all policies and allowed stopping sets. $\square$

Proposition 5.14. $\alpha_i(x)$ and $\beta_i(x)$ are (i) increasing functions of x for $x < M_i$, (ii) decreasing functions of x for $x > M_i$.

Proof. Since $\alpha_i(x) = \beta_i(x)$ when $\lambda_1 = \lambda_2 = 0$ it is sufficient to provide a proof for $\alpha_i(x)$. For part (i) this will be done by supposing that the proposition is not true and showing that this leads to a contradiction.

Suppose then that $\exists\, x_1, x_2$ such that $0 \leqslant x_1 < x_2 \leqslant M_i$ and $\alpha_i(x_1) > \alpha_i(x_2)$. From Proposition 5.11, and since $\rho_i(x)$ is increasing for $x < M_i$ and continuous at M_i, it follows that $\exists\, x_3, x_4$ such that $x_1 \leqslant x_3 < x_4 \leqslant x_2$, $x_4 - x_3 = \Delta$, and

$$\alpha_i(x_3) = \alpha_i(x_3, x_3 + \Delta) > \alpha_i(x_2) \geqslant \alpha_i(x_2, x_2 + \Delta), \tag{5.11}$$

$$\rho_i(x_3 + s) < \rho_i(x_2 + s) \qquad (0 < s < \Delta). \tag{5.12}$$

However, from Propositions 5.12 and 5.13 it follows that (5.11) and (5.12) cannot hold simultaneously. This is the required contradiction.

Part (ii) is an immediate consequence of Propositions 5.12 and 5.13. $\square$

Proposition 5.15. If $x_1 \geqslant M_1$, $x_2 \geqslant M_2$, and

$$V_1 \rho_1(x_1) = V_2 \rho_2(x_2), \text{ then } \alpha_1(x_1) = \alpha_2(x_2).$$

Proof. From Propositions 5.14(ii) and 5.11 it follows that

$$\alpha_i(x_i) = \lim_{\Delta \to 0} \alpha_i(x_i, x_i + \Delta) \qquad (i = 1, 2).$$

The maximized reward rate $\alpha_i(x_i, x_i + \Delta)$ satisfies the equation

$$\alpha_i(x_i, x_i + \Delta) = \sup_{\{g, A, h, B\}} \frac{V_i \rho_i(x_i)\Delta + R_{gA}(1, 0)\lambda_1 \Delta + R_{hB}(2, 0)\lambda_2 \Delta}{\Delta + W_{gA}(1, 0)\lambda_1 \Delta + W_{hB}(2, 0)\lambda_2 \Delta} + o(\Delta).$$

Hence

$$\alpha_i(x_i) = \sup_{\{g, A, h, B\}} \frac{V_i \rho_i(x_i) + R_{gA}(1, 0)\lambda_1 + R_{hB}(2, 0)\lambda_2}{1 + W_{gA}(1, 0)\lambda_1 + W_{hB}(2, 0)\lambda_2} \tag{5.13}$$

and the proposition follows. $\square$

Proposition 5.16. $\beta_i(x) = V_i \rho_i(x) \qquad (x > M_i; i = 1, 2).$

Proof. This follows from (5.13) on putting $\lambda_1 = \lambda_2 = 0$. $\square$

Proposition 5.17. $\alpha_1(\infty)=\alpha_2(\infty)<\alpha_2(0)$.

Proof. With $\leqslant$ in place of $<$ the proposition follows from equation (5.13), together with the assumption that $\rho_i(x)\to0$ as $x\to\infty$. We get strict inequality unless $\rho_2(0)=0$.

To obtain strict inequality when $\rho_2(0)=0$, note that it follows from the converse part of Proposition 5.11 that

$$\alpha_2(0)>\lim_{\Delta\to0}\alpha_2(0,\Delta),$$

and from (5.13) that

$$\lim_{\Delta\to0}\alpha_2(0,\Delta)=\lim_{x\to\infty}\alpha_i(x),\qquad(i=1,2).\qquad\square$$

Propositions 5.10, 5.14, 5.15 and 5.17 lead to a nearly complete specification of the optimal policy.

Let $t_1=\inf\{t:\beta_2(t)>\beta_1(0)\}$ and $t_2=\sup\{t:\beta_2(t)>\beta_1(0)\}$. Let $s(x)$ and $t(x)$ $(>M_2)$ be such that

$$\alpha_1(s(x))=\alpha_2(t(x))=\alpha_2(x)\qquad(0\leqslant x\leqslant t_1);$$

these functions are well-defined since it follows from equation (5.13) that $\alpha_i(t)$ is continuous for $t>M_i$ $(i=1,2)$. Note that it follows from Proposition 5.15 that $V_1\rho_1(s(x))=V_2\rho_2(t(x))$. We may therefore write

$$s(x)=\rho_1^{-1}[V_1^{-1}V_2\rho_2(t(x))],\tag{5.14}$$

adopting the convention that the inverse function ρ_1^{-1} only takes values to the right of M_1.

The function $t(x)$, together with t_1, t_2, and Proposition 5.16, completely determines the priorities between a 2-job and a newly arrived 1-job. If the 1-job arrives when the 2-job is of age $x<t_1$ it receives service until it reaches age $t(x)$; thereafter service is shared so as to preserve the equality $V_1\rho_1(x_1)=V_2\rho_2(x_2)$, where x_1 and x_2 are the respective ages of the 1-job and of the 2-job at a given time. If the 1-job arrives when the 2-job is of age $x\in(t_1,t_2)$ the 2-job continues to receive service until it reaches age t_2; the 1-job is then served until it reaches age $s(t_1)$; and service is then shared so that $V_1\rho_1(x_1)=V_2\rho_2(x_2)$. If the 1-job arrives when the 2-job is of age $x>t_2$ it receives service until it is of age y such that $y>M_1$ and $V_1\rho_1(y)=V_2\rho_2(x)$, and service is then shared so that $V_1\rho_1(x_1)=V_2\rho_2(x_2)$. This is all subject to the fact that no job receives service after it has been completed.

Further arrivals may, of course, interrupt service to one or both jobs, but they will do so according to the same rules. During the phases of shared service there may be more than two jobs, which may be 1-jobs or 2-jobs, between which service is shared so that $V_1\rho_1(x_1)=V_2\rho_2(x_2)=$ a (time-dependent) constant, for every such 1-job and 2-job. Jobs receiving shared service will be described as *ageing in step*.

The allocation of processor time determined by an index policy may be set out in similar fashion for the three other cases of interest: a 1-job arriving when another 1-job is being processed; and a 2-job arriving when either a 1-job or a 2-job is being processed. In every case t_1, t_2, and the functions $t(x)$ and $s(x)$ (given by (5.14) in terms of $t(x)$) are sufficient to specify the policy. A newly arrived 1-job receives priority over a 1-job receiving service iff the job receiving service is of age at least $s(t_1)$. A newly arrived 2-job receives priority over a 1-job (or 2-job) receiving service iff the 1-job (2-job) receiving service is of age at least $s(0)(t(0))$. In all cases, after a sufficiently long interval with no further arrivals the uncompleted jobs age in step.

In order, then, to give a complete prescription for the optimal policy it is sufficient to determine t_1, t_2 and $t(x)$. From their definitions, and using equation (5.2) and Propositions 5.11, 5.14 and 5.16, it follows that t_1 and t_2 ($> M_2$) are the unique solutions of the equations

$$\sup_{t>0} \frac{V_1 \int_0^t f_1(s)e^{-\gamma s}\,ds}{\int_0^t [1-F_1(s)]e^{-\gamma s}\,ds} = V_2 \rho_2(t_2) = \frac{V_2 \int_{t_1}^{t_2} f_2(s)e^{-\gamma s}\,ds}{\int_{t_1}^{t_2}[1-F_2(s)]e^{-\gamma s}\,ds}. \qquad (5.15)$$

Numerical evaluation is straightforward. To determine $t(x)$ an interchange argument will be used.

Suppose there are two 2-jobs ready for processing, one of age $x(<t_1)$ and the other of age $y(>t_2)$. Call them jobs A and B. Let $\mathscr{J}(A)$ and $\mathscr{J}(B)$ denote the sub-families initiated by jobs A and B, respectively, together with job-linked arrivals. Let $R(AB)$ denote the expected reward which results from applying an index policy to $\mathscr{J}(A)$ until it reaches the set $A(y)$, and then applying an index policy to $\mathscr{J}(B)$ until it reaches the set $A(y+\Delta)$. Call this policy AB. Let BA denote the policy for which (i) the processing times lasting for $y-x$ on job A and Δ on job B, or until the respective jobs are completed, are interchanged, (ii) the service of job B, followed by job A, is interrupted by, and shared with, the jobs which arrive in such a fashion that unless and until job A is completed before reaching age y these jobs receive service precisely as they would have done under AB, and (iii) if job A is completed before reaching age y an index policy is applied next to the uncompleted jobs in $\mathscr{J}(A)$ until it reaches the set $A(y)$, and then to the uncompleted jobs of $\mathscr{J}(B)$ until it reaches the set $A(y+\Delta)$.

Thus, for both AB and BA, a point is reached when $\mathscr{J}(A)$ is in a state belonging to the set $A(y)$, and $\mathscr{J}(B)$ is in a state belonging to the set $A(y+\Delta)$. For given realizations of $\mathscr{J}(A)$ nd $\mathscr{J}(B)$ these states are the same for AB and BA, as is the time $s_{A(y)}(2,x)+s_{A(y+\Delta)}(2,y)$ at which they are both reached. Let $R(BA)$ be the expected reward from applying the policy BA up to this point of coincidence. Let the continuation of both AB and BA from this point onwards be optimal.

Proposition 5.18. $t(x)$ is the largest value of y for which $R(AB) - R(BA)$ is $O(\Delta^2)$.

Proof. Suppose first that $y > t(x)$, so that $\alpha_2(y) < \alpha_2(x)$, and consequently AB differs from an index policy only when the jobs in $\mathcal{J}(B)$ are being processed and $\mathcal{J}(B)$ has not yet reached the set $A(y + \Delta)$. The expected time for which an index policy is not followed is $O(\Delta)$, and the extent of departures from an index policy in terms of the indices of the jobs processed is also $O(\Delta)$. It follows that $R(AB)$ differs by $O(\Delta^2)$ from the supremum of the expected reward up to time $s_{A(y)}(2, x) + s_{A(y+\Delta)}(2, y)$ over all policies for which the state at that point is the same as under AB. Policy BA, on the other hand, also differs from an index policy for an expected time which is $O(\Delta)$, and does so to an extent which is $O(1)$ in terms of difference of index values of jobs processed from the maximum index over the available jobs, so that $R(BA)$ falls short of the maximum expected reward up to time $s_{A(y)}(2, x) + s_{A(y+\Delta)}(2, y)$ by an amount which is $O(\Delta)$. If $y = t(x)$, then $\alpha_2(x) = \alpha_2(y)$, and both AB and BA differ from an index policy for times of expectation $O(\Delta)$, and to an extent which is $O(\Delta)$, so that both $R(AB)$ and $R(BA)$ are within $O(\Delta^2)$ of optimality, and therefore differ by $O(\Delta^2)$. $\qquad\square$

Let us now turn to the evaluation of $R(AB) - R(BA)$.

Note first that under both AB and BA interruptions to the service of the 2-jobs A and B up to the time $s_{A(y)}(2, x) + s_{A(y+\Delta)}(2, y)$ from which the two policies coincide may occur because of arrivals of 1-jobs, but do not occur because of the arrival of further 2-jobs. This is because the required value of y is $t(x)$, where $\alpha_2(t(x)) = \alpha_2(x) \geqslant \alpha_2(0)$, and it follows from Proposition 5.14 that $\alpha_2(t) \geqslant \alpha_2(0)$ $(0 \leqslant t \leqslant t(x))$. We may therefore without loss of generality assume that $\lambda_2 = 0$, and for simplicity will write λ for λ_1.

The functions $a(s)$, $b(s)$ and $c(s)$ $(0 \leqslant s \leqslant y - x)$, will be used in writing down an expression for $R(AB) - R(BA)$. These functions are defined by the following equations.

$$a(s) = \exp(\gamma s) \mathbf{E}\{\exp[-\gamma s_{A(x+s)}(2, x)] \mid T(2, x) \geqslant s\}.$$

$$b(s) = \exp(\gamma s) \mathbf{E}\{\exp[-\gamma s_{A(y)}(2, x)] \mid T(2, x) = s\}.$$

$V_1 \exp(-\gamma s) c(s) = \mathbf{E}[$total discounted reward from $\mathcal{J}(2, x)$ under an index policy until state $A(y)$ is reached $\mid T(2, x) = s]$
$\quad - \mathbf{E}[$total discounted reward from $\mathcal{J}(2, x)$ under an index policy until state $A(x + s)$ is reached $\mid T(2, x) \geqslant s]$
$\quad - V_2 \exp(-\gamma s) a(s).$

These definitions mean that under the policy AB and conditional on job A requiring a further processing time s, $e^{-\gamma s} a(s)$ is the expected discount factor for job A, $e^{-\gamma s} b(s)$ is the expected discount factor at the commencement of service on job B, and $V_1 e^{-\gamma s} c(s)$ is the expected total discounted reward from the period after the completion of job A and before the start of job B, when 1-jobs are served until all of them have index values no greater than $\alpha_2(y)$. Explicit expressions for $a(s)$, $b(s)$ and $c(s)$ are derived in the next section. They depend, of course, on x and y as well as on s.

The only rewards which make an $O(\Delta)$ contribution to $R(AB) - R(BA)$ are those from jobs A and B themselves, and those from 1-jobs completed when job A has been completed and job B has not been started under AB. Thus we have

$$
\begin{aligned}
V_2^{-1} & [R(AB) - R(BA)][1 - F_2(x)] \\
& = \left[\int_0^{y-x} e^{-\gamma s}(a(s) + V_1 V_2^{-1} c(s) + \Delta\rho_2(y)b(s)) f_2(x+s)\,\mathrm{d}s \right. \\
& \qquad \left. + e^{-\gamma(y-x)} \Delta f_2(y)b(y-x) \right] \\
& \quad - \left[\Delta\rho_2(y)(1 - F_2(x)) + \int_0^{y-x} e^{-\gamma(s+\Delta)}(a(s+\Delta) \right. \\
& \qquad \left. + V_1 V_2^{-1} c(s+\Delta)) f_2(x+s)\,\mathrm{d}s \right] + O(\Delta^2) \\
& = \Delta\rho_2(y) \left[\int_0^{y-x} e^{-\gamma s} b(s) f_2(x+s)\,\mathrm{d}s \right. \\
& \qquad \left. + (1 - F_2(y))e^{-\gamma(y-x)} b(y-x) - 1 + F_2(x) \right] \\
& \quad - \Delta \int_0^{y-x} e^{-\gamma s} [-\gamma(a(s) + V_1 V_2^{-1} c(s)) + a'(s) \\
& \qquad + V_1 V_2^{-1} c'(s)] f_2(x+s)\,\mathrm{d}s + O(\Delta^2),
\end{aligned}
\tag{5.16}
$$

where primes denote derivatives. Integration by parts shows that the expression in square brackets multiplying $\Delta\rho_2(y)$ on the right-hand side of equation (5.16) may be written as

$$
\int_0^{y-x} [1 - F_2(x+s)]\,\mathrm{d}[e^{-\gamma s} b(s)].
$$

A little further manipulation of the other terms on the right-hand side of equation (5.16), and application of Proposition 5.18, leads to the following result.

Proposition 5.19. $t(x)$ is the largest value of y for which

$$
\rho_2(y) = \frac{\displaystyle\int_0^{y-x} f_2(x+s)\,\mathrm{d}[e^{-\gamma s} a(s)] + V_1 V_2^{-1} \int_0^{y-x} f_2(x+s)\,\mathrm{d}[e^{-\gamma s} c(s)]}{\displaystyle\int_0^{y-x} [1 - F_2(x+s)]\,\mathrm{d}[e^{-\gamma s} b(s)]}.
\tag{5.17}
$$

Equation (5.17) reduces to the form

$$\rho_2(y) = \frac{\displaystyle\int_0^{y-x} f_2(x+s)e^{-\gamma s}\,ds}{\displaystyle\int_0^{y-x} [1-F_2(x+s)]e^{-\gamma s}\,ds} \tag{5.18}$$

when there are no arrivals, since in this case $a(s)=b(s)=1$ and we may assume $V_1=0$. The expression (5.18) also follows directly from equation (5.2) and Propositions 5.11 and 5.16.

5.4.3 Expressions for $a(s)$, $b(s)$ and $c(s)$

The three functions $a(s)$, $b(s)$ and $c(s)$ may be determined by a little judicious use of generating functions. For simplicity consider first the case where the initial age x of job A is zero.

Let $B(u)$ be the busy period initiated by a single newly arrived 1-job if the server works only on uncompleted 1-jobs of age not more than u. Let $N(u)$ be the number of uncompleted 1-jobs of age u at the end of a busy period defined in this way. Let $\Gamma(u,\gamma,z)$ denote the generating function

$$E[e^{-\gamma B(u)} z^{N(u)}].$$

Suppose now that 1-jobs arrive in a Poisson process at the rate λ (no suffix is needed as arrivals of 2-jobs may be disregarded) starting at time 0, that none of them is processed until time t, and starting at time t successive busy periods of the type described in the previous paragraph occur, one for each of the arrivals in the interval $(0,t)$. Let $u(w)$ be the maximum age in the busy period associated with an arrival at time w $(0<w<t)$, which may therefore be written as $B(u(w))$. Suppose $u(w)$ to be a continuous function of w. Let $w_1, w_2, \ldots, w_n$ be the times of arrival in $(0,t)$, so that the total busy period starting at time t may be written

$$B = \sum_{i=1}^n B(u(w_i)).$$

Proposition 5.20. $Ee^{-\gamma B} = \exp\left\{\lambda \int_0^t \Gamma(u(w),\gamma,1)\,dw - \lambda t\right\}.$

Proof. Write $t_{ij} = [m(i-1)+j-1]t/km$ $(i=1,2,\ldots; j=1,2,\ldots)$, where $k, m \in \mathbf{Z}$,

N_{ij} = number of 1-job arrivals in the interval (t_{ij}, t_{ij+1}),

B_{ij} = total of the busy periods initiated by arrivals in the interval (t_{ij}, t_{ij+1}).

Thus $B = \displaystyle\sum_{i=1}^k \sum_{j=1}^m B_{ij}.$

For a Poisson process the numbers of arrivals in disjoint intervals are independent. It follows that the N_{ij}'s and B_{ij}'s are independent, except when they refer to the same interval, so that

$$\mathbf{E}\exp\{-\gamma B\} = \prod_{i=1}^{k} \prod_{j=1}^{m} \mathbf{E}\exp\{-\gamma B_{ij}\}.$$

Now

$$\begin{aligned}
\mathbf{E}\exp\{-\gamma B_{ij}\} &= \mathbf{E}(\exp\{-\gamma B_{ij}\}|N_{ij}=0)\mathbf{P}(N_{ij}=0) \\
&\quad + \mathbf{E}(\exp\{-\gamma B_{ij}\}|N_{ij}=1)\mathbf{P}(N_{ij}=1) \\
&\quad + \mathbf{E}(\exp\{-\gamma B_{ij}\}|N_{ij}>1)\mathbf{P}(N_{ij}>1) \\
&= 1[1 - \lambda t(km)^{-1} + o((km)^{-1})] \\
&\quad + [\mathbf{E}\exp\{-\gamma B(t_{i1})\} + \delta_{ij}][\lambda t(km)^{-1} + o((km)^{-1})] \\
&\quad + o((km)^{-1}),
\end{aligned}$$

where, using the uniform continuity of continuous functions on compact intervals, $\delta_{ij} \to 0$ uniformly over i and j as $k \to \infty$. Thus

$$\mathbf{E}\exp\{-\gamma B\} = \prod_{i=1}^{k} \prod_{j=1}^{m} \{1 - \lambda t(km)^{-1}[1 - \Gamma(x(t_{i1}), \gamma, 1) + \delta_{ij}] + o((km)^{-1})\}.$$

Letting $m \to \infty$, and since

$$\lim_{m \to \infty} (1 + xm^{-1} + o(m^{-1}))^{m} = e^{x} \quad (x \in \mathbf{R}),$$

it follows that

$$\begin{aligned}
\mathbf{E}\exp\{-\gamma B\} &= \prod_{i=1}^{k} \exp\{-\lambda t k^{-1}[1 - \Gamma(x(t_{i1}), \gamma, 1) + \delta_{i}]\} \\
&= \exp\left\{-\lambda t\left[1 - k^{-1}\sum_{i=1}^{k}[\Gamma(x(t_{i1}), \gamma, 1) + \delta_{i}]\right]\right\},
\end{aligned}$$

where $\delta_{i} \to 0$ uniformly over i as $k \to \infty$. Now letting $k \to \infty$ and using the dominated convergence theorem the proposition follows. $\quad\square$

The next proposition uses the functions $s(w)$ and $t(w)$ $(0 \leqslant w \leqslant t_{1})$, defined in the discussion following Proposition 5.17.

Proposition 5.21

$$a(t) = \begin{cases} \exp\left\{\lambda \int_0^t \Gamma(s(w), \gamma, 1)dw - \lambda t\right\} & (0 < t < t_1), \\[2mm] \exp\left\{\lambda \int_0^{t_1} \Gamma(s(w), \gamma, 1)dw - \lambda t_1\right\} & (t_1 < t < t_2), \\[2mm] \exp\left\{\lambda \int_0^{t^{-1}(t)} \Gamma(s(w), \gamma, 1)dw + \lambda(t - t^{-1}(t))\Gamma(s(t^{-1}(t)), \gamma, 1) - \lambda t\right\} & (t_2 < t). \end{cases}$$

Proof. $(0 < t < t_1)$. If job A is completed after processing for a time t, and this processing time is interrupted by 1-jobs which arrive at process times $w_1, w_2, \ldots, w_n$, the time of completion under policy AB has the same distribution as

$$t + \sum_{i=1}^{n} B(s(w_i)).$$

This follows since (i) the arrival times w_i are governed by a Poisson process, and therefore have the same joint distribution whether the busy periods associated with arrivals at each of these times occur immediately, as under policy AB, or are deferred to time t, as described in the preamble to Proposition 5.20, and (ii) the busy periods $B(s(w_i))$ remain independent, and with the same distributions, whenever they occur. Thus using Proposition 5.20,

$$a(t) = \mathbf{E} \exp\left\{-\gamma \sum_{i=1}^{n} B(s(w_i))\right\} = \exp\left\{\lambda \int_0^t \Gamma(s(w), \gamma, 1)dw - \lambda t\right\}.$$

$(t_1 < t < t_2)$. The proof is similar to when $t < t_1$. The difference is that 1-jobs which arrive after process time t_1 do not interrupt job A. Note that since $t < t_2$ no 1-job is processed before process time t except during the 1-job busy period during (or at the beginning of) which it arrives.

$(t_2 < t)$. If job A is completed at process time t, and 1-jobs arrive at process times $w_1, w_2, \ldots, w_n$, for job A, where $0 < w_1 < w_2 < \cdots < w_r < t^{-1}(t) < w_{r+1} < w_{r+2} < \cdots < w_n < t$, then the time at which job A is completed under policy AB has the same distribution as

$$t + \sum_{i=1}^{r} B(s(w_i)) + \sum_{j=1}^{n-r} B_j(s(t^{-1}(t))),$$

where $B_j(s(t^{-1}(t)))$ $(j = 1, 2, \ldots, n-r)$ is the busy period initiated by a 1-job arriving at w_{r+j} when $u(w) = s(t^{-1}(t))$ $(t^{-1}(t) < w < t)$ in the setup of Proposition 5.20. This is because (i) the w_is have the same joint distribution as if there were no interruptions to the service of job A, (ii) the busy periods $B(s(w_i))$ $(i = 1, 2, \ldots, r)$ remain independent and with the same distributions whenever they occur, and are simply shifted in time relative to the setup of Proposition 5.20, and (iii) the total time for which job A and 1-jobs receive service while job A ages from $t^{-1}(t)$ to t and all 1-job arrivals are served until they either reach the age $s(t^{-1}(t))$ or are completed if this happens first, is

$$t - t^{-1}(t) + \sum_{j=1}^{n-r} B_j(s(t^{-1}(t)))$$

if the $n-r$ 1-job busy periods are taken in sequence when job A reaches the age t, must be the same irrespective of the scheduling rule used, and therefore in particular under policy AB, and is independent of the busy periods $B(s(w_i))$ $(i = 1, 2, \ldots, r)$. A further application of Proposition 5.20 gives the form of $a(t)$ for $t_2 < t$. □

Proposition 5.22. $b(t) = \exp\{\lambda t \Gamma(s(0), \gamma, 1) - \lambda t\}$.

Proof. The total time taken for job A to reach age t and all uncompleted 1-job arrivals to reach the age $s(0)$ is, in the notation used for the last part of the previous proof,

$$t + \sum_{i=1}^{n} B_i(s(0)),$$

under the policy which postpones all service to 1-jobs until job A is of age t, and then takes the busy periods initiated by each 1-job in turn. This must therefore be the time taken to reach this state for any other policy which does so while keeping the processor fully occupied, and in particular under the policy AB, under which, if A is completed at age t, this state is reached at the point when service to job B begins. Thus Proposition 5.20 gives

$$b(t) = \mathbf{E} \exp\left\{-\gamma \sum_{i=1}^{n} B_i(s(0))\right\} = \exp\{\lambda t \Gamma(s(0), \gamma, 1) - \lambda t\}. \qquad \square$$

If job A is completed at age $t \in (t_1, t_2)$, then under policy AB the next job to be processed is any 1-job which arrived while job A was being processed between the ages t_1 and t. Any such job will have received no service before job A is completed, since

$$\alpha_1(0) < \alpha_2(t) \qquad (t_1 < t < t_2).$$

Also

$$\alpha_1(0) > \alpha_1(s(s)) = \alpha_2(s) \qquad (0 < s < t_1),$$

so that when job A is completed it has an index value greater than that of any other job which is not in the same category. Since priorities at this point are governed by index values under AB, the first 1-job to be processed will receive service either until it reaches the age $s(t_1)$, or until it is completed if this is earlier; if other 1-jobs of age 0 are available, one of them will then be processed with the same rule for stopping, and so on until the supply of brand new 1-jobs is exhausted.

Let $U_n(\xi)$, or simply U_n, denote V_1^{-1} multiplied by the expected total reward from a 1-job busy period initiated by n age-0 jobs, in which these jobs, and any later arrivals, are processed in turn up to age ξ, or until completion if this is earlier. To evaluate U_n we may suppose that n separate 1-job busy periods, $B_1, B_2, \ldots, B_n$, each initiated by one of the n jobs present at time 0, are taken in turn, since the actual order of processing clearly does not affect U_n. Let $R_1, R_2, \ldots, R_n$ denote the total reward resulting from these successive busy periods, in each case discounted to the beginning of the particular busy period concerned. Thus

$$V_1 U_n = E[R_1 + \exp\{-\gamma B_1\}R_2 + \exp\{-\gamma(B_1 + B_2)\}R_2 + \cdots$$
$$+ \exp\{-\gamma(B_1 + B_2 + \cdots + B_{n-1})\}R_n].$$

Since the pairs (B_i, R_i) $(i = 1, 2, \ldots, n)$, are independently and identically distributed, $ER_i = U_1$, and $E[\exp\{-\gamma B_i\}] = \Gamma(\xi, \gamma, 1)$, it follows that

$$U_n = (1 - \Gamma^n)(1 - \Gamma)^{-1} U_1, \tag{5.19}$$

writing Γ for $\Gamma(\xi, \gamma, 1)$.

Now

$$V_1 U_1 = ER_1 = E[E(R_1|T)], \tag{5.20}$$

where T is the processing time required to complete the initial job. Also

$$E(R_1|T = t) = E\{E[R_1|T = t, M(t)]|T = t\}, \tag{5.21}$$

where $M(t)$ is the number of 1-job arrivals up to time t, and therefore has a Poisson distribution with the parameter λt.

$$E[R_1|T = t, M(t) = m] = V_1 e^{-\gamma t}[1 + U_m],$$

so that substituting back into (5.21), and then into (5.20), we have

$$U_1 = \int_0^\xi e^{-\gamma t}\left[1 + \sum_{m=1}^\infty \frac{(\lambda t)^m}{m!} e^{-\lambda t} U_m\right] f_1(t)\,dt$$
$$+ e^{-\gamma\xi} \sum_{m=1}^\infty \frac{(\lambda\xi)^m}{m!} e^{-\lambda\xi} U_m[1 - F_1(\xi)].$$

Using (5.19), this becomes

$$U_1 = [1 + (1 - \Gamma)^{-1} U_1] \int_0^\xi e^{-\gamma t} f_1(t)\,dt - (1 - \Gamma)^{-1} U_1 \int_0^\xi e^{-\gamma t} e^{-\lambda t(1-\Gamma)} f_1(t)\,dt$$
$$+ (1 - \Gamma)^{-1} U_1 e^{-\gamma\xi}[1 - \exp\{-\lambda\xi(\Gamma - 1)\}][1 - F_1(\xi)].$$

Thus, writing $\psi(\xi, \gamma)$, or simply ψ, for

$$\int_0^\xi e^{-\gamma t} f_1(t)\,dt + e^{-\gamma\xi}[1 - F_1(\xi)],$$

and $I(\xi,\gamma)$, or simply I, for

$$\int_0^\xi e^{-\gamma t} f_1(t)\, dt,$$

we have

$$U_1 = I + (1-\Gamma)^{-1} U_1 [\psi(\xi,\gamma) - \psi(\xi,\gamma+\lambda-\lambda\Gamma)].$$

Now it is a well known result in queueing theory (see for example, Prabhu, 1965, p. 75), and one that is easily proved by a further conditioning argument, that the moment generating functions $\psi(\xi,\gamma)$ and Γ for the service time and for a busy period are related by the identity

$$\psi(\xi,\gamma+\lambda-\lambda\Gamma)=\Gamma.$$

Thus

$$U_1 = (1-\Gamma)(1-\psi(\xi))^{-1} I. \tag{5.22}$$

Substituting for U_1 in (5.19) therefore gives

$$U_n(\xi) = [1 - (\Gamma(\xi,\gamma,1))^n][1-\psi(\xi,\gamma)]^{-1} I(\xi,\gamma). \tag{5.23}$$

Proposition 5.23. If under the policy AB job A is completed at age $t \in (t_1, t_2)$, the contribution $c_1(t)$ to $c(t)$ arising from the following busy period in which 1-jobs are served in turn up to age $s(t_1)$ is

$$e^{-\gamma t} a(t) \frac{1 - \exp\{-\lambda(t-t_1)[1-\Gamma(s(t_1),\gamma,1)]\}}{1-\psi(s(t_1),\gamma)} I(s(t_1),\gamma).$$

Proof. Suppose that the number of age zero 1-jobs queueing when job A is completed at age $t \in (t_1, t_2)$ is N, which therefore has a Poisson distribution with the parameter $\lambda(t-t_1)$. We have, using (5.23),

$$EU_N = \sum_{n=1}^\infty \frac{\lambda^n (t-t_1)^n}{n!} \exp\{-\lambda(t-t_1)\} U_n = \frac{1 - \exp\{-\lambda(t-t_1)(t-\Gamma)\}}{1-\psi} I, \tag{5.24}$$

where the suppressed value of ξ is equal to $s(t_1)$ throughout. This must now be discounted back to time zero by the factor $e^{-\gamma t} a(t)$, which applies to the reward from job A on completion at process time t. $\qquad\square$

If job A is completed at process time t, let $S(t, w)$ $[0 < w < \max(t, t_1)]$ be such that $t + S(t, w)$ is the time at which, under policy AB, all uncompleted 1-jobs are for the first time of age $s(w)$ or more. Let $S(t, w)$ be defined for values of t and w such that

$$0 < t < t(0),$$
$$0 < w < \min(t, t_1) \qquad (t < t_2),$$
$$0 < w < t^{-1}(t) \qquad (t > t_2).$$

Let

$$d(t, w) = \mathbf{E}[\exp\{-\gamma S(t, w)\}]] \tag{5.25}$$

Proposition 5.24.

$$d(t, w) = \exp\left\{ \lambda \int_0^w \Gamma(s(u), \gamma, 1)\, du + \lambda(t - s)\Gamma(s(w), \gamma, 1) - \lambda t \right\}.$$

The proof is very similar to that of Proposition 5.21.

Proposition 5.25. In a 1-job busy period initiated by a single job of age 0, the expected reward which accrues under an index policy during the interval in which the minimum age of all uncompleted jobs increases from u to $u + \Delta$ is

$$V_1 \frac{\partial \Gamma(u, \gamma, 1)}{\partial z} (\rho_1(u) + \lambda U_1(u))\Delta + o(\Delta) \qquad (u > s(t_1)). \tag{5.26}$$

Proof. $N(u)$ is the number of 1-jobs reaching age u. The probability that any one of them is completed while they age in step from u to $u + \Delta$, conditional on $N(u)$, is $V_1 \rho_1(u)N(u)\Delta + o(\Delta)$. The probability of a further arrival during this interval is $\lambda N(u)\Delta + o(\Delta)$. The expected reward, discounted to the time of arrival, of the busy period up to age u generated by a further arrival is $V_1 U_1(u)$. Thus the required expression is

$$V_1 \mathbf{E}[N(u)e^{-\gamma B(u)}](\rho_1(u) + \lambda U_1(u))\Delta + o(\Delta).$$

Clearly

$$\frac{\partial \Gamma(u, \gamma, 1)}{\partial z} = \mathbf{E}[N(u)e^{-\gamma B(u)}],$$

and the proposition follows. □

The next proposition gives expressions for $c(t)$ for different ranges of values of t. For notational simplicity we have already assumed that the initial age x of job A is zero. For the same reason we now also assume that y, the initial age of job B, is equal to $t(0)$.

Proposition 5.26

$$
c(t) = \begin{cases}
-\displaystyle\int_0^t \lambda(t-w)d(t,w)\frac{\partial\Gamma(s(w),\gamma,1)}{\partial z}[\rho_1(s(w))+\lambda U_1(s(w))]s'(w)dw \\
\qquad\qquad\qquad\qquad (0<t<t_1), \\[2mm]
-\displaystyle\int_0^t \lambda(t-w)d(t,w)\frac{\partial\Gamma(s(w),\gamma,1)}{\partial z}[\rho_1(s(w))+\lambda U_1(s(w))]s'(w)dw + c_1(t) \\
\qquad\qquad\qquad\qquad (t_1<t<t_2), \\[2mm]
-\displaystyle\int_0^{t^{-1}(t)} \lambda(t-w)d(t,w)\frac{\partial\Gamma(s(w),\gamma,1)}{\partial z}[\rho_1(s(w))+\lambda U_1(s(w))]s'(w)dw \\
\qquad\qquad\qquad\qquad (t_2<t).
\end{cases}
$$

Proof. The 1-jobs which arrive before work starts on job B under the policy AB all belong to the sub-family $\mathscr{J}(A)$ of job-linked (i.e. bandit-process-linked, see §3.12) Poisson arrivals at the rate λ initiated by job A. It is convenient to think in terms of the subsidiary sub-families of 1-jobs initiated by each of the 1-jobs which as members of $\mathscr{J}(A)$ are directly linked to job A, rather than linked to A via one or more intermediate 1-jobs. Let w_i be the process time of job A when the ith of these initial 1-jobs arrives, and let $\mathscr{J}_i$ be the sub-family of 1-jobs which it initiates. Each of the 1-jobs in $\mathscr{J}(A)$ belongs to one of the sub-families $\mathscr{J}_i$.

The three different expressions for $c(t)$ correspond to the different forms which an index policy takes according as $t<t_1, t_1<t<t_2$, or $t_2<t$. Suppose first that $t<t_1$.

The 1-jobs in $\mathscr{J}_i$ are all processed, starting with the arrival of the initial 1-job, until either they are finished or are of age $s(w_i)$, before service is resumed on job A. Since $\alpha_2(s)$ is increasing for $s<t_1$ no further work is done on $\mathscr{J}_i$ before job A is finished at process time t, and every uncompleted job in $\mathscr{J}_i$ then has an index value of $\alpha_1(s(w_i))=\alpha_2(w_i)$. It follows that the index policy for $\mathscr{J}(A)$ when job A is finished starts by allocating service to the jobs in $\mathscr{J}_k$, where

$$k=\max\{i: w_i<t\}.$$

The jobs in $\mathscr{J}_k$ then receive service so that they age in step, with interruptions when further 1-jobs arrive until these too are either completed or of the same age, until every unfinished job in $\mathscr{J}_k$ is of age $s(w_{k-1})$. At this point service is shared with the jobs in $\mathscr{J}_{k-1}$ which, together with the survivors from $\mathscr{J}_k$, age in step, again with interruptions from further arrivals, until every unfinished job in $\mathscr{J}_k\cup\mathscr{J}_{k-1}$ is of age $s(w_{k-2})$. Service is now also shared with the jobs in $\mathscr{J}_{k-2}$, and ageing in step continues as before. This process continues until every unfinished 1-job is of age $s(0)$ (more generally of age $\rho_1^{-1}(V_1^{-1}V_2\rho_2(y))$ (cf. equation (5.14)), but remember we are assuming $y=t(0)$).

Consider now the sub-family $\mathcal{J}$ initiated by a 1-job which arrives at time zero, and consisting of that job together with job-linked Poisson arrivals of further 1-jobs at the rate λ. The expression (5.26) may be regarded as the contribution to the expected total reward from $\mathcal{J}$ which accrues as the unfinished jobs in $\mathcal{J}$ age from u to $u + \Delta$, when priorities between jobs in $\mathcal{J}$ are determined by the index rule and $\mathcal{J}$ is given priority over all other jobs.

The argument leading to Proposition 5.25 may be adapted to give the contribution to the expected discounted reward $V_1 e^{-\gamma t} c(t)$ attributable to the sub-family $\mathcal{J}_i$ as the minimum age of uncompleted jobs increases from $s(w)$ to $s(w) + \Delta$ and for $t < t_1$. Note first that there is no contribution if $w_i < w$, as in that case the unfinished jobs in $\mathcal{J}_i$ are already of age $s(w_i) > s(w)$ when job A is completed. For $w < w_i$, since an index policy is used both for the jobs in $\mathcal{J}_i$ and in $\mathcal{J}$, the expression (5.26) gives the required contribution with $u = s(w)$, and with appropriate new discount factors. The new discount factors are $e^{-\gamma t}$, to take account of the time t spent serving job A, and $d(t, w)$. The factor $d(t, w)$ is actually an expected discount factor which takes account of the time spent by the server on 1-jobs belonging to sub-families other than $\mathcal{J}_k$ (see equation (5.25)). It follows that the expected discounted reward from $\mathcal{J}_i$ as its jobs age from $s(w)$ to $s(w) + \Delta$, and conditional upon a 1-job arriving when job A is of age w_i, is

$$d(t, w) \frac{\partial \Gamma(s(w), \gamma, 1)}{\partial x} [\rho_1(s(w)) + \lambda U_1(s(w))] \Delta + o(\Delta) \quad (w_i < w). \qquad (5.27)$$

To obtain the expected discounted reward from the sub-family initiated by any 1-job arriving when the age of job A is between w_i and $w_i + \delta$ as its jobs age from $s(w)$ to $s(w) + \Delta$, we must multiply by the probability $\lambda \delta + o(\delta)$, noting that the probability of two or more such sub-families is $o(\delta)$, and hence may be ignored. The total expected discounted reward from all the sub-families of 1-jobs initiated by arrivals as job A ages from w to t, and which accrues as the minimum age of the jobs in these sub-families increases from $s(w)$ to $s(w) + \Delta$, may be obtained by dividing the interval (w, t) into δ-intervals, adding the contributions from the sub-families initiated by arrivals in eac of the δ-intervals, an letting $\delta \to 0$. This has the effect of multiplying the expression (5.27) by $\lambda(t - w)$.

To obtain the total expected discounted reward $V_1 e^{-\gamma t} c(t)$ from $\mathcal{J}(A)$ in the interval between completion of job A at time t, and the time when the minimum age of unfinished 1-jobs is $s(0)$, we must now add the contributions corresponding to intervals in which successive small increases in the minimum age occur. Writing $-s'(w)\delta w + o(\delta w)$ for Δ, where δw is the change (< 0) in w corresponding to an increase Δ in $s(w)$, and letting $\Delta \to 0$, this gives $V_1 e^{-\gamma t}$ multiplied by the first integral expression for $c(t)$ given in the statement of the proposition, as required.

For $t_1 < t < t_2$ the index policy for $\mathcal{J}(A)$ immediately after completion of job A at age t leads to age-zero 1-jobs being processed in turn to age $s(t_1)$, or earlier completion, until there are no more age-zero 1-jobs. From this point service is divided between the set of 'youngest' 1-jobs so that they age in step, until every unfinished 1-job is of age $s(0)$, as when $t < t_1$. The contribution to $c(t)$ which arises from the first of these phases is $c_1(t)$, an expression for which is given by Proposition 5.23. The integral part of the expression for $c(t)$ follows on the lines given for $t < t_1$.

The argument leading to the expression for $t_2 < t$ is similar and is left as an exercise. $\square$

The function $t(x)$ may be calculated iteratively for values of x decreasing from t_1 to 0, using Proposition 5.19 and the generalizations of the expressions for $a(t)$, $b(t)$ and $c(t)$ given in Propositions 5.21, 5.22 and 5.26 which are needed to allow for $x \neq 0$ and $y \neq t(0)$. These generalizations are not difficult to obtain, and the reader is invited to derive one or two of them for himself as an exercise. It must be conceded, however, that the resulting calculations, while tractable, are disappointingly complicated.

EXERCISES

5.1. Establish the assertion following the inequality (5.4) that for this to hold for all allocation plans is equivalent to the inequality (1.2) with $c = \lambda$.

5.2. Before the statement of Theorem 5.6 the definition of job-linked Poisson arrivals for a continuous-time job was given. It was also asserted that an independent stream of Poisson arrivals is equivalent to having job-linked Poisson arrivals at the same rate for every job. Explain carefully why this is so, and how individual arrivals in an independent Poisson stream may be assigned to separate streams so as to form job-linked Poisson arrival streams at the same rate for every job, in particular at those times when service is shared between more than one job.

5.3. Write down the extensions of Propositions 5.21 and 5.22 for $0 < x < t_1$, where x is the initial age of job A.

5.4. Complete the proof of Proposition 5.26 for $x = 0$ and $y = t(0) =$ initial age of job B, and satisfy yourself that you know how to modify the expressions for $c(t)$ when $x \neq 0$ and $y \neq t(0)$.

CHAPTER 6

Multi-population Random Sampling (Theory)

6.1 INTRODUCTION

An arm of the multi-armed bandit problem described as Problem 5 in Chapter 1 generates a sequence $X_1, X_2, \ldots$ of independent and identically distributed random variables taking the values 1 and 0 with probabilities θ and $1 - \theta$. The reward associated with the value X is $a^{t-1}X$ if this is the outcome on the tth pull, counting pulls on every arm. This is one of the problems discussed in the present chapter, which is characterized by bandit processes defined by sequences of independent identically distributed random variables $X_1, X_2, \ldots$. When, as in the multi-armed bandit, these random variables themselves constitute a sequence of rewards the resulting bandit process will be termed a *reward* process. Any function of the X_is may also be taken to define a sequence of rewards, and thereby a bandit process. All bandit processes defined in this way will be termed bandit *sampling* processes. The bandit sampling processes defined by the function which takes the value 1 for the first X_i to be sampled with a value $\geqslant T$, and 0 for all other X_is, will in particular be investigated. In the following section it will be shown that a family of alternative bandit processes (in this chapter we shall be concerned only with simple families) of this type serves as a model for the situation where a number of populations are sampled sequentially with the aim of identifying as rapidly as possible an individual that achieves a target level T on some scale of interest. This type of bandit process will thus be termed a *target* process.

The X_is for a given sampling process we suppose to be drawn from a distribution with an unknown, possibly vector-valued, parameter θ belonging to a family $\mathscr{D}$ of distributions for which a density $f(\cdot | \theta)$ exists with respect to some fixed probability space $(\mathbf{R}, \mathscr{F}_1, \mu_1)$. Adopting the standard Bayesian setup (e.g. see Barra, 1981) θ has a prior density π_0 with respect to a probability space $(\mathbf{R}^r, \mathscr{F}_2, \mu_2)$, and $f(x | \cdot)$ is $\mathscr{F}_2$-measurable for all $x \in \mathbf{R}$. Thus the joint density for θ and X_1 is $\pi_0(\theta) f(x | \theta)$, and it follows that θ has a density with respect to $(\mathbf{R}^r, \mathscr{F}_2, \mu_2)$ conditional on X_1 taking the value x_1, which may be written as

$$\pi_1(\theta | x_1) = \frac{\pi_0(\theta) f(x_1 | \theta)}{\displaystyle\int \pi_0(\phi) f(x_1 | \phi) \, d\mu_2(\phi)}.$$

125

The density π_1 is termed the *posterior* density for θ after observing X_1 to take the value x_1.

Similarly, after observing X_i to take the values $x_i(i = 1, 2, \ldots, n)$ the posterior density for θ is

$$\pi_n(\theta \,|\, x_1, x_2, \ldots, x_n) = \frac{\pi_{n-1}(\theta \,|\, x_1, x_2, \ldots, x_{n-1}) f(x_n \,|\, \theta)}{\int \pi_{n-1}(\phi \,|\, x_1, x_2, \ldots, x_{n-1}) f(x_n \,|\, \phi) d\mu_2(\phi)}$$

$$= \frac{\pi_0(\theta) \prod\limits_{i=1}^{n} f(x_i \,|\, \theta)}{\int \pi_0(\phi) \prod\limits_{i=1}^{n} f(x_i \,|\, \phi) d\mu_2(\phi)}. \tag{6.1}$$

The reader should not be alarmed by the appearance of the measures μ_1 and μ_2 in this account. We shall be concerned only with densities with respect to Lebesgue measure, counting measure, and simple combinations of these two measures. These are the usual densities for continuous, discrete, or mixed, distributions respectively.

Equation (6.1) is sometimes written in the form

$$\pi_n(\theta \,|\, \mathbf{x}_n) \propto \pi_0(\theta) \prod_{i=1}^{n} f(x_i \,|\, \theta),$$

the proportionality sign $\propto$ indicating that an $\mathscr{F}_1^n$-measurable function of $\mathbf{x}_n[= x_1, x_2, \ldots, x_n)]$ has been omitted, one which may be determined from the fact that $\int \pi_n d\mu_2 = 1$. This result is sometimes called the *generalized Bayes' Theorem*.

Two assumptions will be made for the sake of computational tractability. They concern the important statistical concept of sufficiency.

For present purposes a *statistic* is any real-valued function defined for every positive integer n as a $\mathscr{F}_1^n$-measurable function of $\mathbf{X}_n = (X_1, X_2, \ldots, X_n)$. The statistics $u_1, u_2, \ldots, u_m$ are a *sufficient* set of statistics for random samples of size n if the distribution of X_n conditional on $u_1(\mathbf{X}_n), u_2(\mathbf{X}_n), \ldots, u_m(\mathbf{X}_n)$ does not depend on θ. The following equivalent condition for sufficiency is due to Neyman (e.g. see Barra, 1981).

Theorem 6.1. The statistics $u_1, u_2, \ldots, u_m$ are sufficient for random samples of size n iff, for almost all $\mathbf{x}$ and for suitably chosen functions d and h,

$$\prod_{i=1}^{n} f(x_i \,|\, \theta) = d(u_1(\mathbf{x}_n), u_2(\mathbf{x}_n), \ldots, u_m(\mathbf{x}_n), \theta) h(\mathbf{x}_n),$$

where d is a Borel-measurable function defined in $\mathbf{R}^{m+r}$, h is $\mathscr{F}_1^n$-measurable, and $\mathbf{x}_n = (x_1, x_2, \ldots, x_n)$.

The two simplifying assumptions are as follows.

Assumption 1. The family of distributions $\mathscr{D}$ is an *exponential* family. This means (e.g. see Ferguson, 1967) that the density $f(x|\theta)$ may be written in the form

$$f(x|\theta) = c(\theta)\exp\left\{\sum_{j=1}^{m} q_j(\theta)t_j(x)\right\}h(x).$$

Assumption 2. The prior density $\pi_0(\theta)$ may be written in the form

$$\pi_0(\theta) = k_0[c(\theta)]^a \exp\left\{\sum_{j=1}^{m} q_j(\theta)b_j\right\}.$$

Most of the standard distributions do belong to exponential families, so Assumption 1 is not a serious restriction. It means, as Theorem 6.1 shows, that the statistics

$$\sum_{i=1}^{n} t_j(x_i) \qquad (j = 1, 2, \ldots, m)$$

form a sufficient set. From equation (6.1) it therefore follows that the posterior density $\pi_n(\theta|\mathbf{x}_n)$ depends on the observation vector $\mathbf{x}_n$ only as a function of the sufficient statistics. This property gives the clue to the term *sufficient*, the point being that for the purpose of calculating π_n all the information derived from the observations is given by the sufficient statistics.

Assumption 2 means that $\pi_n(\theta|\mathbf{x}_n)$ may be written as

$$\pi_n(\theta|\mathbf{x}_n) = k_n(\mathbf{x}_n)[c(\theta)]^{a+n} \exp\left\{\sum_{j=1}^{m} q_j(\theta)\left(b_j + \sum_{i=1}^{n} t_j(x_i)\right)\right\}, \qquad (6.2)$$

where $k_n(\mathbf{x}_n)$ is chosen so that $\int \pi_n(\theta|\mathbf{x}_n)d\mu_2(\theta) = 1$. Thus, for any n and $\mathbf{x}_n$, π_n is the probability density for a distribution defined by the parameters

$$a+m \quad \text{and} \quad b_j + \sum_{i=1}^{n} t_j(x_i) \qquad (j = 1, 2, \ldots, m).$$

The family $\mathscr{P}$ of distributions with densities of the form (6.2) is said to be *conjugate* to $\mathscr{D}$, a concept discussed in detail by Raiffa and Schlaifer (1961). To restrict the prior distribution to $\mathscr{P}$ is often acceptable, since it typically allows the prior mean and variance of a parameter to be given arbitrary values, and is virtually essential for computational tractability.

This sequential Bayesian setup fits into the bandit process framework if the state of a bandit sampling process is identified as the current (posterior) distribution for θ. For a target process we also need a completion state C, as in the jobs of §2.5, attainment of which means that an X_i above the target has been observed, and that all further rewards are zero rewards. The undiscounted reward $r(S, \pi)$ from continuing the bandit sampling process when it is in state π is now the expected undiscounted reward yielded by the next X_i when π is the current density for θ. Thus

$$r(S, \pi) = \int g(x)f(x|\pi)d\mu_1(x), \tag{6.3}$$

where for a reward process $g(x)=x$, for a target process $g(x)=1$, if $x \geqslant T$ and $g(x)=0$ if $x<T$, and $f(\cdot|\pi)=\int f(\cdot|\theta)\pi(\theta)d\mu_2(\theta)$. The restriction of the prior distribution to a parametric conjugate family means that the state may be specified by the parameters of the current distribution for θ. Note that, for a target process, $r(S, \pi)$ is the probability in state π that the next observation reaches the target. It will be referred to as the *current probability of success (CPS)*.

To refer to the current distribution of the unknown parameter θ as the *state* of a bandit process may seem a little odd. Certainly decision theorists would be more likely to reserve this description for θ itself. The reader may find it helpful to regard θ as the *true* or *absolute* state from an all-knowing divine standpoint, and π as the *existential* state summarizing the incomplete information about θ available to an earthly decision-maker.

It follows from the definition (2.5), and dividing by the discrete-time correction factor $\gamma(1-a)^{-1}$ defined at the end of §2.3, which will be standard for all Markov sampling processes, that for any Markov bandit process B in state x

$$v(B, x) = \sup_{\tau>0} v_\tau(B, x) \geqslant v_1(B, x) = r(B, x). \tag{6.4}$$

This corresponds to the facts that a reward rate per unit time of $r(B, x)$ may be achieved by setting $\tau = 1$, and that higher average reward rates may sometimes be achieved by sequentially adjusting τ so as to be higher or lower according as the prospects for high future rewards improve or deteriorate.

For a bandit sampling process S the relationship between $v(S, \pi)$ and $r(S, \pi)$ is particularly instructive. Consider the limiting case when π represents precise knowledge of the true state θ. Thus the existential state coincides with the true state, and does not change as further X_i's are sampled (except when state C is reached in the case of a target process). The expected reward on each occasion that an observation is made, except in state C, is therefore

$$r(S, \theta) = \int g(x)f(x|\theta)d\mu_1(x).$$

Thus, writing $\sigma = \min\{i: X_i \geqslant T\}$ and applying the discrete-time correction factor, it follows from (2.5) that

$$v_\tau(S, \theta) \begin{cases} = r(S, \theta) & \text{for a reward process if } 0<\tau \\ = r(S, \theta) & \text{for a target process if } 0<\tau\leqslant\sigma \\ < r(S, \theta) & \text{for a target process if } \mathbf{P}(\tau>\sigma)>0. \end{cases}$$

It follows that for either a reward or a target process

$$v(S, \theta) = \sup_{\tau > 0} v_\tau(S, \theta) = r(S, \theta).$$

Thus when the parameter θ is known precisely (6.4) holds with equality. This reflects the fact that there is no further information about θ to be learned from the process of sampling. Except in this limiting case π expresses some uncertainty about θ, so that more reliable information about θ may be obtained by sampling, and this information then used to adjust τ so as to achieve a higher expected average reward rate than $r(S, \pi)$. Thus the difference $v(S, \pi) - r(S, \pi)$ is caused by uncertainty about θ, and may be expected to be large when this uncertainty is large.

Theorem 4.1 tells us that $v(S, \pi)$ increases with a, and so therefore does $v(S, \pi) - r(S, \pi)$, since $r(S, \pi)$ is independent of a. This is because the larger the discount factor the more important are future rewards in comparison with the immediate expected reward $r(S, \pi)$, and consequently the more important does it become to resolve any uncertainty about θ. The difference $v(S, \pi) - r(S, \pi)$ is one measure of the importance of sampling to obtain information, rather than simply to obtain the reward $r(S, \pi)$.

Kelly (1981) has shown that for the Bernoulli reward process (see Problem 5 and §§6.4 and 6.7) with a uniform prior distribution the indices for any two states are ordered, for values of a sufficiently close to one, by the corresponding numbers of zero rewards sampled, provided these are different, irrespective of the numbers of successes. In contrast, the expected immediate reward $r(S, \pi)$ is equal to the proportion of successes. This phenomenon of $r(S, \pi)$ bearing virtually no relationship to the index $v(S, \pi)$, and hence to the policy which should be followed when more than one sampling process is available, means that the prospects of immediate rewards are dominated by the longer term requirement to gather information which may be used to improve the rewards obtained later on.

The remaining sections of this chapter develop the theory of sampling processes. In §6.2 a (possibly new) result on single-processor scheduling is given which establishes that a simple family of alternative target processes is an appropriate model when the aim is to achieve a given target as soon as possible, irrespective of which process it comes from. Section 6.3 explores the circumstances in which the monotonicity condition for Proposition 4.5 is satisfied by a target process, thus leading to a simple expression for index values. Section 6.4 sets out two general methods for calculating index values when no simplifying monotonicity property holds, taking advantage of any location or scale invariance properties and assuming a conjugate prior distribution. This methodology is extended in §6.5 to semi-Markov sampling processes for which the sampling time for each observation is negative-exponential. Section 6.6 describes the Brownian reward process, quoting asymptotic results due to Bather (1983), and Chang and Lai (1987), and showing the asymptotic relationship both of the process itself and of the asymptotic results just mentioned to the normal reward process. Section 6.7 shows that a result similar to Bather's holds for other sampling processes, including target processes, if these are asymptotically normal in an appropriate sense.

In both §§6.5 and 6.6 sampling processes are defined which allow decisions to be taken at any time. This strictly speaking takes us outside the scope of the theorems of Chapter 3. Chapter 5 also does not apply, as we are not dealing with jobs as defined in §2.5. An index theorem which is sufficiently general to cover these cases should not be too difficult to prove, but does not seem as yet to be available. In the implementation of an index policy effort may have to be shared as in §5.4.

Chapter 7 describes in detail the methods and results of calculations of dynamic allocation indices for the normal, Bernoulli and exponential reward processes, and for the exponential target process. The results are tabulated at the back of the book. These cases were selected mainly for their general interest, though the exponential target process is also worthy of note as the original bandit process which arose in the practical problem whose consideration led to the index theorem.

This problem is that of selecting chemical compounds for screening as potential drugs from a number of families of similar compounds. The sampled values represent measurements of some form of potentially therapeutic activity. The target is a level of activity at which much more thorough testing is warranted. A negative exponential distribution of activity is often a reasonable assumption. Index values, which are closely related to the upper tail areas of distributions of activity, are, however, very sensitive to the form of the distribution, particularly if the target is higher than the activities already sampled. A procedure was therefore developed in which observations are first transformed to achieve as good a fit to the negative exponential form as possible, and this type of distribution is then only assumed for the upper part of the distribution, all values below some threshold being treated as equivalent. This led to a model involving target processes based on a distribution with an unknown atom of probability at the origin, and a negative exponential distribution with an unknown parameter over R^+. Calculations of index values for target process of this type, which will be described as *Bernoulli/exponential* target processes, have been reported by Jones (1975), and are described in general terms in §7.7. A brief description of the software package CPSDAI based on these ideas, which has been developed for use by chemists, is given in Bergman and Gittins (1985). Full details may be obtained from the author.

6.2 JOBS AND TARGETS

In the previous section a simple family of alternative target processes was described as a suitable model for a situation where several different populations may be sampled sequentially, and the aim is to find a sample value which exceeds the target as quickly as possible from any of the populations. In fact target processes are jobs (see §2.5), and an index policy therefore maximizes the expected payoff from the set of jobs. This is an apparently different objective. The following theorem shows that the two objectives are actually equivalent.

Theorem 6.2. For a discounted single machine scheduling problem in which each job yields the same reward on completion, any policy that maximizes the expected total reward must also maximize the expected reward from the first job completion.

Proof. Let A be a job which yields a reward of 1 on completion, and B a bandit process which has the same properties as A except that instead of a reward on completion there is a cost of continuation. For an interval of duration h this cost is $h + o(h)$ if the interval occurs before completion, and the cost is zero if the interval occurs after completion. For both A and B the discount parameter is γ. The theorem is a consequence of the following lemma, the superscripts distinguishing the indices (and in the proof other functions) referring to A from those referring to B.

Lemma 6.3. $\dfrac{v^A}{\gamma} + \dfrac{1}{v^B} = -1.$

Proof. Let τ be the stopping time defined by a stopping set Θ_0 which does not include the completion state C. Since A and B pass through the same sequence of states, τ is defined both for A and B. Let T be the (process) time when the process first reaches state C. From our assumptions about the reward for A and the costs for B it follows that to determine v^A only stopping times of the form $\tau_A = \min(\tau, T)$ need be considered, and to determine v^B only stopping times of the form

$$\tau_B = \begin{cases} \tau & \text{if } \tau < T \\ \infty & \text{if } \tau > T \end{cases},$$

need be considered, in both cases for some $\tau > 0$.

We have

$$W^A_{\tau_A} = -R^B_{\tau_B} = \mathbf{E}\left(\int_0^{\tau_A} e^{-\gamma s}\, ds \right) = \gamma^{-1} - \gamma^{-1} \mathbf{E}(e^{-\gamma \tau} A)$$

$$= \gamma^{-1} - \gamma^{-1} \int_{\tau < T} e^{-\gamma \tau}\, dF(\tau, T) - \gamma^{-1} \int_{T < \tau} e^{-\gamma T}\, dF(\tau, T),$$

$$R^A_{\tau_A} = \int_{T < \tau} e^{-\gamma T}\, dF(\tau, T),$$

$$W^B_{\tau_B} = W^A_{\tau_A} + \int_{T < \tau} \int_T^\infty e^{-\gamma s}\, ds\, dF(\tau, T) = W^A_{\tau_A} + \gamma^{-1} R^A_{\tau_A}.$$

It follows that

$$\gamma W^B_{\tau_B} = \gamma W^A_{\tau_A} + R^A_{\tau_A} = -\gamma R^B_{\tau_B} + R^A_{\tau_A},$$

and hence that

$$\gamma^{-1} \frac{R^A_{\tau_A}}{W^A_{\tau_A}} + 1 = -\frac{W^B_{\tau_B}}{R^B_{\tau_B}}.$$

Since

$$\sup_{\tau > 0} \frac{R^A_{\tau_A}}{W^A_{\tau_A}} = v^A, \quad \text{and} \quad \sup_{\tau > 0} \frac{R^B_{\tau_B}}{W^A_{\tau_A}} = v^B, \quad \text{the lemma follows.} \qquad \square$$

The lemma shows that v^A is a strictly increasing function of v^B. Thus given a family $\mathscr{F}_A$ of alternative jobs $A_1, A_2, \ldots$, and a corresponding family $\mathscr{F}_B$ of alternative bandit processes $B_1, B_2, \ldots$, index policies for the two cases are indistinguishable. It therefore follows from Theorem 3.6 or Theorem 5.1 that the same policies are optimal for the reward criterion of $\mathscr{F}_A$ and for the cost criterion of $\mathscr{F}_B$. The second of these criteria amounts to minimizing

$$\mathbf{E}\left(\int_0^t e^{-\gamma s}\, ds \right) = \gamma^{-1} - \gamma^{-1}\, \mathbf{E}(e^{-\gamma t}), \tag{6.5}$$

where t is the time of the first job completion, and this is equivalent to maximizing $\mathbf{E}(e^{-\gamma t})$, the expected reward from the first job completion. This completes the proof when the rewards are all equal to 1. Multiplying all the rewards by a positive constant clearly has no effect on optimality. $\qquad \square$

Thus for $\gamma > 0$ the search for an individual from one of a number of populations for which some measurement reaches a target level may indeed be modelled by a family of alternative target processes with the reward structure of a set of jobs. This is particularly convenient in the limiting case $\gamma = 0$ because then the indices for $\mathscr{F}_B$ are all equal to zero provided each job has a positive probability of reaching the completion state, and it is not immediately clear what the limit of an index policy is as γ tends to zero. This problem does not occur for the indices of the jobs in $\mathscr{F}_A$, and we should find that a policy based on the limits of these indices as γ tends to zero minimizes the limit as γ tends to zero of (6.5), namely the expected time until the target is reached. Strictly speaking some care should be taken to establish conditions under which this assertion is valid. It holds, for example, if the probability of any difference in the selection of processes between an index policy with no discounting and an index policy with $\gamma > 0$ tends to zero as γ tends to zero. This is sufficient for the applications made in this chapter.

6.3 USE OF MONOTONICITY PROPERTIES

In general some values of X_i point to more favourable values of θ than do other values of X_i, and in consequence neither of the sequences $v(S_i, \pi_i)$ or $r(S, \pi_i)$ $(i = 0, 1, 2, \ldots)$ is necessarily monotone. What can happen is that the prior density π_0 is so favourable that, when S is a target process, any sequence of X_is which is capable of making the posterior density π_n even more favourable then π_0 must include a value which exceeds the target T. This consideration leads us to look for prior densities ρ such that any sequence $\{X_i < T; i = 1, 2, \ldots\}$ generates a corresponding sequence of posterior densities $\{\pi_i\}$, $\pi_0 = \pi$, such that

$$r(S, \pi) \geqslant r(S, \pi_i) \qquad (i \geqslant 0).$$

Let us define a probability density π with this property to be *favourable*. From (6.3) it follows that, for a target process, $r(S, \pi) \geqslant 0$ for any state defined by a density π. By definition of the completion state C, $r(S, C) = 0$. Thus if π is favourable the sequence of rewards (or current probabilities of success) associated with the sequence of existential or completion states starting from π are all no greater than $r(S, \pi)$. This is the condition for Proposition 4.5, and we have established the following result.

Proposition 6.4. If S is a target process and the probability density π for the parameter θ is favourable then $v(S, \pi) = r(S, \pi)$.

Usually at least some prior probability distributions are favourable. For example this is true for the Bernoulli, normal, and negative exponential target processes, a brief account of each of which follows. The last two are discussed in considerable detail by Jones (1975). The values assumed for the target T are not restrictive, in the Bernoulli case because there is no other non-trivial possibility, and in the other cases because of the invariance of the processes concerned under changes of scale, and in the normal case also of location. These invariance properties are discussed in §6.4.

Example 6.5 (Bernoulli Target Process). $T = 1$. $\mathbf{P}(X_i = 1) = 1 - \mathbf{P}(X_i = 0) = \theta$. The family of conjugate prior distributions for θ are beta distributions with densities of the form

$$\frac{\Gamma(\alpha + \beta)}{\Gamma(\alpha)\Gamma(\beta)} \theta^{\alpha - 1}(1 - \theta)^{\beta - 1} \qquad (0 \leqslant \theta \leqslant 1),$$

where $\alpha > 0$ and $\beta > 0$.

Thus any state π apart from the completion state may be represented by the parameters α and β, and it follows from (6.3), dropping S from the notation, that

$$r(\alpha, \beta) = \int_0^1 \theta \, \frac{\Gamma(\alpha + \beta)}{\Gamma(\alpha)\Gamma(\beta)} \theta^{\alpha - 1}(1 - \theta)^{\beta - 1} \, d\theta = \frac{\alpha}{\alpha + \beta}.$$

Thus if $N = \min\{n: X_n = 1\}$ the sequence of states starting from (α, β) is $\{(\alpha, \beta), (\alpha, \beta+1), (\alpha, \beta+2), \ldots, (\alpha, \beta+N-1), C\}$. Since

$$\frac{\alpha}{\alpha+\beta} \geqslant \frac{\alpha}{\alpha+\beta+n} \qquad (n \geqslant 0),$$

it follows that every prior distribution state (α, β) is favourable, and hence that

$$v(\alpha, \beta) = r(\alpha, \beta) = \frac{\alpha}{\alpha+\beta}.$$

Example 6.6 (Normal Target Process, Known Variance). $T = 0$, variance $= 1$,

$$f(x|\theta) = \phi(x-\theta) = (2\pi)^{-1/2} \exp\{-\tfrac{1}{2}(x-\theta)^2\} \ (x \in \mathbf{R}).$$

If the prior density for θ is the improper uniform density over the real line, it follows from equations (6.1) and (6.2) that the posterior density for θ after n values have been sampled with mean $\bar{x}$ may be written as

$$\pi_n(\theta|\bar{x}) = (2\pi n^{-1})^{-1/2} \exp\left\{-\frac{n}{2}(\bar{x}-\theta)^2\right\} \qquad (\theta \in \mathbf{R}).$$

Conjugate prior densities for θ are of this form, with n taking any (not necessarily integer) positive value. Thus an arbitrary state $\pi(\neq C)$ may be represented by the parameters $\bar{x}$ and n.

We have

$$f(x|\pi) = f(x|\bar{x}, n) = \int_{-\infty}^{\infty} f(x|\theta)\pi_n(\theta|\bar{x}) \, d\theta$$

$$= \int_{-\infty}^{\infty} (2\pi)^{-1/2} \exp\{-\tfrac{1}{2}(x-\theta)^2\} (2\pi n^{-1})^{-1/2}$$

$$\exp\left\{-\frac{n}{2}(\bar{x}-\theta)^2\right\} d\theta$$

$$= \left(\frac{n}{2\pi(n+1)}\right)^{1/2} \exp\left\{-\frac{n}{2(n+1)}(x-\bar{x})^2\right\},$$

and from (6.3)

$$r(\bar{x}, n) = \int_0^{\infty} f(x|\bar{x}, n) \, dx = \Phi(\bar{x}(1+n^{-1})^{-1/2}). \tag{6.6}$$

For a given initial state $(\bar{x}, n)$ the state after sampling $\{X_i < 0 : i = 1, 2, \ldots, m\}$ with mean $\bar{X}$ is $((n\bar{x}+m\bar{X})(n+m)^{-1}, n+m)$. From (6.6) it follows that if $\bar{x} \geqslant 0$ then

$$r(\bar{x}, n) \geqslant r((n\bar{x}+m\bar{X})(n+m)^{-1}, n+m),$$

and hence that the state $(\bar{x}, n)$ is favourable, so that

$$v(\bar{x}, n) = r(\bar{x}, n).$$

Example 6.7 (Normal Target Process, Unknown Mean and Variance). $T = 0$, $\theta = (\mu, \sigma)$.

$$f(x|\mu, \sigma) = \sigma^{-1} \phi((x - \mu)/\sigma) \ (x \in \mathbf{R}).$$

Conjugate prior densities for μ and σ may be written in the form

$$\pi_n(\mu, \sigma | \bar{x}, s) \propto \sigma^{-n-1} \exp\left\{ -\frac{1}{2\sigma^2} [(n-1)s^2 + n(\bar{x} - \mu)^2 + (x - \mu)^2] \right\}, (\mu \in \mathbf{R}, \sigma > 0)$$

which is the posterior density after observations x_i $(i = 1, 2, \ldots, n)$ with mean $\bar{x}$ and $(n-1)^{-1} \Sigma (x_i - \bar{x})^2 = s^2$, starting from an improper prior proportional to σ^{-1}.

$$f(x|\pi_n) = f(x|\bar{x}, s, n) \propto \left[1 + \frac{n}{n+1} \cdot \frac{(x - \bar{x})^2}{(n-1)s^2} \right]^{-n/2} \quad \text{(see §7.3 for intermediate steps)},$$

thus $u = (n/n + 1)^{1/2} (x - \bar{x})s^{-1}$ has Student's t-distribution with $n - 1$ degrees of freedom.

Thus, as in the known variance case, the density $f(x, \pi_n)$ is symmetrical about $\bar{x}$. The variance of this density, however, now behaves rather differently. For small values of n and fixed s it decreases more rapidly than for known σ^2 as n increases, because of the increase in the degrees of freedom of the t-distribution. This alters the circumstances under which $v(\pi_n) = r(\pi_n)$ in a quite interesting way. For known σ^2 if $\bar{x} > 0$ the modified value of $\bar{x}$ after observing x_{n+1} is reduced unless $x_{n+1} > 0$, and this results in $r(\pi_{n+1}) < r(\pi_n)$, since $\bar{x}$ is the mean of the density $f(x|\pi_n)$. Admittedly the variance of $f(x|\pi_{n+1})$ is smaller than that of $f(x|\pi_n)$, which tends to increase the total probability above the target, 0, but not by enough to offset the decrease caused by the decrease in mean value. For the unknown variance case, on the other hand, the reduction in the variance from $f(x|\pi_n)$ to $f(x|\pi_{n+1})$ may, for small n, be sufficient to cause an increase in CPS although $\bar{x} > 0$ and $x_{n+1} < 0$. For similar reasons it may be impossible to increase CPS with a negative x_{n+1} although $\bar{x} < 0$. Jones has carefully examined this phenomenon and concludes that, for the undiscounted case, if $n > 2.75$

$$v(\bar{x}, s, n) > r(\bar{x}, s, n) \text{ iff } \bar{x} > 0.$$

If $n < 2.75$ there are values of $\bar{x}$ which serve as counter-examples to both parts of this assertion.

Example 6.8 (Exponential Target Process). $T = 1$.

$$f(x|\theta) = \theta e^{-\theta x} \quad (x > 0).$$

Conjugate prior densities may be written in the form

$$\pi_n(\theta|\Sigma) = \Sigma^n \theta^{n-1} e^{-\theta\Sigma}/\Gamma(n) \qquad (\Sigma > 0, n > 0),$$

which is the posterior density after observations x_i $(i = 1, 2, \ldots, n)$ with total Σ, starting from an improper prior density proportional to θ^{-1}.

Thus

$$f(x|\pi_n) = f(x|\Sigma, n) = \int_0^\infty f(x|\theta)\pi_n(\theta|\Sigma)\, d\theta = \frac{\Sigma^n}{\Gamma(n)} \int_0^\infty \theta^n \exp\{-\theta(\Sigma + x)\}\, dx$$

$$= \frac{n\Sigma^n}{(\Sigma + x)^{n+1}}.$$

$$r(\Sigma, n) = \int_1^\infty f(x|\Sigma, n)\, dx = \left(\frac{\Sigma}{\Sigma + 1}\right)^n.$$

Thus $r(\Sigma, n)$ is an increasing function of Σ, and the condition

$$\mathbf{P}\left\{ r\left(\Sigma + \sum_{i=1}^m X_i, n+m\right) \leqslant r(\Sigma, n) \,\middle|\, X_i < 1; i \leqslant m \right\} = 1$$

is therefore equivalent to

$$\Sigma^n(\Sigma + m + 1)^{n+m} \geqslant (\Sigma + 1)^n(\Sigma + m)^{n+m}, \tag{6.7}$$

since $(\Sigma + m)^{n+m}/(\Sigma + m + 1)^{n+m}$ is the supremum of the set of values of $r(\Sigma + \Sigma_{i=1}^m X_i, m+1)$ which are consistent with the condition $X_i < 1 (i \leqslant m)$. Jones shows that if (6.7) holds for $m = 1$ it also holds for $m > 1$, and that (6.7) holds iff $r(\Sigma, n) \geqslant p_n$, where, for all n, $p_n \simeq e^{-1}$ and $p_n \to e^{-1}$ as $n \to \infty$. Thus, using Proposition 6.4,

$$v(\Sigma, n) = r(\Sigma, n) \quad \text{iff } r(\Sigma, n) \geqslant p_n. \tag{6.8}$$

Jones goes on to discuss the Bernoulli/exponential target process for which $f(x|p, \theta) = p\theta e^{-\theta x}$, and $\mathbf{P}\{X = 0\} = 1 - p$, where both p and θ are unknown parameters with conjugate prior distributions (i.e. beta and gamma distributions respectively). A joint conjugate prior for p and θ has parameters m, n and Σ and has the form of the posterior distribution for p and θ, starting from an improper prior density proportional to $p^{-1}(1-p)^{-1}\theta^{-1}$, after observing $m + n$ X_is with m zeros and a total of Σ. It turns out that

$$v(m, n, \Sigma) = r(m, n, \Sigma) \quad \text{iff } r(m, n, \Sigma) \geqslant p_{mn}$$

where $p_{mn} \to e^{-1}$ as $n \to \infty$, for all m.

The results of calculations of $v(\Sigma, n)$ for the undiscounted exponential target process are described in §7.6. Jones reports calculations for the undiscounted Bernoulli/exponential target process. The method of calculation is set out in §7.7.

6.4 GENERAL METHODS OF CALCULATION; USE OF INVARIANCE PROPERTIES

Very often none of the simplifying monotonicity properties of §4.3 holds for a bandit sampling process. This leaves us with two main approaches to the calculation of index values. The first approach, which we might call the *direct* approach, is via the defining equation for the index of a sampling process S in state π,

$$v(S, \pi) = \sup_{\tau > 0} v_\tau(S, \pi) = \sup_{\tau > 0} \frac{R_\tau(S, \pi)}{W_\tau(S, \pi)}. \tag{6.9}$$

The second, or *calibration*, approach uses the standard bandit processes as a measuring device as in the proof of Theorem 3.1. The functional equation (2.2) for the optimal payoff from a SFABP $\{S, \Lambda\}$, consisting of the sampling process S in state π and a standard bandit process Λ with parameter λ, takes the form

$$R(\lambda, \pi) = \max\left[\lambda + aR(\lambda, \pi); r(\pi) + a \int R(\lambda, \pi_x) f(x|\pi) d\mu_1(x) \right]$$

$$= \max\left[\frac{\lambda}{1-a}, r(\pi) + a \int R(\lambda, \pi_x) f(x|\pi) d\mu_1(x) \right], \tag{6.10}$$

where $R(\lambda, \pi)$ and $r(\pi)$ are abbreviated notations for $R(\{S, \Lambda\}, \pi)$ and $r(S, \pi)$, and π_x denotes the image of π under the mapping defined by Bayes' theorem when x is the next value sampled. Note that π defines the state of $\{S, \Lambda\}$ as well as the state of S, since the state of Λ does not change. From Theorem 3.6 it follows that $v(s, \pi) = \lambda$ if and only if it is optimal to select either S or Λ for continuation; that is, if and only if the two expressions inside the square brackets on the right-hand side of equation (6.10) are equal. The calibration approach consists of solving equation (6.10) for different values of λ, and hence finding the values of λ for which the two expressions inside the square brackets are equal as π varies throughout the family of distributions $\mathscr{P}$.

For a target process the range of possible index values is from 0 to 1, and it is convenient to calibrate in terms of a standard target process for which the probability ϕ of reaching the target at each trial is known. For target processes our discussion will focus largely on the undiscounted case (i.e. $a = 1$ or $\gamma = 0$). The recurrence corresponding to (6.10) is

$$M(\phi, \pi, T) = \min\left[\phi^{-1}, 1 + \int_{-\infty}^{T} M(\phi, \pi_x, T) f(x|\pi) d\mu_1(x) \right], \tag{6.11}$$

where $M(\phi, \pi, T)$ is the expected number of values needing to be sampled under an optimal policy to reach the target T, given two target processes, one in state π and the other a standard target process with known success probability ϕ.

The direct approach involves maximizing with respect to a stopping set defined on an infinite state-space. A complete enumeration of all the possibilities is therefore impossible, and some means of restricting the class of stopping sets to be considered must be found. For the Bernoulli reward process (i.e. an arm of the classical multi-armed bandit of Problem 5) this has been achieved by allowing the stopping set $\mathcal{P}_0(\subset\mathcal{P})$ defining a stopping time τ to depend on the process time, and by imposing the restriction $\tau \leqslant T$, where T is a non-random integer. Defining

$$v^T(S, \pi) = \sup_{0 < \tau \leqslant T} v_\tau(S, \pi)$$

we then have $v^T(S, \pi) \leqslant v^\infty(S, \pi) = v(S, \pi)$, and $v^T(S, \pi) \to v(S, \pi)$ as $T \to \infty$. Note that it follows from Lemma 3.2 that the definition (6.9) is unaffected by allowing $\mathcal{P}_0$ to depend on process time. Thus arbitrarily close approximations to $v(S, \pi)$ may be obtained by calculating $v^T(S, \pi)$ for a sequence of increasing values of T.

Thron (1984) has shown that the error caused by the finite horizon T is $O(a^T)$.

Calculations of $v^T(S, \pi)$ were based on the observation that the stopping set $\mathcal{P}_0(t)$ at process time t (measured from the initial state $\pi = \pi_0$) may be assumed to take the form

$$\mathcal{P}_0(t) = \{\pi_t \in \mathcal{P} : v^{T-t}(S, \pi_t) < \phi\} \tag{6.12}$$

for some ϕ independent of t. This is a consequence of the fact that $\mathcal{P}_0(t)$ may be defined as $\{\pi_t \in \mathcal{P} : v^{T-t}(S, \pi_t) < v^T(S, \pi)\}$. That this holds for $T = \infty$ follows immediately from Theorem 3.4 (iii) on putting $M = S$. Now let S^* be a bandit process which (i) is in the state (π_t, t) at process time t, (ii) has rewards identical to those of S up to process time T, and (iii) has large negative rewards after process time T. Thus $v^{T-t}(S, \pi_t) = v(S^*, (\pi_t, t))$, and (6.12) follows from Theorem 3.4 (iii) on putting $M = S^*$.

The conjugate prior distributions for the parameter θ of a Bernoulli process are the family of beta distributions which have densities on the interval $[0, 1]$ of the form

$$\frac{\Gamma(\alpha + \beta)}{\Gamma(\alpha)\Gamma(\beta)} \theta^{\alpha-1}(1 - \theta)^{\beta-1},$$

where $\alpha > 0$ and $\beta > 0$. Thus the state π may be represented by the parameter values α and β, and it follows from (6.3) that for a Bernoulli reward process (dropping S from the notation)

$$r(\alpha, \beta) = \int_0^1 \theta \frac{\Gamma(\alpha + \beta)}{\Gamma(\alpha)\Gamma(\beta)} \theta^{\alpha-1}(1 - \theta)^{\beta-1} d\theta = \frac{\alpha}{\alpha + \beta}.$$

Restricting $\mathcal{P}_0(t)$ to be of the form (6.12), we now have, for positive integers α,

β and N, and $N > \alpha + \beta$,

$$v^{N-\alpha-\beta}(\alpha, \beta) = \sup_{\phi} \frac{r(\alpha, \beta) + \sum\limits_{t=1}^{N-\alpha-\beta-1} a^t \sum\limits_{k=0}^{t} Q(\alpha, \beta, t, k, \phi) r(\alpha+k, \beta+t-k)}{1 + \sum\limits_{t=1}^{N-\alpha-\beta-1} a^t \sum\limits_{k=0}^{t} Q(\alpha, \beta, t, k, \phi)}, \quad (6.13)$$

where

$$Q(\alpha, \beta, t, k, \phi) = \mathbf{P}\{\pi_t = (\alpha+k, \beta+t-k) \cap v^{N-\alpha-\beta-s}(\pi_s) \geq \phi, 1 \leq s \leq t | \pi_0 = (\alpha, \beta)\}.$$

For a given N the expression (6.13) leads to an algorithm along the following lines for calculating $v^{N-\alpha-\beta}(\alpha, \beta)$ for all non-negative integers α and β such that $\alpha + \beta < N$.

(1) If $\alpha + \beta = N - 1$, the stopping time τ in the definition of $v^{N-\alpha-\beta}(\alpha, \beta)$ must be equal to one. Thus

$$v^{N-\alpha-\beta}(\alpha, \beta) = r(\alpha, \beta) = \alpha/(\alpha+\beta). \quad (6.14)$$

(2) Using (6.14), $Q(\alpha, \beta, t, k, \phi)$ may be calculated for $\alpha + \beta = N - 2$, $t = 1$ and $k = 0,1$. We have

$$Q(\alpha, \beta, 1, 1, \phi) = \mathbf{P}\{X_1 = 1 | \pi_0 = (\alpha, \beta)\} \times \begin{cases} 1 & \text{if } v^{N-\alpha-\beta-1}(\alpha+1, \beta) \geq \phi \\ 0 & \text{if } v^{N-\alpha-\beta-1}(\alpha+1, \beta) < \phi, \end{cases}$$

$$Q(\alpha, \beta, 1, 0, \phi) = \mathbf{P}\{X_1 = 0 | \pi_0 = (\alpha, \beta)\} \times \begin{cases} 1 & \text{if } v^{N-\alpha-\beta-1}(\alpha, \beta+1) \geq \phi \\ 0 & \text{if } v^{N-\alpha-\beta-1}(\alpha, \beta+1) < \phi, \end{cases}$$

and

$$\mathbf{P}\{X_1 = 1 | \pi_0 = (\alpha, \beta)\} = \alpha/(\alpha+\beta).$$

Values of $v^{N-\alpha-\beta}(\alpha, \beta)$ for $\alpha + \beta = N - 2$ may now be calculated from (6.13) using these quantities together with (6.14).

(3) Now knowing the function $v^{N-\alpha-\beta}(\alpha, \beta)$ for $\alpha + \beta = N - 1$ and $\alpha + \beta = N - 2$, calculations similar to those described in stage (2) of the algorithm give values of $Q(\alpha, \beta, t, k, \phi)$ for $\alpha + \beta = N - 3$, $t = 1, 2$, and $k \leq t$. These, together with (6.14), may be substituted into (6.13) to give $v^{N-\alpha-\beta}(\alpha, \beta)$ for $\alpha + \beta = N - 3$.

(4) Similar calculations give in turn values $v^{N-\alpha-\beta}(\alpha, \beta)$ for $\alpha + \beta = N - 4$, $N - 5$, and so on, the final quantity to be calculated being $v^N(1, 1)$.

To carry out this entire calculation for a given N takes $O(N^4)$ elementary steps, and the storage requirement is $O(N^2)$. This may be compared with the calibration method for this problem described in Chapter 1, for which the number of elementary steps for a single value of λ (called p in Chapter 1) is $O(N^2)$ and the storage requirement is $O(N)$. Thus the calibration method is better for large values of N. In fact for large values of N both methods of calculation can be made more economical by storing the functions v and Q (for the direct method) and R (for the calibration method) on reduced grids of values, and interpolating as necessary. This reduces the number of elementary steps to $O(N^2)$ and the storage requirement to $O(N)$ for the direct method, and for the calibration method to $O(N)$ and $O(1)$, respectively. Thus the calibration method retains its advantage for large N.

A similar comparison between the two approaches, showing the calibration approach to be preferable for large N, may be carried out for other bandit sampling processes. The only large-scale application of the direct approach to date has in fact been for the Bernouilli reward process, with the results described in §7.4

The calibration method can often be simplified by exploiting the presence of a location or a scale parameter. A parametric family $\mathscr{D}$ of distributions for a random variable X is said (e.g. see Ferguson, 1967) to have (i) a *location* parameter μ if the distribution of $X - \mu$ does not depend on the parameter μ, (ii) a *scale* parameter σ if the distribution of X/σ does not depend on the positive parameter σ, or (iii) joint location and scale parameters μ and σ if the distribution of $(X - \mu)/\sigma$ does not depend on the parameters μ and $\sigma(>0)$. Equivalently, for families $\mathscr{D}$ with densities $f(x|\theta)$ with respect to Lebesgue measure, μ is a location parameter if $f(x|\mu) = g(x - \mu)$ for some g, σ is a scale parameter if $f(x|\sigma) = \sigma^{-1}g(x\sigma^{-1})$ for some g, and μ and σ are joint location and scale parameters if $f(x|\mu, \sigma) = \sigma^{-1}g([x - \mu]\sigma^{-1})$ for some g. For the remainder of this section densities with respect to Lebesgue measure will be assumed to exist.

Location and scale parameters may be used to reduce the complexity of the functions $v(\pi)$, $R(\lambda, \pi)$ and $M(\phi, \pi, T)$. Note that

$$R(\lambda, \pi) = \sup_{\tau} \left\{ R_\tau(\pi) + \lambda(1-a)^{-1}\mathbf{E}(a^\tau|\pi) \right\} \tag{6.15}$$

and

$$M(\phi, \pi, T) = \inf_{\tau} \left\{ C_\tau(\pi, T) + \phi^{-1}Q_\tau(\pi, T) \right\}, \tag{6.16}$$

where the stopping time τ is in both cases the time at which the standard bandit process is first selected, $C_\tau(\pi, T)$ is the expected number of samples drawn from the target process S before either time τ or the target T is reached, and

$$Q_\tau(\pi, T) = \mathbf{P}(\tau \text{ is reached before the target } T),$$

in both cases starting from the state π. Since there are no state-changes when a standard bandit process is selected we may suppose that it is always selected after time τ.

As the following theorems and their corollaries show, considerable simplification results from the presence of location and/or scale parameters, sufficient statistics, and a conjugate prior distribution.

The possible sequences of states through which a bandit sampling process may pass, starting from a given state, are in 1–1 correspondence with the sequences of X_is. Thus stopping times which start when the process is in the given state may be defined as functions of the X_i sequences. In fact it will prove convenient in the proofs to refer to stopping times which occur after a given sequence of X_i values by the same symbol, whether the starting states are the same or not. If τ is a stopping time defined in terms of the X_i sequence let $\tau(b, c)$ $(b > 0)$ denote the stopping time for which $bX_1 + c$, $bX_2 + c, \ldots bX_\tau + c$ is the sequence sampled before stopping. Lower case letters denote the values taken by the random variables represented by the corresponding upper case letters.

Theorem 6.9. If the parameter μ for the family of distributions for a reward process is a location parameter, the sample mean $\bar{X}$ is a sufficient statistic, and π_0 is the (improper) uniform distribution over the real line, then π_n may be identified by its parameters $\bar{x}$ (the value taken by $\bar{X}$) and n,

$$R(\lambda + c, \bar{x} + c, n) = c(1 - a)^{-1} + R(\lambda, \bar{x}, n) \qquad (c \in \mathbf{R}),$$

and

$$v(\bar{x}, n) = \bar{x} + v(0, n).$$

Proof. Since $\bar{X}$ is a sufficient statistic it follows from Theorem 6.1 and equation (6.1) that π_n depends on the vector of sampled values $\mathbf{x}$ only as a function of $\bar{x}$, and may therefore be written as $\pi_n(\mu|\bar{x})$. We have

$$\pi_n(\mu|\bar{x}) \propto [f(\bar{x}|\mu)]^n = [g(\bar{x} - \mu)]^n = [f(\bar{x} + c|\mu + c)]^n \propto \pi_n(\mu + c|\bar{x} + c), \qquad (6.17)$$

and hence

$$\pi_n(\mu|\bar{x}) = \pi_n(\mu + c|\bar{x} + c). \qquad (6.18)$$

In (6.17) the proportionality sign covers multiplication by a positive function of $\mathbf{x}$ which is independent of μ. The first proportionality follows from (6.1) on putting $\pi_0(\mu) = a$ constant, and since $\Pi_{i=1}^{n} f(x_i|\mu)/[f(\bar{x}|\mu)]^n = h(x_1, x_2, \ldots, x_n)/h(\bar{x}, \bar{x}, \ldots, \bar{x})$, and the justification of the second proportionality is similar. The two equalities in (6.17) are consequences of the fact that μ is a location parameter.

Equation (6.18) and the fact that μ is a location parameter give

$$\pi_n(\mu|\bar{x}) \prod_{i=n+1}^{n+m} f(x_i|\mu) = \pi_n(\mu + c|\bar{x} + c) \prod_{i=n+1}^{n+m} f(x_i + c|\mu + c). \qquad (6.19)$$

Thus, writing M for the random variable taking the value μ, the joint distribution of $M, X_{n+1}, X_{n+2}, \ldots, X_{n+m}$ conditional on $\bar{X} = \bar{x}$ is the same as the joint distribution of $M - c, X_{n+1} - c, X_{n+2} - c, \ldots, X_{n+m} - c$ conditional on $\bar{X} = \bar{x} + c$.

It follows that

$$\mathbf{E}(a^\tau | \bar{x}, n) = \mathbf{E}(a^{\tau(1,c)} | \bar{x} + c, n),$$

and

$$R_\tau(\bar{x}, n) = R_{\tau(1,c)}(\bar{x} + c, n) - c[1 - \mathbf{E}(a^\tau | \bar{x}, n)]/(1 - a).$$

Thus

$$R_{\tau(1,c)}(\bar{x} + c, n) + (\lambda + c)(1 - a)^{-1} \mathbf{E}(a^{\tau(1,c)} | \bar{x} + c, n)$$
$$= c(1 - a)^{-1} + R_\tau(\bar{x}, n) + \lambda(1 - a)^{-1} \mathbf{E}(a^\tau | \bar{x}, n),$$

and hence, taking the supremum over τ (or equivalently over $\tau(1, c)$), it follows from (6.15) that

$$R(\lambda + c, \bar{x} + c, n) = c(1 - a)^{-1} + R(\lambda, \bar{x}, n). \qquad (6.20)$$

The notion of calibration now provides the proof of the second part of the theorem. From (6.10) and Theorem 3.8, $v(\bar{x}, n)$ is the unique λ for which

$$\frac{\lambda}{1 - a} = r(\bar{x}, n) + a \iint R\left(\lambda, \frac{n\bar{x} + x}{n + 1}, n + 1\right) f(x | \mu) \pi_n(\mu | \bar{x}) \, dx \, d\mu. \qquad (6.21)$$

From (6.3) and (6.18), and since $f(x + c | \mu + c) = f(x | \mu)$, we have $r(\bar{x} + c, n) = c + r(\bar{x}, n)$. It therefore follows from (6.20) and (6.21), again using (6.18) and $f(x + c | \mu + c) = f(x | \mu)$, this time with $c = -\bar{x}$, that

$$v(\bar{x}, n) = \bar{x} + v(0, n). \qquad \square$$

Corollary 6.10 (and 6.18, see later comment). The conclusions of the theorem hold if μ is a location parameter having a conjugate prior distribution with parameters $\bar{x}$, a location parameter, and $n \, (\geq 0)$, and if these parameters take the values $(n\bar{x} + x)/(n + 1)$ and $n + 1$ when x is the next value sampled.

Proof. The theorem follows from equation (6.18), (6.19) and (6.21), plus the fact that μ is a location parameter. These equations hold under the conditions of the corollary. $\qquad \square$

Note that in Corollary 6.10, and the similar corollaries to Theorems 6.11 and 6.13, the parameter n is not restricted to integer values.

Theorem 6.11. If the parameter $\sigma \, (> 0)$ for the family of distributions for a reward process is a scale parameter, the sample mean X is a sufficient statistic, and $\pi_0(\sigma) \propto \sigma^{-d}$ for some $d > 0$, then π_n may be identified by its parameters $\bar{x}$ and n,

$$R(b\lambda, b\bar{x}, n) = bR(\lambda, \bar{x}, n) \qquad (b > 0),$$

and

$$v(\bar{x}, n) = \bar{x}v(1, n).$$

Proof. This runs along similar lines to the proof of Theorem 6.9. We have

$$\pi_n(\sigma|\bar{x}) \propto \sigma^{-d}[f(\bar{x}|\sigma)]^n = \sigma^{-d-n}[g(\bar{x}\sigma^{-1})]^n = b^{-n}\sigma^{-d}[f(b\bar{x}|b\sigma)]^n \propto \pi_n(b\sigma|b\bar{x}),$$

and hence

$$\pi_n(\sigma|\bar{x}) = b\pi_n(b\sigma|b\bar{x}). \tag{6.22}$$

Thus, since σ is a scale parameter

$$\pi_n(\sigma|\bar{x}) \prod_{i=n+1}^{n+m} f(x_i|\sigma) = b^{m+1}\pi_n(b\sigma|b\bar{x}) \prod_{i=n+1}^{n+m} f(bx_i|b\sigma). \tag{6.23}$$

Hence, writing Σ for the random variable taking the value σ, the joint distribution of $\Sigma, X_{n+1}, X_{n+2}, \ldots, X_{n+m}$ conditional on $\bar{X} = \bar{x}$ is the same as the joint distribution of $b^{-1}\Sigma, b^{-1}X_{n+1}, b^{-1}X_{n+2}, \ldots, b^{-1}X_{n+m}$ conditional on $\bar{X} = b\bar{x}$. It follows that

$$E(a^\tau|\bar{x}, n) = E(a^{\tau(b, 0)}|b\bar{x}, n),$$

and

$$R_\tau(\bar{x}, n) = b^{-1}R_{\tau(b, 0)}(b\bar{x}, n).$$

Thus, using (6.15),

$$R(b\lambda, b\bar{x}, n) = bR(\lambda, \bar{x}, n). \tag{6.24}$$

Calibration again provides a proof of the second part of the theorem. The unique λ satisfying equation (6.21) with σ in place of μ is $v(\bar{x}, n)$. The equations (6.3), (6.22) and $f(b\bar{x}|b\sigma) = b^{-1}f(\bar{x}|\sigma)$ now give $r(b\bar{x}, n) = br(\bar{x}, n)$, and, again using (6.21), the parallel argument to that given for Theorem 6.9 now gives

$$v(\bar{x}, n) = \bar{x}v(1, n). \qquad \square$$

Corollary 6.12 (and 6.20, see later comment). The conclusions of the theorem hold if σ is a scale parameter having a conjugate prior distribution with parameter $\bar{x}$, a scale parameter, and n (≥ 0), and if these parameters take the values $(n\bar{x} + x)/(n+1)$ and $n+1$ when x is the next value samples.

The proof parallels that of Corollary 6.10, using equations (6.22) and (6.23) in place of equations (6.18) and (6.19), plus the fact that σ is a scale parameter.

The proofs of the next two theorems and their respective corollaries are along similar lines to those of the last four results and are left as exercises.

Theorem 6.13. If the parameters μ and σ (>0) for the family of distributions for a reward process are joint location and scale parameters, σ is known, the sample mean $\bar{X}$ is a sufficient statistic, and $\pi_0(\mu)$ is the improper uniform distribution over $\mathbf{R}$, then π_n may be identified by its parameters $\bar{x}$ and n,

$$R(b\lambda + c, b\bar{x} + c, n, b\sigma) = c(1-a)^{-1} + bR(\lambda, \bar{x}, n, \sigma) \qquad (b > 0, c \in \mathbf{R}),$$

and

$$v(\bar{x}, n, \sigma) = \bar{x} + \sigma v(0, n, 1).$$

In the proof the place of equations (6.18) and (6.22) is taken by the equation

$$\pi_n(\mu | \bar{x}, \sigma) = b\pi_n(b\mu + c | b\bar{x} + c, b\sigma).$$

Equations (6.19) and (6.23) are replaced by the equation

$$\pi_n(\mu | \bar{x}, \sigma) \prod_{i=n+1}^{n+m} f(x_i | \mu, \sigma) = b^{m+1}\pi_n(b\mu + c | b\bar{x} + c, b\sigma) \prod_{i=n+1}^{n+m} f(bx_i + c | b\mu + c, b\sigma).$$

Corollary 6.14 (and 6.22, see later comment). The conclusions of the theorem hold if μ and σ are joint location and scale parameters, with σ known and μ having a conjugate prior distribution with parameters $\bar{x}$, σ and n ($\geqslant 0$), for which $\bar{x}$ and σ are joint location and scale parameters, and $\bar{x}$ and n take the values $(n\bar{x} + x)/(n+1)$ and $n+1$ when x is the next value sampled.

Theorem 6.15. If the parameters μ and σ (> 0) for the family of distributions for a reward process are joint location and scale parameters, the sample mean $\bar{X}$ and variance $S^2 = (n-1)^{-1}\Sigma(X_i - \bar{X})^2$ are sufficient statistics, and $\pi_0(\mu, \sigma) \propto \sigma^{-d}$ for some $d > 0$, then π_n may be identified by the parameters $\bar{x}$, s and n,

$$R(b\lambda + c, b\bar{x} + c, bs, n) = c(1-a)^{-1} + bR(\lambda, \bar{x}, s, n) \qquad (b > 0, c \in \mathbf{R}),$$

and

$$v(\bar{x}, s, n) = \bar{x} + sv(0, 1, n).$$

In the proof the place of equations (6.18) and (6.22) is taken by the equation

$$\pi_n(\mu, \sigma | \bar{x}, s) = b\pi_n(b\mu + c, b\sigma | b\bar{x} + c, bs). \qquad (6.25)$$

Equations (6.19) and (6.23) are replaced by the equation

$$\pi_n(\mu, \sigma | \bar{x}, s) \sum_{i=n+1}^{n+m} f(x_i | \mu, \sigma) = b^{m+1}\pi_n(b\mu + c, b\sigma | b\bar{x} + c, bs)$$

$$\times \prod_{i=n+1}^{n+m} f(bx_1 + c | b\mu + c, b\sigma).$$

Corollary 6.16 (and 6.24, see later comment). The conclusions of the theorem hold if μ and σ are joint location and scale parameters, having a conjugate prior distribution with parameters $\bar{x}$, s (> 0) and n ($\geqslant 0$) which satisfies equation (6.25), and if these parameters change by the same rules as the sample mean, variance, and number of observations, as further values are sampled.

The next four theorems are the counterparts of Theorems 6.9, 6.11, 6.13, and 6.15 for target processes, and have similar proofs. Each theorem has a corollary which may be written out in the words of the corollary of its counterpart, and proved along similar lines. These corollaries will be referred to as Corollaries 6.18, 6.20, 6.22, and 6.24, respectively.

Theorem 6.17. If the parameter μ for the family of distributions for a target process is a location parameter, the sample mean $\bar{X}$ is a sufficient statistic, and π_0 is the (improper) uniform distribution over the real line, then

$$M(\phi, \bar{x}, n, T) = M(\phi, \bar{x} + c, n, T + c) \qquad (c \in \mathbf{R}),$$

and

$$v(\bar{x}, n, T) = v(\bar{x} - T, n, 0).$$

Theorem 6.19. If the parameter σ (>0) for the family of distributions for a target process is a scale parameter, the sample mean $\bar{X}$ is a sufficient statistic, and $\pi_0(\sigma) \propto \sigma^{-d}$ for some $d > 0$, then

$$M(\phi, \bar{x}, n, T) = M(\phi, b\bar{x}, n, bT) \qquad (b > 0),$$

and

$$v(\bar{x}, n, T) = v(\bar{x}/T, n, 1).$$

Theorem 6.21. If the parameters μ and σ (>0) for a target process are joint location and scale parameters, σ is known, the sample mean is a sufficient statistic, and $\pi_0(\mu)$ is the improper uniform distribution over $\mathbf{R}$, then

$$M(\phi, \bar{x}, n, \sigma, T) = M(\phi, b\bar{x} + c, n, b\sigma, bT + c) \qquad (b > 0, c \in \mathbf{R}),$$

and

$$v(\bar{x}, n, \sigma, T) = v\left(\frac{\bar{x} - T}{\sigma}, n, 1, 0\right).$$

Theorem 6.23. If the parameters μ and σ (>0) for a target process are joint location and scale parameters, the sample mean and variance are sufficient statistics, and $\pi_0(\mu, \sigma) \propto \sigma^{-d}$ for some $d > 0$, then

$$M(\phi, \bar{x}, s, n, T) = M(\phi, b\bar{x} + c, bs, n, bT + c) \qquad (b > 0, c \in \mathbf{R}),$$

and

$$v(\bar{x}, s, n, T) = v\left(\frac{\bar{x} - T}{s}, 1, n, 0\right).$$

6.5 RANDOM SAMPLING TIMES

A sampling process for which the times $U_1, U_2, \ldots$ needed to sample a value are independent identically distributed random variables is an example of a semi-Markov bandit process. Thus (Theorem 3.6) optimal policies for sampling sequentially from a family of such processes are again index policies. Semi-Markov sampling processes are of interest as models for whole areas of research in a pharmaceutical or agro-chemical laboratory. A family of two semi-Markov reward processes, for example, serves as a model for the herbicide and fungicide programmes. Each sampled value is, in accountants' language, the expected net present value of a potential herbicide or fungicide at the time of its discovery. The aim of the model would be to indicate the relative priorities of the two research programmes.

No index values for semi-Markov reward processes have as yet been calculated, though this would be feasible, particularly when the problem may be reduced by exploiting invariance properties. Provided the distributions of the U_is and of the X_is are independent, and the prior distributions for their parameters are also independent, the theorems of §6.4 continue to hold, with the obvious modification that the state-space must now specify a posterior distribution for the parameters of the U-distribution. We now also have an analogue of Theorem 6.9, referring to the U-distribution instead of the X-distribution. To state this theorem the notation is extended by letting ρ_n be the posterior density for the parameter ($\in \mathbf{R}^+$) of the U-distribution, and including the discount parameter γ in the notation. Since the prior densities ρ_0 and π_0 are by assumption independent, it follows that ρ_n and π_n are also independent.

Theorem 6.25. If the parameter $\sigma(>0)$ for the family of distributions for the sampling time of a semi-Markov reward process is a scale parameter, the sample mean $\bar{U}$ is a sufficient statistic, and $\rho_0(\sigma) \propto \sigma^{-d}$ for some $d > 0$, then ρ_n may be identified by its parameters $\bar{u}$ and n, and for $b > 0$,

$$R(b^{-1}\lambda, b\bar{u}, n, \pi_n, b^{-1}\gamma) = R(\lambda, \bar{u}, n, \pi_n, \gamma),$$

and

$$v(\bar{u}, n, \Pi_n, \gamma) = \gamma v(\bar{u}\gamma, n, \pi_n, 1).$$

Proof. The two parts of the theorem are direct consequences of changing the time unit, for the first part by the factor b, and for the second part by the factor γ. ☐

A further modification is to allow a switch from a sampling process to occur at any time, rather than restricting switches to the times when a new value is obtained. This works out quite neatly when the U-distribution is negative exponential, which means that the times when new values are obtained form a Poisson process in process time.

Let θ be the negative exponential parameter, so that the probability of observing a new value in any time interval $(t, t + \delta t)$ is $\theta \delta t + o(\delta t)$. Let θ have the improper prior density $\rho(\theta | 0, 0) \propto \theta^{-1}$. Suppose that after a time t, n values have been sampled at times $t_1, t_1 + t_2, \ldots, t_1 + t_2 + \cdots + t_n$. The probability density for this outcome is

$$\prod_{i=1}^{n} (\theta e^{-\theta t_i}) \exp[-\theta(t - \Sigma t_i)] = \theta^n e^{-\theta t},$$

$\exp[-\theta(t - \Sigma t_i)]$ being the probability that the $(n+1)$th sampling time is greater than $t - \Sigma t_i$. Thus it follows from the generalized Bayes' Theorem that the posterior density for θ may be written as

$$\rho(\theta | t, n) \propto \theta^{-1} \theta^n e^{-\theta t} \propto t^n \theta^{n-1} e^{-\theta t} / \Gamma(n),$$

which is a gamma density with parameters n and t. Thus the probability density for the further time up to the next observed value is

$$f(u | t, n) = (\Gamma(n))^{-1} t^n \int_0^\infty \theta^{-\theta u} \theta^{n-1} e^{-\theta t} d\theta = nt^n (t+u)^{-n-1}.$$

Theorem 6.26. For a reward process with unrestricted switching, a negative exponential distribution with parameter θ for the sampling time, and an improper prior density $\rho(\theta | 0, 0) \propto \theta^{-1}$, the posterior density for θ after n observations in a time t may be identified by its parameters n and t, and for $b > 0$,

$$R(b^{-1}\lambda, bt, n, \pi_n, b^{-1}\gamma) = R(\lambda, t, n, \pi_n, \gamma),$$

and

$$v(t, n, \pi_n, \gamma) = \gamma v(t\gamma, n, \pi_n, 1).$$

Proof. As for Theorem 6.25.   □

The next theorem states some fairly obvious properties of the function $v(t, n, \pi_n, \gamma)$ which are needed when calculating it.

Theorem 6.27. If a sampling process with unrestricted switching (i) yields a reward $g(x)$ when the value x is sampled, and $g(x)$ increases with x, (ii) has a negative exponential sampling-time distribution with a parameter θ which has an (improper) prior density proportional to θ^{-1}, and (iii) has a positive discount parameter γ, then $v(t, n, \pi_n, \gamma)$ is a decreasing function of t. Moreover, if $g(x)$ is constant, so that the distribution of the sampled values is irrelevant and $v(t, n, \pi_n, \gamma)$ reduces to $v(t, n, \gamma)$, $v(t, n, \gamma)$ is an increasing function of n.

Proof. In state (t, n, π_n) the distribution function $F(u|t, n)$ for the time U until the next observed value is $1 - t^n(t+u)^{-n}$. This is decreasing in t and increasing in n.

Since $F(u|t, n)$ is a continuous function of u, for a given initial state (t_1, m, π_m) a realization of the sampling process may be defined by the sequences $P_1, P_2, \ldots$ and $X_1, X_2, \ldots$ instead of $U_1, U_2, \ldots$ and $X_1, X_2, \ldots$, where

$$P_i = F\left(U_i | t_1 + \sum_{j=1}^{i-1} U_j, m+i-1\right).$$

Note that this definition means that P_i has a uniform distribution on the interval $[0, 1]$. For the given initial state a stopping time τ may be defined by the non-negative functions

$$w_i\{[0, 1]^{i-1} \times [0, \infty]^{i-1} \cap \{w_j > 0, j < i\} \rightarrow$$

$$w_i(P_1, P_2, \ldots, P_{i-1}, X_1, X_2, \ldots, X_{i-1})\} \quad (i = 1, 2, \ldots),$$

where $\tau = \min_i\{\sum_{j=1}^{i-1} U_j + w_i : w_i < U_i\}$, or equivalently by the functions taking values in the interval $[0, 1]$

$$p_i\{[0, 1]^{i-1} \times [0, \infty]^{i-1} \cap \{p_j > 0, j < i\} \rightarrow$$

$$p_i(P_1, P_2, \ldots, P_{i-1}, X_1, X_2, \ldots, X_{i-1})\} \quad (i = 1, 2, \ldots),$$

where $p_i = F(w_i | t_1 + \sum_{j=1}^{i-1} U_j, m+i-1)$. Thus w_i is the maximum time allowed between sampling X_{i-1} and sampling X_i before stopping occurs, and p_i is the conditional probability that X_i is sampled before stopping occurs. Both w_i and p_i depend on the history and state of the process when X_{i-1} is sampled.

If the functions p_i $(i = 1, 2, \ldots)$ define τ starting from the state (t_1, m, π_m), now let σ be the stopping time for the sampling process starting from the state (t_2, m, π_m) which is defined by the same functions p_i. Since m and π_m are unchanged, the joint distributions of the P_is and X_is are the same for both starting states. The definition of σ means that the same is true for the joint distributions of the P_is, X_is and Y_is, where $Y_i = 1$ if X_i is observed before stopping occurs, and $Y_i = 0$ otherwise. Corresponding realizations starting from (t_1, m, π_m) and using τ, and starting from (t_2, m, π_m) and using σ, differ only in the values, U_{i1} and U_{i2} respectively, of the sampling times of the X_is in the two cases. If $t_2 < t_1$ then, for given values of $P_i (i \geqslant 1)$, $U_{i1} > U_{i2}$. This is a consequence of the fact that $F(s|t, n)$ is a decreasing function of t. Thus

$$R_\tau(t_1, m, \pi_m, \gamma) = \mathbf{E} \sum_{i=1}^{\infty} g(X_i) Y_i \exp\left\{-\gamma \sum_{j=1}^{i} U_{j1}\right\}$$

$$< \mathbf{E} \sum_{i=1}^{\infty} g(X_i) Y_i \exp\left\{-\gamma \sum_{j=1}^{i} U_{j2}\right\} = R_\sigma(t_2, m, \pi_m, \gamma).$$

Also $\sigma < \tau$, since as well as $U_{i1} > U_{i2}$ we have, in obvious notation, $w_{i1} > w_{i2}$, so that

$$W_\tau(t_1, m, \pi_m, \gamma) = \gamma^{-1} \mathbf{E}(1 - e^{-\gamma\tau}) > \gamma^{-1} \mathbf{E}(1 - e^{-\gamma\sigma}) = W_\sigma(t_2, m, \pi_m, \gamma).$$

Thus

$$v(t_1, m, \pi_m, \gamma) = \sup_{\tau > 0} \frac{R_\tau(t_1, m, \pi_m, \gamma)}{W_\tau(t_1, m, \pi_m, \gamma)} < \sup_{\sigma > 0} \frac{R_\sigma(t_2, m, \pi_m, \gamma)}{W_\sigma(t_2, m, \pi_m, \gamma)} = v(t_2, m, \pi_m, \gamma),$$

as required.

For the case $g(x) = $ a constant, a virtually identical argument, which now compares realizations starting from states (t, n_1) and (t, n_2), and uses the fact that $F(u|t, n)$ is increasing in n, shows that if $n_2 < n_1$, then

$$v(t, n_1, \gamma) > v(t, n_2, \gamma). \qquad \square$$

As a first example, suppose the rewards are all known to be equal to 1, so that the only uncertainty concerns the intervals between them. Thus π_n may be dropped from the notation, as may γ, being a constant throughout a set of calculations. Index values for other values of γ then follow from Theorem 6.26. Both parts of Theorem 6.27 hold. Thus, given λ (> 0) and n, for some s_n

$$v(t, n) \left\{ \begin{matrix} \geq \\ = \\ < \end{matrix} \right\} \lambda \text{ according as } t \left\{ \begin{matrix} \leq \\ = \\ > \end{matrix} \right\} s_n,$$

and s_n increases with n. The index $v(t, n)$ may be determined by calculating s_n ($n > 0$) for different values of λ. For a given λ this may be done by the calibration procedure using the functional equation (2.4) for the optimal payoff from a SFABP consisting of the reward processes in th state (t, n) and a standard bandit process Λ with parameter λ. This takes the form

$$R(\lambda, t, n) = \sup_{w \geq t} \left\{ \int_t^w e^{-\gamma(u-t)} [1 + R(\lambda, u, n+1)] \frac{nt^n}{u^{n+1}} du + e^{-\gamma(w-t)} \frac{\lambda t^n}{\gamma w^n} \right\}.$$

The supremum occurs at $w = \max[t, s_n]$. Thus, differentiating with respect to w to find the maximum,

$$1 + R(\lambda, s_n, n+1) = \lambda \left(\frac{s_n}{n} + \frac{1}{\gamma} \right).$$

Suppose now that the rules of the SFABP are modified so that no switching is allowed for $n > N$. Thus, if $R(\lambda, t, n, N)$ is the corresponding optimal payoff function,

$$R(\lambda, t, n, N) = \max[\text{Payoff if } \Lambda \text{ is permanently selected, Payoff if}$$
$$S \text{ is permanently selected}]$$

$$= \max[\lambda, nt^{-1}]\gamma^{-1} \qquad (n \geq N). \qquad (6.26)$$

This follows since

$$\text{Payoff if } \Lambda \text{ is permanently selected} = \int_0^\infty e^{-\gamma s} \lambda ds = \lambda \gamma^{-1},$$

and

Payoff if S is permanently selected

$$= \int_0^\infty (\text{Payoff if } S \text{ is permanently selected} | \theta) \rho(\theta | t, n) d\theta$$

$$= \int_0^\infty \frac{\theta}{\gamma} \cdot \frac{t^n \theta^{n-1} e^{-\theta t}}{\Gamma(n)} d\theta = \frac{n}{t\gamma}.$$

The corresponding functional equation is, for $0 < n \leqslant N - 1$,

$$R(\lambda, t, n, N) = \sup_{w \geqslant t} \left\{ \int_t^w e^{-\gamma(u-t)} [1 + R(\lambda, u, n+1, N)] \frac{nt^n}{u^{n+1}} du + e^{-\gamma(w-t)} \frac{\lambda t^n}{\gamma w^n} \right\},$$

(6.27)

the supremum being at $w = \max[t, s_{nN}]$, and, differentiating for the maximum,

$$1 + R(\lambda, s_{nN}, n+1) = \lambda \left(\frac{s_{nN}}{n} + \frac{1}{\gamma} \right).$$

(6.28)

From (6.26) we have $s_{nN} = n\lambda^{-1}$ $(n > N - 1)$. Values of s_{nN} and of the function $R(\lambda, \cdot, n, N)$ may be calculated successively for $n = N - 1, N - 2, \ldots, 1$ from (6.28), and then substituting for $R(\lambda, \cdot, n+1, N)$ in the right-hand side of equation (6.27), using (6.26) to start the process. Clearly s_{nN} increases with N and $\rightarrow s_n$ as $N \rightarrow \infty$. These calculations have not actually been carried out, but would be along similar lines to those described in Chapter 7.

As a second example of a reward process with a negative exponential distribution of sampling times and unrestricted switching times, suppose that the rewards sampled also have a negative exponential distribution. Let θ_1 and θ_2 be the respective parameters of the two negative exponential distributions, each with (improper) prior densities proportional to $\theta_i^{-1} (i = 1, 2)$. The generalized Bayes' theorem shows that π_n is a gamma density with parameters n and Σ, the sum of the first n sampled values. Thus the optimal payoff function for the SFABP formed by the reward process S and a standard bandit process Λ with the parameter λ may be written as $R(\lambda, t, n, \Sigma, \gamma)$. Theorem 6.26 gives $v(t, n, \Sigma, \gamma) = \gamma v(t\gamma, n, \Sigma, 1)$, so calculations are required only for one value of γ, which will be supposed equal to 1 and dropped from the notation. The obvious adaptation of Theorem 6.11 gives $v(t, n, \Sigma, \gamma) = \Sigma v(t, n, 1, \gamma)$; this means that calculations are required only for one value of λ, which will also be supposed equal to 1 and dropped from the notation, so that $R(\lambda, t, n, \Sigma, \gamma)$ becomes $R(t, n, \Sigma)$. The functional equation is

$$R(t, n, \Sigma) = \sup_{w \geqslant t} \left\{ \int_t^w e^{-(u-t)} \int_0^\infty [x + R(u, n+1, \Sigma + x)] \right.$$

$$\left. \times \frac{n\Sigma^n}{(\Sigma + x)^{n+1}} dx \frac{nt^n}{u^{n+1}} du + e^{-(w-t)} \frac{t^n}{w^n} \right\}.$$

(6.29)

From Theorems 6.27 and 6.11 (adapted) it follows that $v(t, n, \Sigma)$ is a decreasing function of t and an increasing function of Σ (its dependence on n is not so clear as this is a parameter in both posterior distributions). Thus, given n and Σ,

$$v(t, n, \Sigma) \left\{ \begin{matrix} \geq \\ = \\ < \end{matrix} \right\} 1 \text{ according as } t \left\{ \begin{matrix} \leq \\ = \\ > \end{matrix} \right\} s_n(\Sigma), \tag{6.30}$$

for some $s_n(\Sigma)$ which increases with Σ. Since $v(t, n, \Sigma) = \Sigma v(t, n, 1)$, the index function may be easily determined from $s_n(\Sigma)$. It follows from (6.30) that the supremum in (6.29) is attained when $w = \max[t, s_n(\Sigma)]$. Differentiation with respect to w shows that

$$\frac{\Sigma}{n-1} + Q(s_n(\Sigma), n, \Sigma) = 1 + \frac{s_n(\Sigma)}{n} \qquad (n > 1), \tag{6.31}$$

where

$$Q(u, n, \Sigma) = \int_0^\infty R(u, n+1, \Sigma + x) \frac{n\Sigma^n}{(\Sigma + x)^{n+1}} \, dx.$$

As in the previous example, arbitrarily close approximations to $s_n(\Sigma)$ may be obtained by backwards induction from N, when the rules of the SFABP are modified to prohibit switching for $n > N$. If $R(t, n, \Sigma, N)$ is the corresponding optimal payoff we now have

$$R(t, n, \Sigma, N) = \max[1, \Sigma t^{-1}]. \tag{6.32}$$

The corresponding functional equation is identical to (6.29), apart from the additional argument N on which the functions R and Q now also depend. The calculations for $s_n(\Sigma)$ now follows the pattern of those for s_n in the previous example, with repeated use of the finite switching-horizon versions of (6.31) and (6.29) in turn, for $n = N-1, N-2, \ldots, 2$. Note that this means iterating with a function of two variables, $R(\cdot, n, \cdot, N)$, rather than of one variable, as in the case of rewards all equal to 1.

6.6 BROWNIAN REWARD PROCESS

No physical particle carries out Brownian motion, though the large number of discrete, randomly directed, shocks to which a gas molecule is subject mean that over any time-interval which is not extremely short its displacement is well approximated by Brownian motion. Similarly, we should certainly have to be ingenious to construct an actual Brownian (or Wiener) reward process. The value of the concept is rather as an approximation to reward processes for which rewards accrue in discrete quanta, and in particular as an approximation to the normal reward process with known variance.

For the normal reward process with known variance σ^2 we may in the first instance assume σ^2 to be 1, since if this is not true for X it is true for the transformed observation $X' = \sigma^{-1} X$. The single unknown parameter μ is thus the mean of the distribution and π_0 is taken to be the improper uniform density over the real line.

$$f(x|\mu)=(2\pi)^{-1/2}\exp\{-\tfrac{1}{2}(x-\mu)^2\},$$

hence μ is a location parameter.

$$\prod_{i=1}^{n}f(x_i|\mu)=\exp\left\{-\frac{n}{2}(\bar{x}-\mu)^2\right\}(2\pi)^{-n/2}\exp\{-\tfrac{1}{2}(\Sigma x_i^2-n\bar{x}^2)\},$$

hence $\bar{X}$ is a sufficient statistic by Theorem 6.1.

$$\pi_n(\mu|\bar{x})\propto\pi_0(\mu)\prod_{i=1}^{n}f(x_i|\mu)\propto(2\pi n^{-1})^{-1/2}\exp\left\{-\frac{n}{2}(\bar{x}-\mu)^2\right\},$$

using Bayes' theorem. Thus $\pi_n(\mu|\bar{x})$ is normal with mean $\bar{x}$ and variance n^{-1}, and conjugate prior densities for μ are of this form, with n not restricted to integer values.

$$v(\bar{x},n)=\bar{x}+v(0,n)$$

by Theorem 6.9. If the observations are now rescaled by multiplying by σ this has the effect of multiplying both $\bar{x}$ and $v(\bar{x},n)$ by σ, so that, extending the notation to indicate the dependence on the variance σ^2,

$$v(\bar{x},n,\sigma)=\bar{x}+\sigma v(0,n,1).$$

This has already been noted as part of Theorem 6.13. Note that σ is a scale parameter.

Consider now a Brownian reward process $\Sigma(t)$, $t\in\mathbf{R}^+$, for which $\Sigma(0)=0$, $E\Sigma(t)=\theta t$, $\mathrm{Var}\,\Sigma(t)=s^2 t$, θ has the improper uniform prior distribution on $\mathbf{R}$, s is known, and with the discount parameter γ. The total discounted reward up to time T is $\int_0^T e^{-\gamma t}\,d\Sigma(t)$,

$$\mathbf{E}\int_0^T e^{-\gamma t}\,d\Sigma(t)=\int_0^T e^{-\gamma t}\,\mathbf{E}d\Sigma(t)=\int_0^T e^{-\gamma t}\,\theta dt=\theta\gamma^{-1}(1-e^{-\gamma T}),$$

and

$$\mathrm{Var}\int_0^T e^{-\gamma t}\,d\Sigma(t)=\int_0^T e^{-\gamma t}\,\mathrm{Var}\,d\Sigma(t)=\int_0^T e^{-\gamma t}\,s^2 dt=s^2\gamma^{-1}(1-e^{-\gamma T}).$$

Note that $\Sigma(T)$ is a sufficient statistic for the process up to time T. This follows from Theorem 6.1 if we write out the joint density function for any $\Sigma(t_i)$ $(1\leqslant i\leqslant r)$, where $t_1<t_2<\cdots<t_r=T$. The state of the process at time T may therefore be expressed as $(\bar{\Sigma},T)$, where $\bar{\Sigma}=\Sigma(T)/T$. The index for a Brownian reward process with parameters s^2 and γ in the state $(\bar{\Sigma},T)$ will be written as $v_B(\bar{\Sigma},T,s,\gamma)$. If we decrease the unit of time by the factor γ this changes $\bar{\Sigma}$, T, s^2 and γ to $\gamma^{-1}\bar{\Sigma}$, γT, $\gamma^{-1}s^2$ and 1. The maximized reward rate v_B is also decreased by the factor γ,

so that

$$v_B(\bar{\Sigma}, T, s, \gamma) = \gamma v_B(\gamma^{-1}\bar{\Sigma}, \gamma T, \gamma^{-1/2}s, 1). \tag{6.33}$$

Within a Brownian reward process we can embed a normal reward process. Let

$$\Delta = T/n, \quad \mu = \int_0^\Delta \theta e^{-\gamma t}\,dt = \theta\gamma^{-1}(1-e^{-\gamma\Delta}),$$

$$X_i = \int_{(i-1)\Delta}^{i\Delta} \Sigma(t)\exp\{-\gamma[t-(i-1)\Delta]\}\,dt \quad (i=1,2,\ldots,n),$$

$$\sigma^2 = \text{Var}\,X_i = s^2\gamma^{-1}(1-e^{-\gamma\Delta}).$$

The X_is form a normal reward process with the discount factor $a = e^{-\gamma\Delta}$. The posterior distribution for μ at time T if only the X_is are observed is therefore $N(\bar{x}, \sigma^2/n)$, so that the posterior distribution for θ is $N(\bar{x}\gamma(1-a)^{-1}, s^2\gamma(1-a)^{-1}n^{-1})$. The posterior distribution for θ at time T when the whole process $\Sigma(t)$ is observed is obtained by letting $n\to\infty$, and is therefore $N(\Sigma(T)/T, s^2/T)$. It follows that the posterior distribution for θ after observing the first nX_is to have the mean $\bar{x}$ is the same as for the fully observed Brownian motion at time $T^* = n\gamma^{-1}(1-a)$ with $\Sigma(T^*) = n\bar{x}$. The index for the normal reward process in the state $(\bar{x}, \sigma^2/n)$ must, then, be less than the index for the Brownian reward process in the state $(\Sigma(T^*)/T^*, s^2/T^*)$, since the first of these processes differs from the second only in (i) providing incomplete information about $\Sigma(t)$, and (ii) restricting the allowed values of the stopping time τ with respect to which the index maximizes the expected reward rate. Thus, using subscripts N and B to distinguish the normal and Brownian reward processes, and multiplying v_N by the discrete-time correction factor $\gamma(1-a)^{-1}$ (see §2.3),

$$\frac{\gamma}{1-a}v_N(\bar{x}, n, \sigma, a) < v_B(\Sigma(T^*)/T^*, T^*, s, \gamma) = v_B\left(\frac{\bar{x}\gamma}{1-a}, \frac{n(1-a)}{\gamma}, \frac{\sigma\gamma^{1/2}}{(1-a)^{1/2}}, \gamma\right). \tag{6.34}$$

We have shown that

$$v_N(\bar{x}, n, \sigma, a) = \bar{x} + \sigma v_N(0, n, 1, a), \tag{6.35}$$

and it follows on letting $n\to\infty$ that

$$v_B(\Sigma(T)/T, T, s, \gamma) = \Sigma(T)/T + s v_B(0, T, 1, \gamma). \tag{6.36}$$

Thus, using (6.33), (6.35) and (6.36), the relationship (6.34) between v_N and v_B may be written in the form

$$v_N(0, n, 1, a) < (1-a)^{1/2}v_B(0, n(1-a), 1, 1). \tag{6.37}$$

The upper bound which (6.37) provides for v_N is close when the set of allowed stopping times for the Brownian motion in which the normal process is embedded are close together. For a fixed value of $T = n(1-a)$ this means that the bound improves as $a\to 1$, since $\Delta = -\gamma^{-1}\log_e a$ is the interval between these times and, as noted in the derivation of (6.33), the factor γ^{-1} may be removed by changing the units in which time is measured. Thus, writing $u(\cdot)=v_B(0,\cdot,1,1)$, we have established, admittedly rather informally, the following theorem.

Theorem 6.28. $(1-a)^{-1/2}v_N(0,\,T/1-a,\,1,\,a)<u(T)$ and $\to u(T)$ as $a\to1$.

The function u has the properties set out in the following theorem.

Theorem 6.29.   (i) $u(T)=[-T^{-1}\log_e(-16\pi T^2\log_e T)]^{1/2}+o(1)$ as $T\to0$.
            (ii) $T^{1/2}u(T)$ is decreasing in T.
            (iii) $Tu(T)\leqslant2^{-1/2}$ and $\to2^{-1/2}$ as $T\to\infty$.

Part (i) of the theorem is due to Chang and Lai (1987), and parts (ii) and (iii) to Bather (1983). The methods which they used are discussed briefly in §9.4. Calculations of the function u are reported in §7.2.

6.7 ASYMPTOTICALLY NORMAL REWARD PROCESSES

The approximation (6.37) for the index of a normal reward process with known variance may be extended to the large class of reward processes that, essentially because of the central limit theorem, may be regarded as being approximately normal. For any reward process the index depends only on the random process defined by the mean of the current distribution for the expected reward. Thus the conditions for a good approximation based on the normal reward process are those which ensure that the change in $\bar{x}$ as n increases to $n+m$ has a distribution which for large n is close to its distribution for the normal case. The main instrument for checking whether these conditions hold is Theorem 6.32. Two well-known results also play a role, and these are stated first.

Theorem 6.30 (Lindberg–Levy Central Limit Theorem). If $Y_1,\,Y_2,\ldots$ are independent identically distributed random variables with mean μ and variance $\sigma^2(\neq0)$, then

$$(n\sigma)^{-1/2}\sum_{i=1}^{n}(Y_i-\mu)\xrightarrow{d}\Phi$$

as $n\to\infty$, and this convergence is uniform.

Theorem 6.31 (Chebyshev's Inequality). If the random variable Y has mean μ and variance σ^2 then

$$\mathbf{P}(|Y-\mu|\geqslant t)\leqslant\sigma^2/t^2\qquad(t>0).$$

Now reverting to the set-up for a sampling process introduced in §6.1, let Θ be the parameter space for θ, and P_n be the joint probability distribution over $\Theta\times R^\infty$ for $\theta,\,X_{n+1},\,X_{n+2},\ldots$ defined by the posterior distribution Π_n for θ and the distribution L_θ for each of the X_is. Let $h_n = (x_1, x_2, \ldots, x_n)$, the history of the process up to time n. The mean θ_1 and variance θ_2 of the distribution defined by $f(x|\theta)$ are both assumed to exist. Our key asymptotic result involves the following random variables which, together with θ_1 and θ_2, are defined in terms of P_n; s is a positive constant, $\bar{x}_n = n^{-1}(x_1 + x_2 + \cdots + x_n)$.

$$U = m^{-1/2}s^{-1}\left(\sum_{i=n+1}^{n+m} X_i - m\theta_1\right), \quad U' = m^{-1/2}\theta_2^{-1/2}\left(\sum_{i=n+1}^{n+m} X_i - m\theta_1\right),$$

$$V = n^{1/2}s^{-1}(\theta_1 - \bar{x}_n), \quad\quad\quad V' = n^{1/2}\theta_2^{-1/2}(\theta_1 - \bar{x}_n),$$

$$W = (m^{-1} + n^{-1})^{-1/2}(m^{-1/2}U + n^{-1/2}V).$$

$F_U, F_{U'}, F_V, F_{V'}$ and F_W are their distribution functions; F_1 and F_2 are those of θ_1 and θ_2. Conditional distribution functions are denoted by $F_{U'|\theta}, F_{2|V'}$ etc.

The following conditions are required.

(i) There is a sequence of Baire functions s_n defined on $h_n(n = 0, 1, 2, \ldots)$ such that, given ε and s (both > 0), $\Pi_n(|\theta_2 - s^2| > \varepsilon) \to 0$ as $n \to \infty$ for any sequence $h_n(n = 0, 1, 2, \ldots)$ on which $s_n(h_n)$ takes the fixed value s for every n. A sequence of h_ns with this property will be termed a *fixed-s sequence*.

(ii) $F_{V'}(v) \to \Phi(v)$ on any fixed-s sequence, $v \in \mathbf{R}$.

(iii) For any bounded interval $I \subset \mathbf{R}$ and $\varepsilon > 0$,

$$\Pi_n(|F_{U'|\theta}(u) - \Phi(u)| < \varepsilon, u \in I) \to 1$$

on any fixed-s sequence as m and $n \to \infty$ in such a fashion that $mn^{-1} \geqslant$ some fixed positive c.

Note that Theorem 6.30 tells us that $F_{U'|\theta}(u) \to \Phi(u)$ uniformly in $u \in \mathbf{R}$ as $m \to \infty$. Condition (iii) is a related, but not equivalent, property.

When the X_is are a reward sequence and $\mathbf{E}\theta_1 = \bar{x}_n$, our interest, as remarked at the beginning of this section, is in showing that for large n the difference $\Delta\bar{x} = \bar{x}_{n+m} - \bar{x}_n$ has a distribution close to its distribution for the normal case. This is the meaning of the theorem which follows. Note that $\Delta\bar{x} = sm^{1/2}n^{-1/2}(m + n)^{-1/2}W$.

Theorem 6.32 (Normal Limit Theorem). (1) If L_θ is $N(\theta_1, \sigma^2)$, where $\sigma^2(= s^2)$ is known, and Π_0 is the improper uniform distribution on the entire real line, then $F_W(w) = \phi(w)(w \in \mathbf{R})$. (2) Under conditions (i), (ii) and (iii), $F_W(w) \to \phi(w)(w \in \mathbf{R})$, on any fixed-s sequence as m and $n \to \infty$ in such a fashion that $mn^{-1} \geqslant$ some fixed positive c.

Proof. (1) $F_{U|\theta} = \Phi(\theta \in \Theta)$. Thus $F_U = \Phi$ and U is independent of V. Also $F_V = \Phi$, since Π_n is $N(\bar{x}_n, \sigma^2\pi^{-1})$. Part 1 of the theorem follows, since W is a linear combination of U and V with coefficients for which the sum of squares is one.

(2) For simplicity of presentation, and with only an easily removable loss of

generality our proof is for the case $mn^{-1} = c$.

$$F_W(w) = \int F_{U|\theta}[(1+c^2)^{1/2}w - cn^{1/2}s^{-1}(\theta_1 - \bar{x}_n)]d\Pi_n(\theta). \qquad (6.38)$$

For the case of part 1 this reduces to

$$F_W(w) = \int_{-\infty}^{\infty} \Phi[(1+c^2)^{1/2}w - cv]\phi(v)\,dv, \qquad (6.39)$$

which, as just shown, is equal to $\Phi(w)$. The proof of part 2 consists of using conditions (i), (ii) and (iii) to obtain bounds on the difference between expressions (6.38) and (6.39).

We first show that, given $\varepsilon > 0$, M may be chosen so that

$$\left| F_W(w) - \int_{-\infty}^{\infty} \Phi[(1+c^2)^{1/2}w - cv]dF_{V'}(v) \right| < 3\varepsilon/5 \qquad (n > M). \qquad (6.40)$$

Let $\xi = \Phi^{-1}(1 - \varepsilon/20)$, and $\delta(0 < \delta < \tfrac{1}{2}s)$ be such that

$$|\Phi[(1+c^2)^{1/2}ws\theta_2^{-1/2} - cv] - \Phi[(1+c^2)^{1/2}w - cv]| < \varepsilon/5$$

if $|s - \theta_2^{1/2}| < \delta$ and $|v| < \xi$. There must be such a δ because of the uniform continuity of continuous functions on compact intervals. Now choose M so that

$$\Pi_n(\{|V'| \geq \xi\} \cup \{|\theta_2 - s^2| \geq \delta\} \cup \{|F_{U'|\theta}(u) - \Phi(u)| \geq \varepsilon/5$$
$$\text{for some } |u| < 2(1+c^2)^{1/2}|w| + c\xi\}) < \varepsilon/5 \qquad (n > M). \qquad (6.41)$$

This is possible because of conditions (ii), (i) and (iii) in turn.

Next note that

$$F_{U|\theta}[(1+c^2)^{1/2}w - cn^{1/2}s^{-1}(\theta_1 - \bar{x}_n)] =$$
$$= F_{U'|\theta}[(1+c^2)^{1/2}ws\theta_2^{-1/2} - cn^{1/2}\theta_2^{-1/2}(\theta_1 - \bar{x}_n)].$$

It thus follows from (6.41) that the integrand in (6.38) may be replaced by $\Phi[(1+c^2)^{1/2}ws\theta_2^{-1/2} - cn^{1/2}\theta_2^{-1/2}(\theta_1 - \bar{x}_n)]$, and then by $\Phi[(1+c^2)^{1/2}w - cn^{1/2}\theta_2^{-1/2}(\theta_1 - \bar{x}_n)]$, with an error at each stage of no more than $\varepsilon/5$, except on a set of Π_n-measure at most $\varepsilon/5$, on which the total error of the two changes is at most one. Now writing $dF_{2|V'}(\theta_2)dF_{V'}(v)$ in place of $d\Pi_n(\theta)$, which we can do since the integrand now depends on θ only through θ_2 and V', and carrying out the integration with respect to θ_2, the error bound (6.40) follows.

The rest of the proof is along similar lines. The convolution $\int \Phi[(1+c^2)^{1/2}w - cv]\,dF_{V'}(v)$ may be rewritten as $\int F_{V'}[(1+c^2)^{1/2}w - c^{-1}y]\phi(y)\,dy$. Choose N so that $|F_{V'}(v) - \phi(v)| < \varepsilon/5$ for $|v| \leq (1 + c^{-2})^{1/2}|w| + c^{-1}\xi$ and $n > N$, which is possible by (ii). Arguing along the lines of the previous paragraph it follows that

$$\left| \int_{\infty}^{\infty} \Phi[(1+c^2)^{1/2}w - cv]\, dF_{V'}(v) \right.$$

$$\left. - \int_{\infty}^{\infty} \Phi[(1+c^{-2})^{1/2}w - c^{-1}y]\phi(y)\, dy \right| < 2\varepsilon/5 \qquad (n > N). \qquad (6.42)$$

Since $\int \Phi[(1+c^2)^{1/2}w - c^{-1}y]\phi(y)\, dy = \Phi(w)$, we have from (6.40) and (6.42) that

$$|F_W(w) - \Phi(w)| < \varepsilon \qquad (n > \max(M, N)),$$

and the theorem is proved. □

From the inequality (6.37) and the comments after it, it follows that, when for each Π_n the expected value of the next reward in the sequence is $\bar{x}_n$, and the conditions of the theorem hold,

$$[v_a(\Pi_n, n) - \bar{x}_n]s^{-1}n(1-a)^{1/2} \to 2^{-1/2} \qquad (6.43)$$

as $n(1-a) \to \infty$ on a fixed-s sequence and $a \to 1$. The fact that condition (iii) requires $mn^{-1} \geqslant c$ might at first sight be expected to invalidate this conclusion, since the calculation of $v_a(\Pi_n, n)$ involves values of $\bar{x}_{n+m}$ for all $m > 0$. The reason it does not do so is that we are considering values of a close to 1. This means that many members of the sequence $\bar{x}_{n+m}$ ($m = 0, 1, 2, \ldots$) have an appreciable effect in the calculation of $v_a(\Pi_n, n)$, and a may be chosen near enough to 1 for the higher values of m, for which the normal approximation holds, to swamp the effect of the lower values, for which it does not.

Applications of the theorem may be extended by considering a function $G(y, n)$ on $\mathbf{R}^+ \times \mathbf{R}$ satisfying appropriate regularity conditions. Suppose $G(y, n)$, $\partial G(y, n)/\partial y$, and $n^{1/2}[G(y, n+m) - G(y, n)]$, as n tends to infinity tend uniformly in bounded y intervals and $m \geqslant 0$ to $G(y)$, $dG(y)/dy$, and 0, respectively, where $G(y)$ is continuously differentiable. Let $\Delta G = G(\bar{x}_{n+m}, n+m) - G(\bar{x}_n, n)$, and a fixed-$\bar{x}$ sequence be a sequence of h_n's on which $\bar{x}_n = \bar{x}$ ($n = 1, 2, \ldots$) for some $\bar{x}$.

Corollary 6.33. Under conditions (i), (ii) and (iii) and the above regularity conditions,

$$P_n\{|\Delta G/\Delta \bar{x} - dG(\bar{x})/dy| > \varepsilon\} \to 0$$

on any fixed-$(\bar{x}, s)$ sequence, as m and $n \to \infty$ in such a fashion that $mn^{-1} \geqslant$ some fixed positive c.

Proof. For brevity write the conclusion of the corollary as $\Delta G/\Delta \bar{x} \xrightarrow{P} dG(\bar{x})/dy$. We have

$$\Delta G = G(\bar{x} + \Delta \bar{x}, n + m) - G(\bar{x} + \Delta \bar{x}, n) + \Delta \bar{x} \frac{\partial}{\partial y} G(\bar{x} + \alpha \Delta \bar{x}, n),$$

for some $\alpha \in (0, 1)$. To establish the corollary it is therefore sufficient to show that

$$\text{(a)} \quad \frac{\partial}{\partial y} G(\bar{x} + \alpha \Delta \bar{x}, n) \xrightarrow{P} \frac{dG(\bar{x})}{dy},$$

and

$$\text{(b)} \quad (\Delta \bar{x})^{-1} [G(\bar{x} + \Delta \bar{x}, n+m) - G(\bar{x} + \Delta \bar{x}, n)] \xrightarrow{P} 0.$$

The limit (a) follows from part 2 of the theorem, the uniform convergence of $\partial G(y, n)/\partial y$ to $dG(y)/dy$ on bounded y-intervals, and continuity of dG/dy at $y = \bar{x}$. From part 2 of the theorem it also follows that

$$\lim_{\delta \to 0} [\liminf_{n \to \infty} P_n(n^{1/2} |\Delta \bar{x}| > \delta)] = 1. \tag{6.44}$$

Now writing the expression in (b) as $(n^{1/2} \Delta \bar{x})^{-1} n^{1/2} [G(\bar{x} + \Delta \bar{x}, n+m) - G(\bar{x} + \Delta \bar{x}, n)]$, the limit (b) follows from (6.44), the uniform convergence to zero of $n^{1/2} [G(y, n+m) - G(y, n)]$, and a further application of part 2 of the theorem.

This corollary may be applied to bandit processes for which the expected value of the next reward in the sequence, when the current distribution of θ is Π_n, is $G(\bar{x}_n, n)$, where $G(y, n)$ satisfies the stated conditions. A difference ΔG between members of the sequence of expected rewards is now well approximated for large n by multiplying the difference $\Delta \bar{x}$, for the situation where this sequence coincides with the $\bar{x}_n$s, by $dG(\bar{x}_n)/dy$, which may be treated as independent of n provided we are considering only large values of n. Asymptotically, then, the relationship between the two expected reward sequences may be regarded as constant and linear. Under these circumstances it follows from the definition (2.5) that the same linear relationship must hold between the corresponding dynamic allocation indices. The asymptotic result corresponding to (6.43) is therefore

$$[v_a(\Pi_n, n) - G(\bar{x}_n, n)] [sdG(\bar{x})/dy]^{-1} n(1-a)^{1/2} \to 2^{-1/2} \tag{6.45}$$

as $n(1-a) \to \infty$ on a fixed-$(\bar{x}, s)$ sequence and $a \to 1$.

EXERCISES

6.1. Verify the forms given in §6.3 for the conjugate families of prior distributions for Bernoulli, normal (with σ^2 first known, then unknown) and exponential sampling processes.

6.2. Write out the proofs of one or two of Theorems 6.13, 6.15, 6.17, 6.19, 6.21 and 6.23. For example, Theorem 6.15 provides a suitably tricky test of understanding of the methods of these proofs. For this case you may find it helpful to note that for odd values of n the statistics $\bar{X}$ and S^2 take the values

$\bar{x}$ and s^2 if one of the X_is takes the value $\bar{x}$, half the remainder of the X_is take the value $\bar{x}+s$, and the other X_is take the value $\bar{x}-s$. For even values of n a slightly different set of X_i-values is needed.

6.3. Verify the inequality $v(t, n_1, \gamma) > v(t, n_2, \gamma)$ at the end of the proof of Theorem 6.27.

6.4. Verify equation (6.32).

CHAPTER 7

Multi-population Random Sampling (Calculations)

7.1 INTRODUCTION

This chapter sets out in some detail the methods by which the general principles described in Chapter 6 have been used to calculate dynamic allocation indices. This is done in turn for reward processes based on a normal distribution with known variance, a normal distribution with unknown variance, a $\{0, 1\}$ distribution and a negative exponential distribution, and for undiscounted target processes based on a negative exponential distribution, and on a negative exponential distribution together with an atom of probability at the origin. The results of these calculations are also described for each case except the Bernoulli/exponential target process, for which some results are given by Jones (1975) and more accurate calculations are planned. Two approximations which should prove useful are given in §7.7.

Tables are given at the back of the book. They seem likely to be accurate to within 1 or 2 in the last digit tabulated. Jones also gives results for the normal target process, both with known variance (1970) and with unknown variance (1975).

7.2 NORMAL REWARD PROCESS (KNOWN VARIANCE)

Because of Theorem 6.13 and Corollary 6.14 we may without loss of generality assume the known variance σ^2 to be equal to 1. The single unknown parameter μ is thus the mean of the distribution, and π_0 is taken to be the improper uniform density over the real line

$$f(x|\mu) = (2\pi)^{-1/2} \exp\{-\tfrac{1}{2}(x-\mu)^2\},$$

hence μ is a location parameter.

$$\prod_{i=1}^{n} f(x_i|\mu) = \exp\left\{-\frac{n}{2}(\bar{x}-\mu)^2\right\}(2\pi)^{-n/2}\exp\{-\tfrac{1}{2}(\Sigma x_i^2 - n\bar{x}^2)\},$$

hence $\bar{X}$ is a sufficient statistic by Theorem 6.1.

$$\pi_n(\mu|\bar{x}) \propto \pi_0(\mu)\prod_{i=1}^{n} f(x_i|\mu) \propto (2\pi n^{-1})^{-1/2}\exp\left\{-\frac{n}{2}(\bar{x}-\mu)^2\right\},$$

using the generalized Bayes' theorem (equation (6.1)). Thus $\pi_n(\mu|\bar{x})$ is normal with mean $\bar{x}$ and variance n^{-1}, and conjugate prior densities for μ are of this form, with n not restricted to integer values.

$$f(x|\pi_n) = f(x|\bar{x}, n) = \int_{-\infty}^{\infty} f(x|\mu)\pi_n(\mu|\bar{x})\,d\mu$$

$$= \int_{-\infty}^{\infty} (2\pi)^{-1/2}\exp\{-\tfrac{1}{2}(x-\mu)^2\}(2\pi n^{-1})^{-1/2}\exp\left\{-\frac{n}{2}(\bar{x}-\mu)^2\right\}d\mu$$

$$= \left(\frac{n}{2\pi(n+1)}\right)^{1/2}\exp\left\{-\frac{n}{2(n+1)}(x-\bar{x})^2\right\}.$$

$$r(\bar{x}, n) = \int_{-\infty}^{\infty} xf(x|\bar{x}, n)\,dx = \bar{x}.$$

Thus the functional equation (6.10) becomes

$$R(\lambda, \bar{x}, n) = \max\left[\frac{\lambda}{1-a}, \bar{x} + a\int_{-\infty}^{\infty} R\left(\lambda, \frac{n\bar{x}+x}{n+1}, n+1\right)f(x|\bar{x}, n)\,dx\right], \quad (7.1)$$

and $v(\bar{x}, n)$ is the value of λ for which the two expressions inside the square brackets are equal.

The conditions of Theorem 6.9 (and of the more general Theorem 6.13) hold, so $v(\bar{x}, n) = \bar{x} + v(0, n)$ and it is sufficient to calculate $v(\bar{x}, n)$ for $\bar{x} = 0$. From (7.1), and since

$$R(\lambda+c, \bar{x}+c, n) = c(1-a)^{-1} + R(\lambda, \bar{x}, n),$$

it follows that

$$R(\lambda, 0, n) = \max\left[\frac{\lambda}{1-a}, a\int_{-\infty}^{\infty} R\left(\lambda - \frac{x}{n+1}, 0, n+1\right)f(x|0, n)\,dx\right]. \quad (7.2)$$

Now defining

$$Q(\lambda, n) = R(\lambda, 0, n) - \lambda(1-a)^{-1},$$

this gives

$$Q(\lambda, n) = \max\left[0, -\lambda + a\int_{-\infty}^{\infty} Q\left(\lambda - \frac{x}{n+1}, n+1\right)f(x|0, n)\,dx\right]. \quad (7.3)$$

Thus, writing ξ_n for $v(0, n)$, ξ_n is the value of λ for which the two expressions inside the square brackets in equation (7.2) (or equivalently in equation (7.3)) are equal. $Q(\lambda, n)$ is the difference between the payoff from an optimal sequential policy using a reward process S starting in the state $\pi_n = (0, n)$, and a standard bandit process Λ with the parameter λ, and the payoff using only the standard bandit process. Thus $Q(\lambda, n) = 0$ if λ is large enough for the optimal policy to start (and therefore permanently continue) with Λ. On the other hand, if λ is sufficiently large and negative for the optimal policy to keep selecting S for a long time with high probability, then $Q(\lambda, n)$ is essentially the difference between always selecting S and always selecting Λ, namely

$$0(1-a)^{-1} - \lambda(1-a)^{-1}.$$

We should find, therefore, that $Q(\lambda, n) = 0$ for $\lambda > \xi_n$, and $Q(\lambda, n) + \lambda(1-a)^{-1} \to 0$ as $\lambda \to -\infty$. This means that (7.3) may be written as

$$Q(\lambda, n) = \max\left[0, \, -\lambda + a\left(\frac{n}{2\pi(n+1)}\right)^{1/2} \int_{(n+1)\xi_{n+1}}^{\infty} Q\left(\lambda - \frac{x}{n+1}, n+1\right)\right.$$

$$\left. \times \exp\left\{-\frac{nx^2}{2(n+1)}\right\} dx\right]$$

$$= \max\left[0, \, -\lambda + a\left(\frac{n(n+1)}{2\pi}\right)^{1/2} \int_{-\infty}^{\xi_{n+1}} Q(y, n+1)\right.$$

$$\left. \times \exp\left\{-\tfrac{1}{2}n(n+1)(y-\lambda)^2\right\} dy\right]. \tag{7.4}$$

The calculation of ξ_n was based on equation (7.4).

An approximation to $R(\lambda, 0, n)$, the payoff from S and Λ under an optimal sequential policy, may be obtained by allowing only policies which select the same bandit process at all times from the point when n reaches the value N onwards. Let $R(\lambda, 0, n, N)$ denote this approximation to $R(\lambda, 0, n)$, and $Q(\lambda, n, N)$ the corresponding approximation to $Q(\lambda, n)$. Thus $R(\lambda, 0, n, N)$ and $Q(\lambda, n, N)$ are increasing in N, since an increase in N means an enlargement of the set of policies over which optimization takes place, and tend to $R(\lambda, 0, n)$ and $Q(\lambda, n)$, respectively, as N tends to infinity. We have, for $n \geqslant N$,

$$R(\lambda, 0, n, N) = \max[\lambda(1-a)^{-1}, 0],$$

$$Q(\lambda, n, N) = R(\lambda, 0, n, N) - \lambda(1-a)^{-1} = \max[0, -\lambda(1-a)^{-1}], \tag{7.5}$$

and, for $1 < n < N+1$,

$$Q(\lambda, n-1, N) = \max\left[0, -\lambda + a\left(\frac{n-1}{2\pi n}\right)^{1/2} \int_{n(\lambda - \xi_{nN})}^{\infty} Q\left(\lambda - \frac{x}{n}, n, N\right)\right.$$

$$\times \exp\left\{-\frac{(n-1)x^2}{2n}\right\} dx\bigg]$$

$$= \max\left[0, -\lambda + a\left(\frac{n(n-1)}{2\pi}\right)^{1/2} \int_{-\infty}^{\xi_{nN}} Q(y, n, N)\right.$$

$$\times \exp\left\{-\tfrac{1}{2}n(n-1)(y-\lambda)^2\right\} dy\bigg], \qquad (7.6)$$

where ξ_{nN} is the corresponding approximation to ξ_n, and $\nearrow \xi_n$ as $N \to \infty$. Thus ξ_{n-1N} is the value of λ for which the expression after the comma in the square brackets in equation (7.6) (which will be written as $Q^*(\lambda, n-1, N)$) is equal to 0.

The method of calculation was a backwards induction using equation (7.6) to obtain in succession values of ξ_{nN} and of the function $Q(\cdot, n, N)$ for $n = N-1, N-2, \ldots, 1$. Equation (7.5) was used to provide values of $Q(\cdot, N, N)$, and clearly $\xi_{NN} = 0$, since ξ_{NN} is the value of λ for which the expected payoff from a permanent choice of S is the same as from a permanent choice of Λ, when S is in the state $(0, N)$. At each stage of the calculation values of $Q^*(\lambda, n, N)$ were obtained by numerical integration, the value ξ_{nN} of λ for which $Q^*(\lambda, n, N) = 0$ was found by the bisection method, and a grid of values of $Q(\lambda, n, N)$ was calculated for use in the next stage of the backwards induction.

The numerical integration for values of $Q^*(\lambda, n, N)$ was carried out using Romberg's method (e.g. see Henrici, 1964) to evaluate separately the contributions to the total integral from those parts of the five intervals with the end-points $\lambda \pm 6n^{-1}, \lambda \pm 4n^{-1}, \lambda \pm 2n^{-1}$, that are below the point ξ_{nN}, contributions from values of y outside the range $\lambda \pm 6n^{-1}$ being negligible. The grid of values of $Q(\cdot, n, N)$ was defined separately for the five intervals with the end-points

$$\xi_{nN} - 5n^{-1/2}, \xi_{nN} - 4n^{-1/2}, \xi_{nN} - 3n^{-1/2}, \xi_{nN} - 2n^{-1/2}, \xi_{nN} - 6n^{-1}, \xi_{nN} \quad (n > 40),$$

and with the end-points

$$\xi_{nN} - 5n^{-1/2}, \ \xi_{nN} - 4n^{-1/2}, \ \xi_{nN} - 3n^{-1/2}, \ \xi_{nN} - 2n^{-1/2}, \ \xi_{nN} - n^{-1/2}, \ m_{nN} \quad (n \leqslant 40).$$

For $n > 40$ the numbers of values calculated in the five successive intervals were 6, 10, 16, 40 and 50, and for $n \leqslant 40$ the numbers were 6, 10, 24, 30 and 50. Within each interval the grid was equally spaced. The values of $Q(\cdot, n, N)$ required for the next set of integrals were obtained by quadratic interpolation for $\xi_{nN} > \lambda > \xi_{nN} - 5n^{-1/2}$. For $\lambda < \xi_{nN} - 5n^{-1/2}$ the approximation $Q(\lambda, n, N) \simeq -\lambda(1-a)^{-1}$ was used. The grid values which most influence the calculation of the sequence of ξ_{nN}'s are those near to the corresponding value of ξ_{nN}. Consequently more grid values were calculated in the intervals close to ξ_{nN} than in those further away, so as to reduce interpolation errors, and the grid values close to ξ_{nN} were also calculated with higher precision by reducing the maximum error allowed in the Romberg procedure.

The actual numbers of grid points were determined by carrying out repeated calculations for different numbers of points. In this way the convergence of the calculated ξ_{nN} values as the number of grid points increased was directly observed. This meant that suitable numbers could be chosen, and also provided a means of estimating bounds for the error due to interpolation between grid points. Similar iterative procedures were used to determine appropriate maximum errors for numerical integration and root-finding, and corresponding bounds for the errors in calculating ξ_{nN}. The loss of accuracy from these two causes turns out to be negligible compared with that caused by the use of a discrete grid of points.

To obtain index values we need values of ξ_n rather than the lower bound for ξ_n given by ξ_{nN}. It is reasonable to conjecture that

$$(\xi_n - \xi_{nN})/\xi_n = O(a^{N-n}), \tag{7.7}$$

since this represents geometric discounting of the effect of the finite horizon using the discount factor a. Repeated calculations with different values of N show that (7.7) is approximately correct, thereby providing a means of bounding the error caused by estimating ξ_n by ξ_{nN}. This source of error can be made negligible by starting the calculations from a suitable value of N.

Reverting now to the fuller notation of §6.6 (apart from the subscript N referring to the fact that we are dealing with a normal process), it follows from Theorem 6.28 and Theorem 6.29(iii) that

$$n(1-a)^{1/2}v(0, n, 1, a) < n(1-a)u(n(1-a)) \leqslant 2^{-1/2}, \tag{7.8}$$

that the difference of the first two of these quantities is small when a is close to 1, and that the difference of the second two is small when $n(1-a)$ is large. Because of the inequalities (7.8), and since, by Theorem 6.13,

$$v(\bar{x}, n, \sigma, a) = \bar{x} + \sigma v(0, n, 1, a),$$

the index $v(\bar{x}, n, \sigma, a)$ is shown in Tables 1 and 2 in the form of values of $n(1-a)^{1/2}v(0, n, 1, a)$.

This normalized index shows the convergence to $2^{-1/2}$ ($=0.70711$) indicated by the inequalities (7.8) and subsequent comments as $a \to 1$ and $n(1-a) \to \infty$. There is also clear evidence of convergence as $n \to \infty$ and a remains fixed. The rate of convergence appears to be $O(n^{-1})$. The values shown in Table 1 for $n = \infty$ were obtained by further calculations for much larger values of n than those appearing elsewhere in the table.

The left-hand inequality in (7.8) means that a lower bound to $Tu(T)$ may be read off from Table 1. This lower bound is shown for a selection of values of T in Table 2. The function u is of some interest in its own right since the indices for any Brownian reward process may be obtained from it by scaling, as shown in §6.6. Values of the lower bound for small T were examined for evidence of the asymptotic behaviour given in Theorem 6.29(i). No very striking agreement was detected, presumably because no calculations were available for small enough value of T, nor perhaps for a sufficiently near 1.

7.3 NORMAL REWARD PROCESS (MEAN AND VARIANCE BOTH UNKNOWN)

The unknown parameters are the mean μ and standard deviation σ; $\pi_0 \propto \sigma^{-1}(\sigma > 0)$.

$$f(x|\mu,\sigma)=(2\pi\sigma^2)^{-1/2}\exp\left\{-\frac{1}{2\sigma^2}(x-\mu)^2\right\},$$

hence μ and σ are joint location and scale parameters.

$$\prod_{i=1}^{n} f(x_i|\mu,\sigma)=(2\pi\sigma^2)^{-n/2}\exp\left\{-\frac{1}{2\sigma^2}[\Sigma(x_i-\bar{x})^2+n(\bar{x}-\mu)^2]\right\},$$

hence $S^2 = (n-1)^{-1}\Sigma(X_i-\bar{X})^2$ and $\bar{X}$ are jointly sufficient statistics by Theorem 6.1. The generalized Bayes theorem gives

$$\pi_n(\mu,\sigma|\bar{x},s)\propto\pi_0(\mu,\sigma)\prod_{i=1}^{n} f(x_i|\mu,\sigma)$$

$$\propto\sigma^{-n-1}\exp\left\{-\frac{1}{2\sigma^2}[(n-1)s^2+n(\bar{x}-\mu)^2+(x-\mu)^2]\right\}.$$

$$f(x|\pi_n)=f(x|\bar{x},s,n)=\int_0^\infty\int_{-\infty}^\infty f(x|\mu,\sigma)\pi_n(\mu,\sigma|\bar{x},s,n)\,\mathrm{d}\mu\,\mathrm{d}\sigma$$

$$\propto\int_0^\infty\int_{-\infty}^\infty\sigma^{-n-2}\exp\left\{-\frac{1}{2\sigma^2}\left[(n-1)s^2+\frac{n}{n+1}(x-\bar{x})^2\right.\right.$$

$$\left.\left.+(n+1)\left(\mu-\frac{n\bar{x}+x}{n+1}\right)^2\right]\right\}\cdot\mathrm{d}\mu\,\mathrm{d}\sigma$$

$$\propto\int_0^\infty\sigma^{-n-1}\exp\left\{-\frac{1}{2\sigma^2}\left[(n-1)s^2+\frac{n}{n+1}(x-\bar{x})^2\right]\right\}\mathrm{d}\sigma$$

$$\propto\left[1+\frac{n}{n+1}\cdot\frac{(x-\bar{x})^2}{(n-1)s^2}\right]^{-n/2}.$$

Writing $t = (n/n+1)^{1/2}(x-\bar{x})s^{-1}$, it follows that t has Student's t-distribution with $n-1$ degrees of freedom, i.e. it has the density

$$g(t, n-1) = \frac{\Gamma\left(\dfrac{n}{2}\right)}{[(n-1)\pi]^{1/2}\Gamma\left(\dfrac{n-1}{2}\right)}\left(1+\frac{t^2}{n-1}\right)^{-\frac{n}{2}}.$$

$$r(\bar{x}, s, n) = \int_{-\infty}^{\infty} xf(x|\bar{x}, s, n) = \bar{x},$$

and the functional equation (6.10) becomes

$$R(\lambda, \bar{x}, s, n) = \max\left[\frac{\lambda}{1-a}, \bar{x}+a\int_{-\infty}^{\infty} R\left(\lambda, \frac{n\bar{x}+x}{n+1}\right),\right.$$

$$\left. s(s, \bar{x}, x, n+1), n+1)f(x|\bar{x}, s, n)\,\mathrm{d}x\right], \qquad (7.9)$$

where

$$s(s, \bar{x}, x, n+1) = (n^{-1}(n-1)s^2 + (n+1)^{-1}(x-\bar{x})^2)^{1/2}.$$

The two expressions inside the square brackets are equal when $\lambda = v(\bar{x}, s, n)$.
  The conditions of Theorem 6.15 hold, so

$$v(\bar{x}, s, n) = \bar{x} + sv(0, 1, n),$$

and it is sufficient to calculate $v(\bar{x}, s, n)$ for $\bar{x}=0$ and $s=1$. From (7.9), and since

$$R(b\lambda+c, b\bar{x}+c, b, s, n) = c(1-a)^{-1} + bR(\lambda, \bar{x}, s, n) \quad (n>1),$$

we have

$$R(\lambda, 0, 1, n) = \max\left[\frac{\lambda}{1-a}, a\int_{-\infty}^{\infty} s(1, 0, x, n+1)\right.$$

$$\left. \times R\left(\frac{\lambda - x(n+1)^{-1}}{s(1, 0, x, n+1)}, 0, 1, n+1\right)f(x|0, 1, n)\,\mathrm{d}x\right].$$

Writing $Q(\lambda, n) = R(\lambda, 0, 1, n) - \lambda(1-a)^{-1}$, this becomes, for $n>1$,

$$Q(\lambda, n) = \max\left[0, -\lambda + a\int_{-\infty}^{\infty} n^{-1/2}(n-1+t^2)^{1/2}\right.$$

$$\left. \times Q\left(\frac{n^{1/2}\lambda - (n+1)^{-1/2}t}{(n-1+t^2)^{1/2}}, n+1\right)g(t, n-1)\,\mathrm{d}t\right]. \qquad (7.10)$$

The second expression inside the square brackets in equation (7.10) will be denoted by $Q^*(\lambda, n)$. Writing $v(0, 1, n) = \zeta_n$, we have $Q^*(\zeta_n, n) = 0$. As for the normal reward process with known variance, calculations were based on the analogue of (7.10) when a switch from S to Λ is allowed only for $n \leq N$. Extending the notation in the obvious way we have

$$Q(\lambda, n-1, N) = \max\left[0, -\lambda + a \int_{-\infty}^{\infty} (n-1)^{-1/2}(n-2+t^2)^{1/2} \right.$$

$$\left. \times Q\left(\frac{(n-1)^{1/2}\lambda - n^{-1/2}t}{(n-2-t^2)^{1/2}}, n, N \right) g(t, n-2) \, dt \right]$$

$$= \max[0, Q^*(\lambda, n-1, N)] \qquad (2 < n < N+1). \qquad (7.11)$$

$$Q(\lambda, n, N) = \max[0, -\lambda(1-a)^{-1}] \qquad (n > N-1), \qquad (7.12)$$

$$Q^*(\zeta_{n-1, N}, n-1, N) = 0 \qquad (2 < n < N+1),$$

$$\zeta_{nN} = 0 \qquad (n \geq N).$$

Also $\zeta_{nN} \nearrow \zeta_n$ as $N \to \infty$.

Equations (7.11) and (7.12) are the counterparts of equations (7.6) and (7.5) for the normal reward process with known variance. In equation (7.6) the integration over x is reduced to the range $n(\lambda - \xi_{nN}) < x$, using the fact that $Q(y, n, N) = 0$ for $y \geq \xi_{nN}$. The integral in equation (7.11) cannot be reduced in the same simple way, as the function $[(n-1)^{1/2}\lambda - n^{-1/2}t](n-2-t^2)^{-1/2}$ is not monotone in t for all values of n and λ. There are in fact three possible cases, depending on whether the equation in t

$$\frac{(n-1)^{1/2}\lambda - n^{-1/2}t}{(n-2-t^2)^{1/2}} = \zeta_{nN} \qquad (7.13)$$

has 0, 1 or 2 real roots. These lead to integrals with 2, 1 or 0 infinite end-points, respectively.

The calculations of $\zeta_{NN}, \zeta_{N-1, N}, \zeta_{N-2, N}, \ldots, \zeta_{2, N}$ was carried out in sequential fashion along the lines of the calculation of $\xi_{NN}, \xi_{N-1, N}, \xi_{N-2, N}, \ldots, \xi_{1, N}$ for the known variance case. The numerical integration for values of $Q^*(\lambda, n-1, N)$ proceeded by first determining the roots of equation (7.13), and hence the set A of values of t for which the integrand is non-zero, and then applying Romberg's method to evaluate separately the contributions to the total integral which arise from the intersections of the set A with the five intervals with the end-points $\pm 6(1 + 200n^{-3})$, $\pm 4(1 + 60n^{-3})$, $\pm 2(1 + 20n^{-3})$. The grid of values of $Q(\cdot, n, N)$ was defined separately for the five intervals with the end-points

$$\zeta_{nN} - 5n^{-1/2}(1 + 100n^{-3}), \quad \zeta_{nN} - 4n^{-1/2}(1 + 60n^{-3}), \quad \zeta_{nN} - 3n^{-\frac{1}{2}}$$

$$\zeta_{nN} - 2n^{-1/2}(1 + 20n^{-3}), \quad \zeta_{nN} - 6n^{-1}(1 + 60n^{-3}), \quad \zeta_{nN}$$

for $n > 40$, and with end-points defined in the same way for $21 \leq n \leq 40$ except that

$\zeta_{nN} - n^{-1/2}(1 + 6n^{-3})$ replaces $\zeta_{nN} - 6n^{-1}(1 + 6n^{-3})$. For $6 \leqslant n \leqslant 20$ two more intervals were added, and a further two intervals for $3 \leqslant n \leqslant 5$. The wider intervals and the additional intervals, compared with the known variance case, were to allow for the greater dispersion of the t-distribution, compared with the standard normal distribution, for small numbers of degrees of freedom.

For a given value of the discount factor a the difference in the definitions of the quantities $\xi_n (= v(0, n, 1))$, defined in §7.2, and $\zeta_n (= v(0, 1, n))$, defined in this section, is that in the first case the variance of the normal reward process is known to be 1, and in the second case the sample variance is 1 and the prior distribution for the true variance expresses ignorance. In both cases the mean value of the first n rewards, and hence the expected value of subsequent rewards, is 0, so that ξ_n and ζ_n represent the extent to which the expected reward rate up to a stopping time may be increased by making the stopping time depend on the reward sequence. The scope for such an increase itself depends on the degree of uncertainty about the mean of the underlying normal distribution. This is greater if both the mean and the variance are unknown than if the variance is known, since the t-distribution, which expresses the uncertainty when the variance is unknown, has fatter tails than the normal distribution, which does so when the variance is known. Since the t-distribution also converges to a $N(0, 1)$ distribution as the number of degrees of freedom tends to infinity, it is not surprising that the calculations show that $\xi_n/\zeta_n > 1$ and $\searrow 1$ as $n \to \infty$. For this reason the calculated values of ξ_n are given in Table 3 in terms of the ratio ξ_n/ζ_n.

7.4 BERNOULLI REWARD PROCESS

A Bernoulli reward process is an arm of the classical multi-armed bandit problem, given as Problem 5 in Chapter 1. This problem was first considered by Bellman (1956). X_i takes values 0 or 1 with probabilities $(1 - \theta)$ and θ. The parameter θ has an improper prior density $\pi_0 \propto \theta^{-1}(1 - \theta)^{-1}$. After n observations, α of them 1's, and $\beta = n - \alpha$ of them 0's, the posterior density $\pi_n \propto \theta^{\alpha - 1}(1 - \theta)^{\beta - 1}$, which is a beta density with parameters α and β, mean αn^{-1}, and variance $\alpha\beta n^{-2}(n + 1)^{-1}$.

In the notation of §6.7, $\theta_1 = \theta$, $\theta_2 = \theta(1 - \theta)$ and $\bar{x}_n = \alpha n^{-1}$. If we now put $s_n^2 = \alpha\beta n^{-2}$, which means that a fixed-s sequence is a fixed-$\bar{x}$ sequence, the conditions of Theorem 6.32 may be checked. Condition (i) follows from Chebyshev's theorem (6.31). Note that here a fixed-s sequence defined for integer values of n involves non-integer values of α. Since the corresponding sequence of beta distributions Π_n is well-defined this causes no difficulty.

The distribution function for a standardized beta variate tends to Φ as the parameters tend to infinity in a fixed ratio (Cramer, 1946). It follows that

$$\Pi_n \left\{ n^{1/2} \frac{\theta_1 - \alpha n^{-1}}{[\alpha\beta n^{-1}(n + 1)^{-1}]^{1/2}} < v \right\} \to \Phi(v) \tag{7.14}$$

as $n \to \infty$ on a fixed-s sequence ($v \in \mathbf{R}$). This, together with condition (i), implies condition (ii). Condition (iii) follows, with a little care, by applying the Lindberg–Levy central limit theorem (6.30) to $F_{U'|\theta}$ as $m \to \infty$ with $\theta = \alpha n^{-1}$, plus the fact that $\Pi_n(|\theta_1 - \alpha n^{-1}| > \varepsilon) \to 0$ as $n \to \infty$ if αn^{-1} is fixed, as it is on a fixed-s sequence. Thus the limit (6.43) holds, and takes the form

$$[v(\alpha, \beta) - \alpha n^{-1}]n^2 \alpha^{-1/2} \beta^{-1/2}(1-a)^{1/2} \to 2^{-1/2} \qquad (7.15)$$

as $n(1-a) \to \infty$ on a fixed-αn^{-1} sequence and $a \to 1$.

Calculations for this case were carried out by the direct method described in §6.4. The limit (7.15) indicates that for large values of $n(1-a)$ and fixed a close to 1, and for values of α and β such that αn^{-1} is constant, $v(\alpha, \beta) - \alpha n^{-1} \propto n$. In fact it turns out that for any a and any constant value of αn^{-1}, $v(\alpha, \beta)$ is well approximated for large values of n by an equation of the form

$$[v(\alpha, \beta) - \alpha n^{-1}]^{-1} = n\alpha^{-1/2}\beta^{-1/2}(A + Bn + Cn^{-1})(1-a)^{1/2}, \qquad (7.16)$$

where A, B and C depend on a and on αn^{-1}.

In this case the calculation of index values involves no approximation of continuous functions by grids of values nor numerical integration, and the only important error arises from the finite value N of n from which the iterations start, and the consequent restriction on the set of allowed stopping times. As for the other calculations of index values for sampling processes, the importance of this source of error was estimated by comparing the index values calculated using different values of N. Such errors were found to be negligible in the first four significant figures of index values for

$$a = 0.99, N = 300, n \leqslant 150, \qquad (7.17)$$

and for

$$a \leqslant 0.95, N = 300, n \leqslant 200. \qquad (7.18)$$

Values of $v(\alpha, n - \alpha)$ for $n (\in \mathbf{Z}) \leqslant 300$ were calculated starting from $N = 300$ for $a = 0.5, 0.6, 0.7, 0.8, 0.9, 0.95$ and 0.99. For each value of a, and for values of $\lambda = \alpha n^{-1}$ ranging between 0.025 and 0.975, a weighted least squares fit of the form $A + Bn + Cn^{-1}$ to the function

$$y(n) = [v(n\lambda, n - n\lambda) - \lambda]^{-1}[\lambda(1 - \lambda)(1 - a)]^{-1/2}$$

was carried out for each λ at values of n such that $n\lambda$ is an integer, using the weight function $[y(n)]^{-2}$. The ranges of n-values used were chosen with a lower end-point high enough for a good fit to be achieved, and with an upper end-point low enough to satisfy (or almost satisfy) the appropriate constraint (7.17) or (7.18). The fitted values of A, B and C are shown in Tables 12, 13 and 14, and the corresponding ranges of n-values in Table 15.

For values of α and $\beta (\in \mathbf{Z})$ up to 20 for $a = 0.5, 0.6, 0.7$, and up to 40 for $a = 0.8$, $0.9, 0.95, 0.99$, the index function $v(\alpha, \beta, a)$ is given in Tables 5 to 11. For larger values of α and β (which need not be integer values) an approximation to $v(\alpha, \beta, a)$ may be obtained by assuming $v(n\lambda, n-n\lambda, a)-\lambda$ to be a linear function of $[\lambda(1-\lambda)]^{1/2}$ and interpolating (or if necessary extrapolating) the set of approximations to $v(n\lambda, n-n\lambda, a)-\lambda$ given by the least squares fits for the required value of n and for the thirteen fitted values of λ. For integer values of α and β and for values of n $(=\alpha+\beta)$ subject to the bounds (7.17) or (7.18) the index values obtained by this approximation procedure have been compared with the directly calculated values. For the most part the discrepancy is at most one in the fourth significant figure, the main exceptions being for values of αn^{-1} outside the range $(0.025, 0.975)$, and for values of n just above the upper bounds of the ranges covered by Tables 5 to 11. More details of the observed accuracy of the approximation are given in Table 4.

Because of the relationship between the indices for the Bernoulli and normal reward processes which Theorem 6.32 establishes for large n, and of the n-dependence of the normal reward process index, described in §7.2, and since for the Bernoulli process $v(\alpha, \beta)-\alpha n^{-1}$ is small for large n, it is fairly certain that the four-figure accuracy noted in Table 4 when n is large and $0.025 \leqslant \alpha n^{-1} \leqslant 0.975$ holds for $n > 200$. Clearly the main conclusions set out in Table 4 for lage n must also hold at non-integer values of α and β. What happens for large n and $|\alpha n^{-1}-0.5| > 0.475$ is much less clear, particularly when $\min(\alpha, \beta) < 1$.

For non-integer α and β, $\min(\alpha, \beta) \geqslant 1$, and

$$1 < (\alpha, \beta) \leqslant 20 \qquad (a = 0.5, 0.6, 0.7),$$
$$1 < (\alpha, \beta) \leqslant 40 \qquad (a = 0.8, 0.9, 0.95, 0.99),$$

it should be possible to interpolate fairly accurately from the values given in Tables 5 to 11, using one of the standard procedures for interpolating functions of more than one variable (see, for example, Fox and Parker, 1968).

Calculations of index values for the Bernoulli reward process have also been reported (briefly) by Gittins and Jones (1979), Robinson (1982), and Katehakis and Derman (1985).

7.5 EXPONENTIAL REWARD PROCESS

There is one unknown parameter θ.

$$f(x|\theta) = \theta e^{-\theta x},$$

so θ^{-1} is a scale parameter, call it σ. $\pi_0(\theta) \propto \theta^{-1}$, so the prior density for the scale parameter σ is proportional to σ^{-1}.

$$\prod_{i=1}^{n} f(x_i|\theta) = \theta^n e^{-\theta \Sigma},$$

where $\Sigma = \Sigma_{i=1}^{n} x_i$, and is a sufficient statistic by Theorem 6.1.

$$\pi_n(\theta|\Sigma) \propto \pi_0(\theta) \prod_{i=1}^{n} f(x_i|\theta) \propto \Sigma^n \theta^{n-1} e^{-\theta\Sigma}/\Gamma(n).$$

Thus $\pi_n(\theta|\Sigma)$ is a gamma density with the parameters n and Σ, and conjugate prior densities for θ are of this form.

$$f(x|\pi_n) = f(x|\Sigma, n) = \int_0^\infty f(x|\theta)\pi_n(\theta|\Sigma)\,d\theta$$

$$= \frac{\Sigma^n}{\Gamma(n)} \int_0^\infty \theta^n \exp\{-\theta(\Sigma + x)\}\,dx = \frac{n\Sigma^n}{(\Sigma + x)^{n+1}}.$$

$$r(\Sigma, n) = \int_0^\infty xf(x|\Sigma, n)\,dx = \frac{\Sigma}{n-1}.$$

The functional equation (6.10) becomes

$$R(\lambda, \Sigma, n) = \max\left[\frac{\lambda}{1-a}, \frac{\Sigma}{n-1} + an\Sigma^n\right.$$

$$\left. \times \int_0^\infty R\left(\lambda, \frac{\Sigma + x}{n+1}, n+1\right)(\Sigma + x)^{-n-1}\,dx\right], \qquad (7.19)$$

and $v(\Sigma, n)$ is the value of λ for which the two expressions inside the square brackets are equal.

The conditions of Theorem 6.11 hold, so $v(b\Sigma, n) = bv(\Sigma, n)$ and it is sufficient to calculate the sequence of values of Σ for which $v(\Sigma, n) = 1 (n = 1, 2, \ldots)$. These will be written as Σ_n. Defining $S(\Sigma, n) = R(1, \Sigma, n)$, and $S^*(\Sigma, n)$ as the optimal payoff from bandit processes S and Λ starting with S in the state (Σ, n) and subject to the constraint that at time zero S must be selected, equation (7.19) reduces to

$$S(\Sigma, n) = \max[(1-a)^{-1}, S^*(\Sigma, n)], \qquad (7.20)$$

where

$$S^*(\Sigma, n) = \Sigma(n-1)^{-1} + an\Sigma^n \int_\Sigma^\infty S(z, n+1)z^{-n-1}\,dz, \qquad (7.21)$$

and

$$S^*(\Sigma_n, n) = (1-a)^{-1}. \qquad (7.22)$$

$S^*(z, n+1)$ is strictly increasing in z, thus $S(z, n+1) = (1-a)^{-1}$ for $z \leqslant \Sigma_{n+1}$, and for $\Sigma < \Sigma_{n+1}$ the integral from Σ to Σ_{n+1} in the expression for $S^*(\Sigma, n)$ may be evaluated explicitly, giving

$$S^*(\Sigma, n) = \Sigma(n-1)^{-1} + \frac{a}{1-a}\left[1 - \left(\frac{\Sigma}{\Sigma_{n+1}}\right)^n\right]$$

$$+ an\Sigma^n \int_{\Sigma_{n+1}}^{\infty} S(z, n+1)z^{-n-1}\, dz. \tag{7.23}$$

As usual, the calculation of Σ_n was based on the analogues of equations (7.20) to (7.23) under the constraint that a switch from the reward process S to the standard bandit process Λ is not allowed for $n > N$. These lead to an approximation Σ_{nN} which decreases as N increases and tends to Σ_n as N tends to infinity. The equations in question, all for $n > 2$, are

$$S(\Sigma, n-1, N) = \max[(1-a)^{-1}, S^*(\Sigma, n-1, N)], \tag{7.24}$$

$$S^*(\Sigma, n-1, N) = \Sigma(n-2)^{-1} + a(n-1)\Sigma^{n-1}$$

$$\times \int_{\Sigma}^{\infty} S(z, n, N)z^{-n}\, dz \qquad (n \leqslant N), \tag{7.25}$$

$$S^*(\Sigma_{n-1,N}, n-1, N) = (1-a)^{-1} \tag{7.26}$$

$$S^*(\Sigma, n-1, N) = \Sigma(n-2)^{-1} + \frac{a}{1-a}\left[1 - \left(\frac{\Sigma}{\Sigma_{nN}}\right)^{n-1}\right]$$

$$+ a(n-1)\Sigma^{n-1}\int_{\Sigma_{nN}}^{\infty} S(z, n, N)z^{-n}\, dz$$

$$(\Sigma < \Sigma_{nN}, n \leqslant N), \tag{7.27}$$

$$S^*(\Sigma, n-1, N) = \Sigma(n-2)^{-1}(1-a)^{-1} \qquad (n > N). \tag{7.28}$$

From (7.26) and (7.28) it follows that $\Sigma_{NN} = N - 1$. $\Sigma_{N-1,N}, \Sigma_{N-2,N}, \ldots, \Sigma_{2,N}$ were calculated in turn using equations (7.24) to (7.27).

Since $\Sigma(n-1)^{-1}$ is the expected reward from the next observation from S when it is in the state (Σ, n), and since for a given value of Σn^{-1} the uncertainty associated with future rewards decreases as n increases, it follows that $\Sigma_{nN}(n-1)^{-1}$ increases with n for $n \leqslant N$. This is because the contribution to the index value arising from the possibility of learning the value of the parameter θ more precisely is less when there is not much more learning to be done, which is to say when n is large. It follows that Σ_{nN} increases with n, and hence that equation (7.27) is valid for values of Σ in the neighbourhood of $\Sigma_{n-1, N}$. A further consequence of the fact that Σ_{nN} increases with n is that if S is in a state (Σ, n) for which $\Sigma \geqslant \Sigma_{NN} = N - 1$ then $\Sigma \geqslant \Sigma_{Nm}$ $(n \leqslant m \leqslant N)$, and hence the optimal policy under the constraint of no switch from S to Λ after n reaches the value N is one which selects S at all times. This policy yields the expected payoff $\Sigma(n-1)^{-1}(1-a)^{-1}$, so that

$$S(\Sigma, n, N) = \Sigma(n-1)^{-1}(1-a)^{-1} \qquad (\Sigma \geqslant \Sigma_{NN}). \qquad (7.29)$$

Each successive value of Σ_{nN} was found by solving for Σ the equation $S^*(\Sigma, n, N) = (1-a)^{-1}$. Equation (7.27) was used to calculate $S^*(\Sigma, n, N)$, so that once the integral has been evaluated we are left with an nth order polynomial to solve for Σ, for which the Newton–Raphson method was used. Equations (7.25) and (7.27) were then used to calculate a grid of values of $S(\Sigma, n, N) = S^*(\Sigma, n, N)$ for $\Sigma > \Sigma_{nN}$, to use in evaluating the integrals required in the next stage of the calculation.

From (7.24) and (7.28) it follows that

$$S(\Sigma, N, N) = \max[1, \Sigma(N-1)^{-1}](1-a)^{-1}, \qquad (7.30)$$

so that $S(\cdot, N, N)$ has a discontinuous first derivative at the point $\Sigma = N - 1 = \Sigma_{NN}$. Now substituting (7.30) in the right-hand side of (7.27) for $n = N$ it follows that $S^*(\cdot, N-1, N)$ has a discontinuous second derivative at $\Sigma = \Sigma_{NN}$, and from (7.24) that $S(\cdot, N-1, N)$ has a discontinuous second derivative at $\Sigma = \Sigma_{NN}$ and a discontinuous first derivative at $\Sigma = \Sigma_{N-1, N}$. Equation (7.27) now shows that these discontinuities are inherited by $S^*(\cdot, N-2, N)$, though now transformed by the integral into discontinuities of the third and second derivatives respectively. Equation (7.24) transmits the properties of $S^*(\cdot, N-2, N)$ to $S(\cdot, N-2, N)$ for $\Sigma > \Sigma_{N-2, N}$, and adds a discontinuity of the first derivative at $\Sigma = \Sigma_{N-2, N}$. This process clearly may be formalized as an inductive argument showing that $S(\cdot, n, N)$ has a discontinuous rth order derivative at $\Sigma_{n+r-1, N}$ $(r \leqslant N-n+1)$.

These discontinuities in the derivatives of $S(\Sigma, n, N)$ mean that errors would arise from interpolating values of this function between a set of points which straddles discontinuity points. In order to minimize this source of error the grid of calculated values $S(\Sigma, n, N)$ was chosen so as to include considerable numbers of discontinuity points. The necessary integrals of the form $\int S(z, n, N) z^{-n} dz$ were evaluated numerically using Romberg's procedure separately over several different intervals and then summing. Those intervals close to the value $z = \Sigma_{nN}$ were chosen so as not to straddle discontinuity points. Quadratic interpolation, requiring three grid points for each interpolated value, was carried out to give the required values of $S(z, n, N)$ between grid points, and wherever possible the three grid points were chosen so as to avoid straddling discontinuity points. The grid of values of $S(\Sigma, n, N)$ was calculated for the following values of Σ, unless $\Sigma \geqslant \Sigma_{NN}$, in which case $S(\Sigma, n, N)$ is given by (7.29). For brevity the second suffix N has been omitted throughout.

$$\Sigma_n, 7, \Sigma_{n+1}, 7, \Sigma_{n+2}, 7, \Sigma_{n+3}, 3, \Sigma_{n+4}, 3, \Sigma_{n+5}, 1, \Sigma_{n+6}, 1,$$

$$\Sigma_{n+7}, 1, \Sigma_{n+8} \quad \text{(all } n < N\text{)}.$$

$$\Sigma_{n+8}, \Sigma_{n+9}, \ldots, \Sigma_{n+16}; 7, n+20 \quad (2 \leqslant n \leqslant 6).$$

$$\Sigma_{n+8}, \Sigma_{n+9}, \ldots, \Sigma_{n+12}; 7, n+20 \quad (7 \leqslant n \leqslant 10).$$

$$\Sigma_{n+8}; 7, n+4n^{1/2}, 3, n+6n^{1/2} \quad (11 \leqslant n \leqslant 16).$$

$$\Sigma_{n+8}, 3, n+2n^{1/2}; 7, n+4n^{1/2}, 3, n+6n^{1/2} \quad (17 \leqslant n \leqslant 36).$$

$$\Sigma_{n+8}, 7, n+2n^{1/2}; 7, n+4n^{1/2}, 3, n+6n^{1/2} \quad (n > 36).$$

The integers in the above specification indicate a corresponding number of equally spaced grid points between the two points separated by each integer; thus there were seven equally spaced grid points between Σ_n and Σ_{n+1} for all n, for example. The semicolons separate grid points at which $S(\Sigma, n, N)$ was calculated more (to the left) or less (to the right) accurately; the reason for this is that values of $S(\Sigma, n, N)$ for Σ close to Σ_{nN} have more influence on the calculation of Σ_{mN} for $m < n$, and it is therefore more important to calculate them accurately. For values of Σ greater than the largest grid point for a given n, $S(\Sigma, n, N)$ was approximated as $\Sigma(n-1)^{-1}(1-a)^{-1}$, corresponding to the assumption that the standard bandit process Λ is never selected after the reward process S reaches the state (Σ, n). This is a good approximation for large Σ, and means that the condition to $\int S(z, n, N)z^{-n} \, \mathrm{d}z$ from this part of the range may be evaluated explicitly.

In the notation of §6.7, $\theta_1 = \theta^{-1}$ and $\theta_2 = \theta^{-2}$. Thus the (improper) prior density $\pi_0(\theta) = $ a constant leads after n observations totalling Σ to a posterior gamma density with parameters $n+1$ and Σ, with respect to which θ_1 has mean $\bar{x}_n$ $(= \Sigma/n)$ and variance $(n-1)^{-1}\bar{x}_n^2$. With this modified prior and $s_n = \bar{x}_n$ the conditions of Theorem 6.32 may be checked.

Condition (i) follows from Chebyshev's theorem (6.31). Condition (ii) follows from the Lindberg–Levy central limit theorem (6.30) when we note that V' is asymptotically a standardized gamma variate whose distribution function may be expressed as the convolution of $n+1$ identical negative exponential distribution functions. U' and θ are independent with respect to P_n, and condition (iii) is thus an immediate consequence of the same theorem.

Thus Theorem 6.32 holds, as therefore does the limit (6.44).

The posterior distribution for θ starting from the prior density $\pi_0(\theta) \propto \theta^{-1}$, and after n observations totalling Σ, is the same as with $\pi_0(\theta) = $ a constant after $n-1$ observations totalling Σ. Thus, with $\pi_0(\theta) \propto \theta^{-1}$, the limit (6.44) holds if we replace $\bar{x}_n$ by $\Sigma/(n-1)$ and n by $n-1$. It then takes the form

$$\frac{v(\Sigma, n) - \Sigma/(n-1)}{\Sigma/(n-1)} (n-1)(1-a)^{1/2} \to 2^{-1/2} \tag{7.31}$$

as $n(1-a)\to\infty$ on a fixed-$(\Sigma/(n-1))$ sequence and $a\to 1$. Since $v(b\Sigma, n) = bv(\Sigma, n)$, and by definition $v(\Sigma_n, n) = 1$, it follows from (7.31) that $\Sigma_n/(n-1)\to 1$ as $n(1-a)\to\infty$ and $a\to 1$, and hence that

$$(n-1-\Sigma_n)(1-a)^{1/2}\to 2^{-1/2} \qquad (7.32)$$

as $n(1-a)\to\infty$ and $a\to 1$.

Because of the limit (7.32), values of $v(\Sigma, n)$ have been tabulated in the form $(n-1-\Sigma_n)(1-a)^{1/2}$, see Table 16. The tabulated values are accurate to the five significant figures shown. As in the normal case, the accuracy was checked by repeated calculations with different numbers of grid points, different error bounds for numerical integration and root-finding, and different starting points for dynamic programming iteration. A more detailed account of the calculations is given by Amaral (1985).

From these results it seems very likely that $(n-1-\Sigma_n)(1-a)^{1/2}$ tends to a limit as n tends to infinity for every fixed a. This is the analogue of the behaviour of $n(1-a)^{1/2}v(0, n, 1, a)$ for the normal reward process with known variance. The fact that the limits in the two cases appear to differ more widely for values of a which are not close to 1 reflects the failure of the asymptotic normality argument associated with the limit (6.43) when a is not near 1.

7.6 EXPONENTIAL TARGET PROCESS

Suppose without loss of generality that the target $T = 1$, as this is just a matter of appropriately scaling the unit of measurement (see Theorem 6.17). The distributional assumptions are the same as for the exponential reward process, and the functional equation (6.11) becomes

$$M(\phi, \Sigma, n, 1) = \min\left[\phi^{-1}, 1 + n\Sigma^n \right.$$

$$\left. \times \int_0^1 M\left(\phi, \frac{\Sigma+x}{n+1}, n+1, 1\right)(\Sigma+x)^{-n-1}\,dx\right]. \qquad (7.33)$$

Defining $N(\Sigma, n) = M(\phi, \Sigma, n, 1)$, and $N^*(\Sigma, n)$ as the minimum expected number of samples from S and the standard target process before reaching the target 1, starting with S in the state (Σ, n) and subject to the constraint that S must be selected at time zero, equation (7.33) reduces to

$$N(\Sigma, n) = \min[\phi^{-1}, N^*(\Sigma, n)], \qquad (7.34)$$

where

$$N^*(\Sigma, n) = 1 + n\Sigma^n \int_\Sigma^{\Sigma+1} N(z, n+1)z^{-n-1}\,dz. \qquad (7.35)$$

If $\Sigma(\phi, n)$ is the solution of the equation in Σ,

$$N^*(\Sigma, n) = \phi^{-1}, \qquad (7.36)$$

we have

$$v(\Sigma(\phi, n)n^{-1}, n) = \phi.$$

For $\Sigma < \Sigma(\phi, n+1) < \Sigma + 1$ part of the integral in (7.35) may be evaluated explicitly, giving

$$N^*(\Sigma, n) = 1 + \phi^{-1}\left(\left[\frac{\Sigma}{\Sigma(\phi, n+1)}\right]^n - 1\right) + n\Sigma^n \int_{\Sigma(\phi, n+1)}^{\Sigma+1} N(z, n+1)z^{-n-1}\,\mathrm{d}z. \quad (7.37)$$

For $\Sigma + 1 < \Sigma(\phi, n+1)$ we have

$$N^*(\Sigma, n) = 1 + \phi^{-1}\left(\left[\frac{\Sigma}{\Sigma+1}\right]^n - 1\right). \qquad (7.38)$$

Values of $\Sigma(\phi, n)$ for a given ϕ may be determined by a backwards induction calculation using equations (7.34), (7.37) and (7.38). Repeated calculations of this type for different values of ϕ establish a set of curves in (Σ, n)-space on which the index values are known. Index values at other points may then be obtained by interpolation. This procedure will now be described rather more fully. Note that the dependence of $N(\Sigma, n)$ and $N^*(\Sigma, n)$ of ϕ has been suppressed from the notation.

The calculation of the quantities $\Sigma(\phi, n)$ was based on the analogues of equations (7.24) to (7.28) under the constraint that a switch from the target process S to the standard bandit process Λ is not allowed for $n > N$. These lead to approximations $\Sigma(\phi, n, N)$ which decrease as N increases and tend to $\Sigma(\phi, n)$ as N tends to infinity. The equations in question, all for $n > 1$, are

$$N(\Sigma, n-1, N) = \min[\phi^{-1}, N^*(\Sigma, n, N)], \qquad (7.39)$$

$$N^*(\Sigma, n-1, N) = 1 + 1 + (n-1)\Sigma^{n-1} \int_{\Sigma}^{\Sigma+1} N(z, n, N)z^{-n}\,\mathrm{d}z \qquad (n \leqslant N), \qquad (7.40)$$

$$N^*(\Sigma(\phi, n-1, N), n-1, N) = \phi^{-1}, \qquad (7.41)$$

$$N^*(\Sigma, n-1, N) = 1 + \phi^{-1}\left(\left[\frac{\Sigma}{\Sigma(\phi, n, N)}\right]^{n-1} - 1\right)$$

$$+ (n-1)\Sigma^{n-1} \int_{\Sigma(\phi, n, N)}^{\Sigma+1} N(z, n, N)z^{-n}\,\mathrm{d}z$$

$$(\Sigma < \Sigma(\phi, n, N) < \Sigma + 1, n \leqslant N), \qquad (7.42)$$

$$N^*(\Sigma, n-1, N) = 1 + \phi^{-1}\left(\left[\frac{\Sigma}{\Sigma+1}\right]^{n-1} - 1\right) \quad (\Sigma + 1 < \Sigma(\phi, n, N), n \leqslant N), \qquad (7.43)$$

$$N^*(\Sigma, n-1, N) = \left(\frac{\Sigma}{\Sigma-1}\right)^{n-1} \qquad (n > N). \qquad (7.44)$$

Equation (7.44) is a consequence of the fact that, for $n > N$, $N^*(\Sigma, n-1, N)$ is the expected number of samples from S required to reach the target starting from the state $(\Sigma, n-1)$, and for a given θ the expected number required is e^θ. The right-hand side of (7.44) is the expectation of e^θ with respect to the gamma distribution with the parameters Σ and $n-1$, which is the current posterior distribution for θ in the given state.

When S is in the state (Σ, n) the probability that the next value sampled from S exceeds the target 1, or *current probability of success (CPS)*, is

$$\int_1^\infty f(x|\Sigma, n)\,dx = \int_1^\infty \frac{n\Sigma^n\,dx}{(\Sigma+x)^{n+1}} = \left(\frac{\Sigma}{\Sigma+1}\right)^n.$$

The index $\nu(\Sigma, n)$ of S in the state (Σ, n) exceeds the CPS by an amount corresponding to uncertainty about the parameter θ, and the consequent possibility of estimating its value more accurately after further sampling from S. Since this uncertainty decreases as n increases, the value of the CPS which leads to a given index-value ϕ increases with n, and tends to ϕ as n tends to infinity. For a given $\bar{x}$ it is easy to show that CPS is a decreasing function of n. Since CPS is an increasing function of $\bar{x}$ it follows that the value of $\bar{x}$ which leads to the index-value ϕ increases with n, which is to say that $\Sigma(\phi, n)n^{-1}$ increases with n. A simple extension of this informal argument shows that $\Sigma(\phi, n, N)n^{-1}$ increases with n for all ϕ and N, and therefore $\Sigma(\phi, n, N)$ certainly increases with n. For the next few paragraphs ϕ and N are fixed, and $\Sigma(\phi, n, N)$ is abbreviated as Σ_n.

From (7.41) and (7.44) it follows that

$$\Sigma_N = (1 - \phi^{1/N})^{-1}.$$

$\Sigma_{N-1}, \Sigma_{N-2}, \ldots, \Sigma_1$ were calculated in turn using equations (7.39) to (7.43). Each successive value of Σ_n was obtained by solving for Σ the equation $N^*(\Sigma, n, N) = \phi^{-1}$ using the bisection method. Equations (7.40), (7.42) and (7.43) were then used to calculate a grid of values of $N(\Sigma, n, N) = N^*(\Sigma, n, N)$ for $\Sigma > \Sigma_n$ to use in evaluating the integrals required in the next stage of the calculation.

The function $S(\cdot, n, N)$ which occurs in the calculation of index values for the exponential reward process has a discontinuous rth order derivative at $\Sigma_{n+r-1, N}(r \leqslant N-n+1)$. The argument used for that case shows, when applied to the exponential target process, that $N(\cdot, n, N)$ has a discontinuous rth derivative at $\Sigma_{n+r-1}(r \leqslant N-n+1)$. These discontinuities all originate as discontinuities of the first derivative of $N(\cdot, n, N)$ at Σ_n, defined by equations (7.39 and (7.41), for all the different values of n, and are then inherited as discontinuities of successively higher order derivatives via equation (7.40) as n decreases.

For $N(\cdot, n, N)$ there are further discontinuities of higher order derivatives generated by the discontinuity of the first order derivative at Σ_n. These are discontinuities transmitted on at least one occasion via the upper limit of the integral in equation (7.49), rather than exclusively via the lower limit. It is not difficult to show that $N(\cdot, n, N)$ has a discontinuous rth derivative at $\Sigma_{n+r-1} - k(r \leqslant N - n + 1)$ for any positive integer k such that $\Sigma_{n+r-1} - \Sigma_n > k$. The most important of these additional discontinuities are those for which $k = 1$, as these include discontinuities of lower order derivatives than those for $k > 1$. Accordingly a scheme of calculation was adopted which avoids interpolating values of $N(\cdot, n, N)$ over ranges which include those points Σ_{n+r-1} and $\Sigma_{n+r-1} - 1$ $(r \leqslant N - n + 1)$ which are close to Σ_n.

Integrals of the form $\int N(z, n, N) z^{-n} \, dz$ were evaluated by Romberg's procedure over sub-intervals chosen so that those close to the value $z = \Sigma_n$ do not straddle discontinuity points of the types just mentioned. Quadratic interpolation to obtain values of $N(z, n, N)$ between grid points was as far as possible carried out using sets of three grid points which do not straddle these discontinuity points. The grid of values of $N(\Sigma, n, N)$ was calculated for values of Σ based on the following pattern, unless $\Sigma \geqslant \Sigma_N$, when $N(\Sigma, n, N) = \Sigma^n(\Sigma - 1)^{-n}$.

$\Sigma_n, 7, \Sigma_{n+1}, 7, \Sigma_{n+2}, 7, \Sigma_{n+2}, 3, \Sigma_{n+4}, 3, \Sigma_{n+5},$

$1, \Sigma_{n+6}, 1, \Sigma_{n+7}, 1, \Sigma_{n+8}$ (all $n < N$).

$\Sigma_{n+8}, \Sigma_{n+9}, \ldots, \Sigma_{n+16}; 7, (n+20)(-\log_e \phi)^{-1}$ $(2 \leqslant n < 6)$.

$\Sigma_{n+8}, \Sigma_{n+9}, \ldots, \Sigma_{n+12}; 7, (n+20)(-\log_e \phi)^{-1}$ $(7 \leqslant n \leqslant 10)$.

$\Sigma_{n+8}; 7, (n+4n^{1/2})(-\log_e \phi)^{-1}, 3, (n+6n^{1/2})(-\log_e \phi)^{-1}$ $(11 \leqslant n \leqslant 16)$.

$\Sigma_{n+8}, 3, (n+2n^{1/2})(-\log_e \phi)^{-1};$

$(n+4n^{1/2})(-\log_e \phi)^{-1}, 3, (n+6n^{1/2})(-\log_e \phi)^{-1}$ $(17 \leqslant n \leqslant 36)$.

In addition those values of $\Sigma_m - 1$ $(m = 1, 2, \ldots)$ for which, for a given n, $\Sigma_n < \Sigma_m - 1 < \Sigma_{n+8}$ were taken as grid points. The integers in the above array indicate numbers of additional equally spaced grid points between the Σ_{n+r} values separated in the list by the integer in question. If the two Σ_{n+r} values are also separated by values of $\Sigma_m - 1$, the additional points were equally spaced in, for example, the interval $(\Sigma_{n+r}, \Sigma_m - 1)$, so that the number of additional points in any such sub-interval was $2^k - 1$ for some integer k, and so that the number of additional points separating two Σ_{n+r} values was always at least equal to the number shown in the array.

The expected reward from the next observation when the target has not yet been reached is $e^{-\theta}$ for a given value of θ. The expectation of $e^{-\theta}$ with respect to the current distribution Π_n with parameters n and $\bar{x} \, (= \Sigma n^{-1})$ is $(n\bar{x})^n (n\bar{x} + 1)^{-n}$, which $\rightarrow \exp(-\bar{x}^{-1})$ as $n \rightarrow \infty$. The functions $(ny)^n(ny+1)^{-1}$ and $\exp(-y^{-1})$ satisfy the conditions for $G(y, n)$ and $G(y)$ of Corollary 6.33, as a little manipulation shows, and conditions (i) to (iii) of Theorem 6.32 were checked in the previous section. However, the situation differs from that considered in the discussion leading to the limit (6.45) because the sequence of expected rewards $(n\bar{x})^n(n\bar{x}+1)^{-n}$ terminates as soon as the target is reached.

In fact (6.45) may be applied directly to a modified target process for which rewards occur for every observation X_i which exceeds the target, rather than just for the first such observation. Writing $s(t)(t=0,1,2,\dots)$ for the sequence of states, and p_t for the probability of exceeding the target with an observation taken when the state is $s(t)$, the index for a general target process with this modification, and with the discount factor a, may be written

$$v_a^M = \sup_{\tau > 0} \frac{E(p_0 + ap_2 + a^2 p_2 + \cdots + a^{\tau-1}p_{\tau-1})}{E(1 + a + a^2 + \cdots + a^{\tau-1})}. \tag{7.45}$$

Now note that the index of a target process without the modification and with no discounting may be written in the same form as (7.45), but with a^r replaced by $q_0 q_1 \cdots q_{r-1}$ $(r=1,2,\dots,\tau-1)$, where $q_r = 1 - p_r$. This follows by calibrating the target process against the class of standard target processes for which the probability of reaching the target is known and the same for each observation. The calibration procedure is similar to that used, for example, in the discussion of Problem 4 in Chapter 1.

This observation suggests a method of obtaining an approximation for $v(\Sigma, n)$ for the exponential target process. Provided q_r only varies slowly with r, which is true for large values of n, we should find that

$$v(\Sigma,\ n) = v_q^M(\Sigma,\ n),$$

where $q = q_0 = 1 - \Sigma^n(\Sigma+1)^{-n} = q(\Sigma, n)$. Plugging all this into (6.45), we may conjecture that

$$[v(\Sigma, n) - 1 + q(\Sigma, n)]\Sigma\exp(-n\Sigma)^{-1}[1 - q(\Sigma, n)]^{1/2} \to 2^{-1/2} \tag{7.46}$$

as $n(1-q) \to \infty$ and $q \to 1$.

Now note that it follows from Chebyshev's theorem (6.31) that

$$\Pi_n(|\bar{x}^{-1} - \theta| > \varepsilon) \to 0 \quad \text{as} \quad n \to \infty\ (\varepsilon > 0).$$

Thus for large n the modified target process behaves like a standard bandit process with the parameter

$$P_\theta(X_i \geqslant 1 | \theta = \bar{x}^{-1}) = \exp(-\bar{x}^{-1}),$$

so that

$$v_q^M(n\bar{x}, n) \to \exp(-\bar{x}^{-1}) \quad \text{as} \quad n \to \infty.$$

We also have

$$1 - q(n\bar{x}, n) = (n\bar{x})^n(n\bar{x}+1)^{-n} \to \exp(-\bar{x}^{-1}) \quad \text{as} \quad n \to \infty.$$

It follows that the limit (7.46) may be rewritten in the form

$$n\Delta(\Sigma, n) = -\frac{n[v(\Sigma, n) - \Sigma^n(\Sigma + 1)^{-n}]}{[v(\Sigma, n)]^{1/2}\log_e[v(\Sigma, n)]} \to 2^{-1/2} \tag{7.47}$$

as $n(1-q) \to \infty$ and $q \to 1$. The limit (7.47) remains a conjecture because we have not shown that it is legitimate to replace q_r by q when q is close to 1, even for large values of n.

The results of calculations of the quantities $\Sigma(v, n)$, where $v(\Sigma(v, n), n) = v$, are shown in Table 17. Because of the conjectured limit (7.47), values of $n\Delta(\Sigma, n)$ are also tabulated in Table 18. The accuracy of the calculations was checked along the lines described for other cases.

The tables are consistent with the conjectured limit and, more importantly, suggest that $n\Delta(\Sigma, n)$ tends to a limit as n tends to infinity for any fixed value v of $v(\Sigma, n)$, and that this limit does not change rapidly with v, thus offering a means of extrapolation and interpolation to find values of $v(\Sigma, n)$ which have not been directly calculated. In fact, for all $n \geqslant 10$, a function of the form

$$A - Bn^{(C + Dn + En^{-1})},$$

where the five constants depend on v, turns out to approximate $n\Delta(\bar{x}, n)$ well for any fixed value v of $v(\Sigma, n)$. Table 19 shows values of these constants for v between 0.01 and 0.3. For $10 \leqslant n \leqslant 850$ the relative error in $\Delta(\Sigma, n)$ is at most 0.6% for the tabulated values of v. Note that $\Delta(\Sigma, n) = 0 (n \in Z)$ for $v \geqslant 0.4$. This is a consequence of (6.8).

7.7 BERNOULLI/EXPONENTIAL TARGET PROCESS

As in the case of the exponential target process, we suppose the observations to be scaled so that the target $T = 1$. The sampled random variable X for this process has a density with respect to the measure μ on the non-negative real line for which $\mu(\{0\}) = 1$, and $\mu(I(a, b)) = b - a \ (b \geqslant a \geqslant 0)$, where $I(a, b)$ is the open interval (a, b). There are two unknown parameters $p \ (0 < p < 1)$ and $\theta \ (> 0)$.

$$f(0|p, \theta) = 1 - p, \qquad f(x|p, \theta) = p\theta e^{-\theta x},$$

so that the distribution of X is made up of an atom of probability $1 - p$ at the origin together with a negative exponential distribution over $\mathbf{R}^+$.

As mentioned in §6.1 the reason for considering this particular process is that a simple family of alternative Bernoulli/exponential target processes is appropriate as a model for the search for a coumpound which shows activity at some target level in a commercial pharmaceutical project. Different target processes correspond to different classes of compounds, any of which may include active compounds. The distribution of activity within a class of compounds typically tails off for higher levels of activity, and the activity scale may fairly easily be transformed so as to produce an approximately negative exponential distribution. The atom of probability at $x = 0$ corresponds to the fact that compounds frequently show no observable activity. It is also frequently desirable to ignore any activity below some threshold level for the purpose of estimating the parameter θ, as extrapolation errors caused by a distribution that is only approximately negative exponential tend to be less serious when the range over which a negative exponential distribution is assumed is reduced. This may be achieved by setting the origin of the activity scale at the threshold level.

Fairly extensive tables of index values for the Bernoulli/exponential target process have been calculated, and their use in new-product chemical research discussed, by Jones (1975). Here we shall simply establish the form of the functional equation and exhibit the asymptotic relationships between the index for this process, and those for the exponential target process, and for a Bernoulli reward process with discounting only when a successful trial occurs.

The prior density $\pi_0(p, \theta) \propto p^{-1}(1-p)^{-1}\theta^{-1}$.

$$\prod_{i=1}^{N} f(x_i|p, \theta) = \binom{N}{n} p^n (1-p)^m \theta^n e^{-\Sigma\theta},$$

where m is the number of zero x_i's, $m+n=N$, and $\Sigma = n\bar{x} = \Sigma_{x_i>0} x_i$. Hence n and Σ are jointly sufficient statistics and

$$\pi_N(p, \theta|\Sigma, n) \propto \pi_0(p, \theta) \prod_{i=1}^{N} f(x_i|p, \theta) \propto \frac{\Gamma(n+m)}{\Gamma(n)\Gamma(m)} p^{n-1}(1-p)^{m-1} \frac{\Sigma^n}{\Gamma(n)} \theta^{n-1} e^{-\Sigma\theta}.$$

Thus the posterior distributions for p and θ are independent and, respectively, beta with parameters m and n, and gamma with parameters n and Σ. Conjugate prior distributions for p and θ are therefore of this form, though we shall widen the class of distributions under consideration by allowing the second parameter n_1 of the beta distribution, and the first parameter n_2 of the gamma distribution, to take different values.

$$f(0|\pi_N) = f(0|m, n_1, \Sigma, n_2) = \int_0^\infty \int_0^1 f(0|p, \theta)\pi_N(p, \theta|m, n_1, \Sigma, n_2)\,dp\,d\theta$$

$$= \int_0^\infty \int_0^1 (1-p)\pi_N(p,\theta|m, n_1, \Sigma, n_2)\,dp\,d\theta$$

$$= \int_0^\infty \frac{\Gamma(n_1+m)}{\Gamma(n_1)\Gamma(m)} p^{n_1-1}(1-p)^m\,dp = \frac{m}{m+n_1}.$$

$$f(x|\pi_N) = f(x|m, n_1, \Sigma, n_2) = \int_0^\infty \int_0^1 f(x|p, \theta)\pi_N(p, \theta|m, n_1, \Sigma, n_2)\,dp\,d\theta$$

$$= \int_0^1 p \frac{\Gamma(n_1+m)}{\Gamma(n_1)\Gamma(m)} p^{n_1-1}(1-p)^m\,dp \int_0^\infty \theta e^{-\theta x} \frac{\Sigma^{n_2}}{\Gamma(n_2)} \theta^{n_2-1} e^{-\Sigma\theta}\,d\theta$$

$$= \frac{n_1}{m+n_1} \cdot \frac{n_2 \Sigma^{n_2}}{(\Sigma+x)^{n_2+1}}(x>0).$$

The functional equation (6.11) becomes

$$M(\phi, m, n_1, \Sigma, n_2, 1)$$

$$= \min\left[\phi^{-1}, 1 + \frac{m}{m+n_1} M(\phi, m+1, n_1, \Sigma, n_2, 1) \right.$$

$$\left. + \frac{n_1}{m+n_1} \int_0^1 M\left(\phi, m, n_1+1, \frac{\Sigma+x}{n_2+1}, n_2, 1 \right) \frac{n_2 \Sigma^{n_2}}{(n_2\bar{x}+x)^{n_2+1}} \, dx \right]. \quad (7.48)$$

When the two expressions inside the square brackets in (7.48) are equal $v(m, n_1, \Sigma, n_2) = \phi$.

Equation (7.48) may be solved by backwards induction along the lines followed for the exponential target process (§7.6). The starting point now is a grid of values of $M(\phi, m, n_1, \Sigma, n_2, 1)$ for a given ϕ, for m and n_1 such that $m+n_1 = N$, and for $n_2 = n_1 + c$, where c is fixed. Successive grids for $m+n_1 = N-1, N-2$ etc. may then be calculated by substituting in the right-hand side of (7.48). This procedure is described in more detail by Jones. The complexity of the calculations is now of order N^2 for a single value of ϕ, rather than of order N as for the exponential target process, and an approximation to $v(m, n_1, \Sigma, n_2)$ based on the index $v(\Sigma, n_2)$ for the exponential process is therefore of interest.

Proposition 7.1. $v(m, n_1, \Sigma, n_2) > \dfrac{n_1 - 1}{m + n_1 - 1} v(\Sigma, n_2).$

Proof. The sequence $X_1, X_2, \ldots$ of values sampled from an exponential process which starts from the state (Σ, n_2) has the same distribution as the sub-sequence of positive X_is sampled from a Bernoulli/exponential process which starts from the state (m, n_1, Σ, n_2)—the sub-sequence, that is to say, obtained by striking out any zero-valued X_is. If τ is a stopping time for the exponential process, let $\sigma(\tau)$ denote the stopping time for the Bernoulli/exponential process defined in terms of the sub-sequence of positive corresponding X_is. We have $R_{\sigma(\tau)}(m, n_1, \Sigma, n_2) = R_\tau(\sigma n_2)$, since zero-valued X_is produce zero-valued rewards, and $W_{\sigma(\tau)}(m, n_1, \Sigma, n_2) = E(T/m, n_1, \Sigma, n_2) W_\tau(\Sigma, n_2)$, where T is the interval between successive positive X_is for the Bernoulli/exponential process.

$$E(T|m, n_1, \Sigma, n_2) = E\{E(T|p)|m, n_1, \Sigma, n_2\} = \int_0^1 p^{-1} \frac{\Gamma(m+n_1)}{\Gamma(m)\Gamma(n_1)} p^{n_1-1}(1-p)^{m-1} \, dp$$

$$= \frac{m+n_1-1}{n_1-1}.$$

Thus

$$v(m, n_1, \Sigma, n_2) > \sup_{\tau > 0} \frac{R_{\sigma(\tau)}(m, n_1, \Sigma, n_2)}{W_{\sigma(\tau)}(m, n_1, \Sigma, n_2)} = \frac{n_1 - 1}{m + n_1 - 1} \cdot \sup_{\tau > 0} \frac{R_\tau(\Sigma, n_2)}{W_\tau(\Sigma, n_2)}$$

$$= \frac{n_1 - 1}{m + n_1 - 1} v(\Sigma, n_2). \qquad (7.49)$$

since the stopping times of the form $\sigma(\tau)$ are a subset of the stopping times for the Bernoulli/exponential process. In fact they are a subset which take no account of changes in the distribution of p as more information becomes available. Fairly obviously a higher rate of return may be achieved by using such information, hence the strict inequality in (7.49). □

The lower bound to $v(m, n_1, \Sigma, n_2)$ given by Proposition 7.1 gives a good approximation when the ignored information mentioned in the proof is relatively unimportant. This is so when p is already known fairly accurately, which is to say when m and n_1 are both large.

When n_2 is large a different approximation comes into play, based on the fact that a large n_2 means that not much remains to be learnt about θ, which means that $P\{X > 1 | X > 0, m, n_1, \Sigma, n_2\}$ does not change much as n_2 increases. In fact we have

$$P\{X > 1 | X > 0, m, n_1, \Sigma, n_2\} = \left(\frac{\Sigma}{\Sigma + 1}\right)^{n_2} \simeq \exp(n\Sigma^{-1}) \simeq e^{-\theta}$$

for large values of n_2.

For a given initial state $\pi(0)$ defined by (m, n_1, Σ, n_2) let $\pi(t)$ denote the state at process time t, $\tau = \min\{t : \pi(t) \in \Theta_0\}$ for some subset Θ_0 of the state-space which does not include the completion state C, and $T = \min\{t : \pi(t) = C\}$. For $t < \min(\tau, T)$ let

$$p(\pi(t)) = P\{X > 1 | \pi(t)\},$$

$$q(\pi(t)) = P\{\pi(t + 1) \in \Theta_0 | \pi(t)\},$$

$$r(\pi(t)) = 1 - p(\pi(t)) - q(\pi(t)).$$

As mentioned in §6.2, in determining the index-value attention may be restricted to stopping times of the form $\tau_A = \min(\tau, T)$ for some Θ_0. Thus

$$v(m, n_1, \Sigma, n_2)$$

$$= \sup_{\tau > 0} \frac{p(\pi(0)) + r(\pi(0))p(\pi(1)) + r(\pi(0))r(\pi(1))p(\pi(2)) + r(\pi(0))r(\pi(1))r(\pi(2))p(\pi(3)) + \cdots}{1 + r(\pi(0)) + r(\pi(0))r(\pi(1)) + r(\pi(0))r(\pi(1))r(\pi(2)) + \cdots}$$

$$(7.50)$$

If $n_2 \gg \max(m, n_1)$ a good approximation to $v(m, n_1, \Sigma, n_2)$ may be obtained by restricting Θ_0 to be of the form

$$\{(m', n'_1, \Sigma', n'_2): (m', n'_1) \in \Theta_0^* \subset \mathbf{R}^+ \times \mathbf{R}^+\}.$$

This restriction amounts to ignoring any further information about θ and making τ depend only on the information about p obtained from the sampling process; this is reasonable if θ is known much more accurately than p, as the assumption $n_2 \gg \max(m, n_1)$ implies. We may also assume that if $(m', n'_1) \in \Theta_0^*$ then $(m' + h, n'_1) \in \Theta_0^* (h > 0)$ and $(m', n'_1 - h) \in \Theta_0^* (0 < h < n'_1)$, as the beta distributions for p with the parameters $(m' + h, n'_1)$ and $(m', n'_1 - h)$ are stochastically less than the distribution with parameters (m', n'_1), and the option of stopping is therefore relatively more attractive. With Θ_0 of this form, and writing $\pi = (m', n'_1, \Sigma', n'_1)$ $(\pi \in \Theta_0 \cup \{C\})$,

$$p(\pi) = \frac{n'_1}{m' + n'_1} \left(\frac{\Sigma'}{\Sigma' + 1}\right)^{n'_2},$$

$$q(\pi) = \begin{cases} \dfrac{m'}{m' + n'_1} & \text{if } (m' + 1, n'_1) \in \Theta_0^* \\ 0 & \text{otherwise} \end{cases},$$

$$r(\pi) = \frac{n'_1}{m' + n'_1}\left[1 - \left(\frac{\Sigma'}{\Sigma' + 1}\right)^{n'_2}\right] + \begin{cases} 0 & \text{if } (m'_1 + 1, n'_1) \in \Theta_0^* \\ \dfrac{m'}{m' + n'_1} & \text{otherwise} \end{cases}.$$

Writing $Q = \Sigma^{n_2}(\Sigma + 1)^{-n_2}$, and ignoring changes in this quantity as n_2 increases, it follows on substituting expressions of the above form for $p(\pi(t))$, $q(\pi(t))$ and $r(\pi(t))$ in (7.50) that, to the stated approximations,

$$v(m, n_1, \Sigma, n_2) = Qv(m, n_1),$$

where $v(m, n_1)$ is the index for a Bernoulli reward process in the state (m, n_1), for which the discount factor between successive trials is 1 if the second trial of the pair is unsuccessful and $1 - Q$ if it is successful. Index values for this modified Bernoulli reward process may be calculated by making the obvious changes to the procedure described in §6.4, again with a big improvement in the complexity of the calculation compared with backwards induction based on equation (7.48).

EXERCISES

7.1. Work through the derivations of some of the equations (7.3), (7.10), (7.19), (7.23), (7.34), (7.37) and (7.48).

7.2. The quantities $Q(\lambda, n, N)$ appearing in equations (7.5) and (7.6), and in equations (7.11) and (7,12), are defined as maxima over sets of policies which increase with N. Explain why it follows from this that the associated quantities ξ_{nN} and ζ_{nN} are increasing with N, and why they tend to the limits ξ_n and ζ_n as $N \to \infty$.

7.3. Chebyshev's Theorem (6.31) was cited as establishing Condition (i) of Theorem 6.32 in the two applications discussed in §§7.4 and 7.5. Show that this claim is justified.

7.4. Derive the expression for the allocation index of a target process described in the paragraph following equation (7.45).

CHAPTER 8

Search Theory

8.1 INTRODUCTION

Iterative procedures for finding the maximum of a function are search processes, as are the processes modelled by families of alternative target processes. Our concern in this chapter is with search in the concrete sense of looking for a hidden object. In a sense this amounts to looking for the maximum of a function which is known to take the value zero everywhere except at the point where the object is hidden, where it takes the value one. The specification of the problem is completed by a prior probability distribution for the location of the object, a function giving the costs of searching different points or regions, and a second function giving the probability of finding the object if a paticular point is searched and the object is at that point.

The theory of optimal search for a physical object was initiated by Koopman, working for the United States Navy during the Second World War. The most complete theory is for a unique stationary object, though there is an obvious need for a theory which extends to multiple objects, including false targets, and to mobile objects, which may or may not wish to be found, depending on the friendliness or otherwise of the searcher's intentions. Developments of the theory have taken place along each of these directions and are still continuing. The key text, though it does not deal with objects which move so as to help or hinder the search process, is the book by Stone (1975). This is complemented by Gal's (1980) book on search games. Koopman (1979) sets the subject in an interesting historical perspective, and Strümpfer (1980) has compiled a useful index of about 400 books and papers. The final section of the book by Ahlswede and Wegener (1987) is also on this subject.

The starting point for this chapter is the discrete search problem given as Problem 3 in Chapter 1. A solution to this problem based on dynamic programming was given by Blackwell (reported by Matula, 1964). As we shall see it also provides a neat example of a simple family of alternative bandit processes, and this formulation leads quickly to the solution of a more general version of the problem. Lehnerdt (1982) gives a useful account of work in this area.

The two-person zero-sum game, in which one of the players hides the object, also has some interesting features. This is discussed in §8.4, after introducing the classical minimax theorem of game theory in §8.3. This version of hide-and-seek is one of those problems for which the solution in continuous time is much neater than in discrete time; §8.5 shows what happens. §§8.4 and 8.5 are based on papers by Roberts and Gittins (1978), Gittins and Roberts (1979), and Gittins (1979).

8.2 A DISCRETE SEARCH PROBLEM

Problem 3 (see also Chapter 1). A stationary object is hidden in one of n boxes. The probability that a search of box i finds the object if it is in box i is q_i. The probability that the object is in box i is p_i, and changes by Bayes' theorem as successive boxes are searched. The cost of a single search of box i is c_i. How should the boxes be sequenced for search so as to minimize the expected cost of finding the object?

This problem is an obvious candidate for modelling as a simple family of alternative bandit processes, the boxes corresponding to bandit processes, box i being in state p_i, and searching a box corresponding to continuation of the bandit process. The difficulty is that each time a box is searched without finding the object the effect of Bayes' theorem is not only to decrease the probability that the object is in that box, but also to increase the probability that it is in any of the other boxes. The state of a bandit process only changes when it is continued, so we must find some other way to proceed. A related SFABP in fact turns out to have an identical cost structure, thereby providing a solution in terms of an index.

Problem 3A. The ith bandit process in a simple family $\mathscr{F}$ of n alternative bandit processes passes in deterministic sequence through states $j = 1, 2, \ldots$ when it is continued ($i = 1, 2, \ldots, n$). It takes a time c_i to pass from one state to the next, and a cost $p_i(1 - q_i)^{j-1} q_i$ accrues per unit time until the jth of these transitions.

The SFABP $\mathscr{F}$ has, then, a cost structure of the type covered by Corollary 3.10. Under a given policy PA let $c(i, j)$ be the time at which bandit process i reaches state j. Thus the total cost under policy PA is

$$\sum_{i=1}^{n} p_i \sum_{j=1}^{\infty} (1 - q_i)^{j-1} q_i c(i, j). \tag{8.1}$$

For any policy PA for Problem 3A there is a corresponding policy P for Problem 3 which searches box i whenever policy PA continues bandit process i, until the point when the object is found. Under policy P the expected cost of finding the object is also given by the expression (8.1), so that corresponding sets of policies are optimal for the two problems.

Optimal policies for the search problem are therefore defined by the index policies for $\mathscr{F}$ given by letting $\beta \to 0$ in the definition (2.5). For box i after j searches in the box i and the object not yet found, the appropriate index is that for bandit process i in state j. This is

$$v_i(j) = \sup_{\{N > j\}} \frac{\sum\limits_{r=j+1}^{N} p_i(1 - q_i)^{r-1} q_i}{(N - j) c_i}. \tag{8.2}$$

Since $(1 - q_i)^{r-1}$ is a decreasing function of r, the supremum in (8.2) occurs at

$N = j+1$, so that

$$v_i(j) = \frac{p_i(1-q_i)^j q_i}{c_i}. \tag{8.3}$$

The expression (8.1) and subsequent calculations are expressed in terms of the initial probability p_i that the object is in box i ($i = 1, 2, \ldots, n$). After $j(i)$ unsuccessful searches of box i, for each i, the effect of Bayes' theorem is to transform these prior probabilities into posterior probabilities p'_i proportional to $p_i(1-q_i)^{j(i)}$. It thus follows from (8.3) that optimal policies are those conforming to the index $p'_i q_i / c_i$. If we adopt the convention that the probabilities p_i change according to Bayes' theorem the following result has therefore been established.

Theorem 8.1. Optimal policies for Problem 3 are those conforming to the index $p_i q_i / c_i$.

This is the result obtained by Blackwell.

The present derivation via the SFABP $\mathscr{F}$, given by Kelly (1979), has the advantage of leading immediately to considerable generalization. A simpler, and more direct, proof is given as an exercise at the end of the chapter.

Suppose now that the probability of finding the object on the jth search of box i, conditional on not finding it before then, and provided it is in box i, is q_{ij}, and that the cost of the jth search of box i is c_{ij}. For the associated SFABP $\mathscr{F}$ now let bandit process i pass through the sequence of states $x_i(0)(=0), x_i(1), x_i(2), \ldots$. Let c_{ij} be the time taken to pass from state $x_i(j-1)$ to $x_i(j)$, and let a cost per unit time of

$$p_i \prod_{r=1}^{j-1} (1-q_{ir}) q_{ij}$$

accrue until the state $x_i(j)$ has been reached. This formulation allows the sequences $(c_{i1}, q_{i1}), (c_{i2}, q_{i2}), \ldots$ to be independent random processes ($i = 1, 2, \ldots, n$). The expected total cost under a given policy PA for $\mathscr{F}$ is now

$$\mathbf{E}\left\{ \sum_{i=1}^{n} p_i \sum_{j=1}^{\infty} \prod_{r=1}^{j-1} (1-q_{ir}) q_{ij} c(i,j) \right\},$$

where $c(i,j)$ is the time at which bandit process i reaches state $x_i(j)$. This is also the expected total time required to find the object under the corresponding policy P for the search problem. Again it follows from Corollary 3.10 that optimal policies for the search problem are those based on an index, which now takes the form

$$v_i(x_i(j)) = \sup_{\{N>j\}} \frac{p_i \mathbf{E}\left\{ \sum_{r=j+1}^{N} \prod_{s=j+1}^{r-1} (1-q_{is}) q_{ir} \mid x_i(j) \right\}}{\mathbf{E}\left\{ \sum_{i=j+1}^{N} c_{ir} \mid x_i(j) \right\}}, \tag{8.4}$$

where N is now an integer-valued stopping variable for bandit process i, and p_i is the current posterior probability that the object is in box i.

8.3 TWO-PERSON ZERO-SUM GAMES

Two-person games have two players, A and B. Each player has an action space, $\mathscr{X}$ for player A and $\mathscr{Y}$ for player B, from which he randomly selects an action, x, for A and y for B, using an arbitrary probability distribution, F for A and G for B, which he chooses without the knowledge of the other player. As a result each player receives a payment which depends on x and y. In the case of a zero-sum game the sum of these payments is zero.

Let $a(x, y)$ be the payment received by A, in effect a payment from B to A, if the actions selected are x and y. The aim of player A is to make the expected payment.

$$a(F, G) = \int_{\mathscr{X}} \int_{\mathscr{Y}} a(x, y) \, dF(x) \, dG(y)$$

large, irrespective of B's choice of G. Conversely, B aims to make this quantity small, irrespective of F.

Under fairly general conditions the following theorem holds.

Theorem 8.2 (The minimax theorem).

$$\inf_{G} \sup_{F} a(F, G) = \sup_{F} \inf_{G} a(F, G).$$

The first proof of this famous result was given by von Neumann and Morgenstern (1928). To establish $\geqslant$ is in fact trivial, since for any F^* and G^* we have

$$\sup_{F} a(F, G^*) \geqslant a(F^*, G^*) \geqslant \inf_{G} a(F^*, G),$$

and this inequality is preserved on taking the infimum of the left-hand side over G^* and the supremum of the right-hand side over F^*. The reverse inequality is more difficult to prove.

The common value of the inf sup and the sup inf is termed the *value* of the game. If V is the value of the game and F^* is such that

$$\inf_{G} a(F^*, G) = V,$$

then F^* is termed a *maximum* strategy for A. If

$$\sup_{F} a(F, G^*) = V,$$

then G^* is *minimax* for B. The importance of maximin and minimax strategies is twofold. In the first place they are conservative or safety-first, since the most effective counter-strategy by the opponent is made as ineffective as possible. Secondly, if either player uses one of these conservative strategies his opponent can do no better than to do the same, since this ensures an expected payment equal to V. It is thus rather difficult to argue in favour of any other strategy, particularly if the other player knows the minimax theorem!

When it can be applied the following theorem tells us how to play a particular game.

Theorem 8.3. If G^* is an optimal counter to a given strategy F^* for A, and

$$a(x, G^*) = \int_Y a(x, y)\, dG^*(y) \leqslant a(F^*, G^*) \quad (x \in \mathscr{X}),$$

then the minimax theorem holds, F^* is maximin and G^* is minimax.

Proof. We have

$$\inf_G \sup_F a(F, G) = \inf_G \sup_x a(x, G) \leqslant \sup_x a(x, G^*) \leqslant a(F^*, G^*)$$

$$\leqslant \inf_y a(F^*, y) \leqslant \sup_F \inf_y a(F, G) = \sup_F \inf_G a(F, G).$$

Since we already know that

$$\inf_G \sup_F a(F, G) \geqslant \sup_F \inf_G a(F, G)$$

all these inequalities must in fact be equalities, and the theorem follows. ☐

8.4 A GAME OF HIDE-AND-SEEK

The seeking, or searching, part of Version 1 of the game is as in Problem 3, except that for simplicity we now suppose that $c_i = 1 (i = 1, 2, \ldots, n)$. The difference is that the initial location probability vector $(p_1, p_2, \ldots, p_n)$ is now chosen by a second player, the hider, with the aim of leaving the object undetected for as long as possible. The hider and the searcher are players A and B in a two-person zero-sum game. The payment received by the hider is the time, or total number of searches of any of the boxes, required for the searcher to find the object. Version 2 of the game is similar, except that now the hider is player B and the searcher player A, the searcher is allowed just one search of one box, and the payment is 1 if he finds the object and 0 otherwise.

For Version 2 the action-spaces $\mathscr{X}$ and $\mathscr{Y}$ both consist of n elements corresponding to the n boxes. Let p_i be the probability that the hider puts the object in box i, and r_i the probability that the searcher looks in box i $(i = 1, 2, \ldots, n)$. Consider strategies P^* and R^* for hider and searcher defined as follows.

$$P^*: p_i = q_i^{-1} \bigg/ \left(\sum_{j=1}^{n} q_j^{-1} \right),$$

$$R^*: r_i = q_i^{-1} \bigg/ \left(\sum_{j=1}^{n} q_j^{-1} \right).$$

P and R are both what are known as *equalizer* strategies, which means that if either is played the expected payment is the same for any opposing strategy. Thus P^* is an optimal counter to R^* and, putting $G^* = P^*$, $F^* = R^*$, the conditions of Theorem 8.3 are satisfied, so that P^* is minimax, R^* is maximin and the value of the game is $(\Sigma q_j^{-1})^{-1}$.

For Version 1, and a given initial location probability vector, Theorem 8.1 tells us that the searcher's optimal policy is always to search next the box for which $p_i q_i$, which is the probability of finding the object if it is in box i, has the largest current value. Since the strategy P^* is minimax for Version 2 of the game we have

$$\min_{\{\Sigma p_j = 1\}} \max_i p_i q_i = \left(\sum_{j=1}^{n} q_j^{-1} \right)^{-1},$$

and the minimax occurs when the current location probability vector is as given by strategy P^*. It follows that to achieve a maximin strategy for Version 1 the hider must choose a strategy P_0, which we can think of as a point in R^n on the hyperplane $\Sigma p_i = 1$, so that the entire sequence of points $P_0, P_1, P_2, \ldots$ into which P_0 is transformed as the searcher looks in successive boxes are all in some average sense as close to P^* as possible.

If the hider could move the object after each successive search he would do so so that the searcher always faced a vector of current location probabilities given by P^*. This would effectively reduce the game to a succession of Version 2 games. However, in Version 1 the object is hidden just once, so the hider's strategy consists just of the initial point P_0, leaving P_1, P_2 and so on to be determined by Bayes' theorem. If P_0 is set equal to P^*, P_1 will differ from P^*, whereas if P_0 is chosen so that $P_1 = P^*$, P_0 will differ from P^*; if P_0 and P_1 are both to be close to P^* some compromise between these conflicting reqirements may be necessary.

In fact the extent of the conflict is often not very great since, if P_0 is close to P^*, P_1 and its successors also tend to be close to P^*. After an unsuccessful search of box i the effect of Bayes' theorem is to transform the location probability vector $(p_1, p_2, \ldots, p_n)$ as follows.

$$p_i \rightarrow p_i(1 - q_i)(1 - p_i q_i)^{-1}, \tag{8.5}$$

$$p_j \rightarrow p_j(1 - p_i q_i)^{-1} \quad (j \neq i).$$

Thus p_i is reduced, and $p_j(j \neq i)$ is increased. If the searcher uses the index policy given by Theorem 8.1 this means that the largest $p_j q_j$ is reduced, and all the other

$p_j q_j$s are increased in the same ratio. Apart from the overshoot which may result from the discrete nature of these changes, this is a procedure which leads to the situation where

$$p_i q_i = \left(\sum_{j=1}^{n} q_j^{-1} \right)^{-1} \qquad (i = 1, 2, \ldots, n),$$

or in other words to the point P^*. When combined with overshoot the effect is that if P_0 is close to P^*, then $P_1, P_2, \ldots$ all remain within some fixed neighbourhood $N(P^*)$ of P^*. This is most easily seen when $n = 2$. We have

$$P^* = (q_2, q_1)(q_1 + q_2)^{-1}.$$

Provided p_1 is initially in the interval

$$\left(\frac{q_2(1-q_1)}{q_1 + q_2 - q_1 q_2}, \frac{q_2}{q_1 + q_2 - q_1 q_2} \right)$$

it remains in that interval throughout the sequence $P_0, P_1, P_2, \ldots$. There is something of a paradox in the facts that the hider would like the sequence to remain near P^*, and that the searcher ensures that it does by following an optimal policy.

One of the consequences of the tendency for the location probability vector to remain near P^* is that if the object may be moved between searches, but at a cost, the cost does not have to be very high for it not to be worth moving. If it is worth moving, movement should be towards P^*. A model of this type has been discussed by Norris (1962).

From this discussion it is fairly clear that the maximin strategy for the hider must be a point in $N(P^*)$. Any point Q outside $N(P^*)$ is eventually transformed to a point R in $N(P^*)$ as the search proceeds in accordance with Theorem 8.1, and it may be shown that the hider does better to start from R than from Q.

The form of the exact minimax and maximin strategies for Version 1 may, as for Version 2, be deduced from Theorem 8.3. For the hider the action space may once again be identified with the n boxes. For the seacher an action is now defined by an infinite sequence of integers between 1 and n specifying the order in which boxes are to be searched until the object is found.

Lemma 8.4. There is a strategy $\bar{P}$ for the hider for which an optimal counter-strategy is an equalizer strategy.

Proof. Let $C(P)$ denote the set of index counter-strategies to the hider's strategy P, which Theorem 8.1 tells us to be optimal. There may be more than one member of $C(P)$ owing to the possibility of ties between index values. Our claim is that $\bar{P}$ may be chosen so that a particular index strategy in $C(\bar{P})$ is an equalizer strategy. For simplicity the following proof is for $n = 2$; the extension to arbitrary finite n is not difficult.

For $n=2$, $C(P)$ and other functions of P may be written as functions of p, where $P=(p, 1-p)$. Let $C_i(p)$ be the strategy in $C(p)$ which always selects box i ($i=1, 2$) whenever the index values $p_1 q_1$ and $p_2 q_2$ are equal (on the assumption that the hider has used strategy p). Let $C(p, \lambda)$ be the strategy which coincides with $C_1(p)$ with probability λ, and with $C_2(p)$ with probability $1-\lambda$. Let

$$V(p) = \mathbf{E} \text{ (number of searches to find the object using}$$
$$\text{a strategy in } C(p) \text{ when the hider plays } p).$$

$$V_i(p, \lambda) = \mathbf{E} \text{ (number of searches to find the object using}$$
$$\text{a strategy } C(p, \lambda)$$
$$\text{|object is in box i)} \qquad (i=1, 2; 0 \leqslant \lambda \leqslant 1).$$

Now note that for any positive integer m, and $0 \leqslant x < y \leqslant 1$, the mth search of box is

(i) no earlier under $C_i(x)$ than under $C_i(y)$ ($i=1$, 2),
(ii) no earlier under $C_1(x)$ than under $C_2(y)$,
(iii) no earlier under $C_2(x)$ than under $C_1(x)$.

From these observations it follows that

$$V_1(x, \lambda) \geqslant V_1(y, \lambda) \qquad (\lambda = 0, 1),$$
$$V_1(x, 1) \geqslant V_1(y, 0), \qquad\qquad\qquad (8.6)$$
$$V_1(x, 0) \geqslant V_1(x, 1).$$

Moreover it is not difficult to show that $V_i(x, 1)$ is right continuous as a function of x, and $V_i(x, 0)$ is left continuous ($i=1, 2$). From the inequalities (8.6) it therefore follows that

$$\lim_{y \searrow x} V_1(y, 0) = \lim_{y \searrow x} V_1(y, 1) = V_1(x, 1) \quad (0 < x < 1),$$
$$(8.7)$$
$$\lim_{y \nearrow x} V_1(y, 0) = \lim_{y \nearrow x} V_1(y, 1) = V_1(x, 0) \quad (0 < x < 1),$$

Similarly we have

$$V_2(x, \lambda) \leqslant V_2(y, \lambda) \quad (\lambda = 0, 1; 0 \leqslant x < y \leqslant 1), \qquad (8.8)$$

and

$$\lim_{y \searrow x} V_2(y, 0) = \lim_{y \searrow x} V_2(y, 1) = V_2(x, 1) \quad (0 < x < 1),$$
$$(8.9)$$
$$\lim_{y \nearrow x} V_2(y, 0) = \lim_{y \nearrow x} V_2(y, 1) = V_2(x, 0) \quad (0 < x < 1),$$

Clearly

$$V_1(0, \lambda) = V_2(1, \lambda) = \infty \quad \text{and} \quad V_1(1, \lambda) = V_2(0, \lambda) = 1. \qquad (8.10)$$

Define

$$D(x, \lambda) = V_1(x, \lambda) - V_2(x, \lambda).$$

Since

$$V_i(x, \lambda) = \lambda V_i(x, 1) + (1 - \lambda)V_i(x, 0) \quad (i = 1, 2), \tag{8.11}$$

it follows from the first of the inequalities (8.6), and from (8.8) and (8.10), that $D(x, \lambda)$ is decreasing in x, and tends to ∞ as x tends to 0, and to $-\infty$ as x tends to 1. Let

$$\bar{p} = \sup\{p : D(p, \lambda) > 0\},$$

a definition which because of (8.7), (8.9) and (8.11) does not depend on λ. From the right continuity of $D(p, 1)$ and the left continuity of $D(p, 0)$ it follows that

$$D(\bar{p}, 0) \geqslant 0 \geqslant D(\bar{p}, 1),$$

and hence, using (8.11), that, for some $\lambda \in [0, 1]$, $D(\bar{p}, \lambda) = 0$. Since the strategy $C(\bar{p}, \lambda)$ belongs to the set of index strategies $C(\bar{p})$ this completes the proof of the lemma for $n = 2$.   □

Proposition 8.5. The strategy $\bar{P}$ for the hider given by Lemma 8.4 is maximin; the searcher's equalizer optimal counter-strategy is minimax, and the minimax theorem holds for Version 1 of the game of hide and seek.

Proof. All this follows immediately from Lemma 8.4 and Theorem 8.3.   □

The remaining task is to calculate $\bar{P}$, $V(\bar{P})$, and the searcher's minimax strategy in particular cases. Theorem 8.1 reduces the calculation of $V(P)$, for any P, to the evaluation of the expectation of a discrete random variable for which the distribution is easily calculated. Since

$$V(\bar{P}) = \sup_P V(P),$$

$\bar{P}$ and $V(\bar{P})$ may then be evaluated by some form of hill-climbing procedure. The function $V(P)$ is concave, so there is no risk of finding a local maximum which is not the global maximum. The concavity follows since $V(P)$ is the infimum over all searcher strategies of the expected number of searches, and for any fixed strategy this is a linear, and hence concave, function of P. A suitable randomization between the deterministic index strategies in the set $C(\bar{P})$ is then required to provide a strategy for the searcher for which the expected number of searches conditional on the box i in which the object is hidden is $V(\bar{P})$ for all i. In most cases good approximations are fairly readily computable.

Exact solutions may be obtained when the ratios

$$\log(1 - q_i)/\log(1 - q_j) \qquad (i, j = 1, 2, \ldots, n),$$

are expressible as the ratios of small integers. These are particularly neat for $n = 2$.

Let

$$\log(1-q_1)/\log(1-q_2)=n_1/n_2 \qquad (n_1,n_2\in \mathbf{Z}). \qquad (8.12)$$

Represent $P=(p,1-p)$ by p. Thus

$$p^*=q_1^{-1}/(q_1^{-1}+q_2^{-1}).$$

Let $T_1^{m_1}T_2^{m_2}p$ be the point to which p is transformed by Bayes' theorem after m_i unsuccessful searches of box i ($i=1,2$). Thus

$$T_1^1 T_2^0 p^* \text{ (or } T_1 p^*)=\frac{q_1^{-1}-1}{q_1^{-1}+q_2^{-1}-1}, \quad T_1^0 T_2^1 p^* \text{ (or } T_2 p^*)=\frac{q_1^{-1}}{q_1^{-1}+q_2^{-1}-1},$$

and the interval $N(p^*)$ within which $\bar{p}$ must lie is $[T_1 p^*, T_2 p^*]$.

The set of points

$$S=\{T_1^{m_1}T_2^{m_2}p^*:0\leqslant m_i\leqslant n_i, i=1,2; T_1 p^*\leqslant T_1^{m_1}T_2^{m_2}p^*\leqslant T_2 p^*\}$$

is of particular significance, since it must include the point $\bar{p}$. This is a consequence of the fact that the optimal index counter-strategy to a hider strategy given by p is the same for all values of p lying between a pair of successive points in S, so that $V(p)$ is piece-wise linear, with discontinuities of slope only at points in S. The periodic nature of index policies when an integer relationship of the type (8.12) holds also means that the calculation of $V(p)$ is simplified.

Details and results of calculations along these lines may be found in Roberts and Gittins (1978) for $n=2$, and in Gittins and Roberts (1979) for $n>2$. The most important practical conclusion is that P^* is often equal to $\bar{P}$, and is usually at least a good approximation. For example, extensive calculations indicate that for $n=2$, and points $(q_1,q_2)\in \mathbf{R}^2$ which lie between the curves $(1-q_1)^2=(1-q_2)$ and $(1-q_1)=(1-q_2)^2$,

$$V(\bar{p})<1.00075\ V(p^*).$$

8.5 HIDE-AND-SEEK IN CONTINUOUS TIME

A continuous-time analogue of Version 1 of the discrete-time hide-and-seek problem is of interest both in its own right and as a good approximation to the discrete-time problem for small values of the q_i's. As in the case of the continuous-time jobs considered in Chapter 5, in continuous time it is natural to allow the searcher to divide his effort at any given time between the n boxes.

Let $u_i(t)$ $(\Sigma u_i(t)=1)$, a Lebesgue-measurable function of t, be the proportion of the searcher's total effort allocated to box i at time t, provided the object is not found before then. Let

$$U_i(t)=\int_0^t u_i(s)\,\mathrm{d}s,$$

and let $exp\{-\lambda_i U_i(t)\}$ be the probability that the object is not found by time t if it is in box i, so that λ_i is the *detection rate* for box i. For a given initial value the location probability vector again changes by Bayes' theorem as the search proceeds. Let

$$P(t) = (p_1(t), p_2(t), \ldots, p_n(t))$$

denote its value at time t.

If $P(0) = P$, the probability that detection does not occur before time t is

$$\sum_{i=1}^{n} p_i \exp\{-\lambda_i U_i(t)\},$$

and the expected time to detection is

$$\int_0^{\infty} \sum_{i=1}^{n} p_i \exp\{-\lambda_i U_i(t)\}\, dt. \qquad (8.13)$$

The searcher's problem is to choose the allocations $u_i(\cdot)$ so as to minimize the expression (8.13). By analogy with Theorem 8.1 we expect to find that at time t effort should be concentrated on boxes for which

$$p_i \lambda_i \exp\{-\lambda_i U_i(t)\} = \max_j p_j \lambda_j \exp\{-\lambda_j U_j(t)\},$$

since

$$p_i(t) = \frac{p_i \exp\{-\lambda_i U_i(t)\}}{\sum_j p_j \exp\{-\lambda_j U_j(t)\}}. \qquad (8.14)$$

Fairly clearly this could be established along the lines of the proof of Theorem 8.1, via a suitable continuous-time index theorem, and extended to allow λ_i to be a random process in the continuous variable $U_i(t)$, thus giving a continuous-time analogue of the index (8.4). Rather than embark on this not entirely trivial programme we shall be content with a less general proof, starting with a well-known result.

Theorem 8.6 (The Strong Lagrangian Principle). F and G real-valued functions defined on the set $\mathscr{X}$, μ a constant. If the minimum over $\mathscr{X}$ of $F - \mu G$ occurs at ξ, and $G(\xi) = c$, then the minimum of F over $\{x : G(x) = c\}$ occurs at ξ.

Proof. Let $G(x) = c$. Thus

$$F(x) - F(\xi) = F(x) - \mu G(x) - [F(\xi) - \mu G(\xi)] \geqslant 0. \qquad \square$$

Lemma 8.7. The minimum with respect to $(U_1, U_2, \ldots, U_n)$ of

$$\sum_{i=1}^{n} p_i \exp\{-\lambda_i U_i\}, \text{ subject to } \Sigma U_i = t, U_j > 0 \quad (j = 1, 2, \ldots, n),$$

occurs at a point of the form $U_i = U_i(t)$, where

$$U_i(t) = \begin{cases} 0 & \text{if } \lambda_i p_i \leqslant \mu \\ \lambda_i^{-1} \log_e \left(\dfrac{\lambda_i p_i}{\mu} \right) & \text{if } \lambda_i p_i > \mu \end{cases}, \quad (i = 1, 2, \ldots, n).$$

The lemma is a simple consequence of Theorem 8.6.

The quantities $U_i(t)$ given by the lemma are such that

$$\mu \begin{cases} = \lambda_i p_i \exp\{ -\lambda_i U_i(t) \} & \text{if } U_i(t) > 0 \\ \geqslant \lambda_i p_i & \text{if } U_i(t) = 0 \end{cases}, \quad (i = 1, 2, \ldots, n). \qquad (8.15)$$

There is only one vector $(U_1(t), U_2(t), \ldots, U_n(t))$ for which (8.15) holds, together with $\Sigma U_i(t) = t$. Moreover, recalling (8.14), these conditions are satisfied simultaneously for all $t > 0$ by setting

$$u_i(s) \begin{cases} = 0 & \text{if } \lambda_i p_i(t) < \max_j \lambda_j p_j(t) \\ \propto \lambda_i^{-1} & \text{if } \lambda_i p_i(t) = \max_j \lambda_j p_j(t) \end{cases}, \quad (i = 1, 2, \ldots, n; s \geqslant 0). \qquad (8.16)$$

Now bringing together (8.13), Lemma 8.7 and (8.16), we have the following theorem.

Theorem 8.8. The searcher's optimal strategy when $P(0)$ is known is at each time t to identify the set of boxes

$$B(t) = \{ i : \lambda_i p_i(t) = \max_j \lambda_j p_j(t) \},$$

and to set

$$u_i(t) = \begin{cases} 0 & \text{if } i \notin B(t) \\ \lambda_i^{-1} \Big/ \displaystyle\sum_{j \in B(t)} \lambda_j^{-1} & \text{if } i \in B(t) \end{cases}, \quad (i = 1, 2, \ldots, n).$$

This theorem is the continuous-time analogue of Theorem 8.1 with $c_i = 1 \ (i = 1, 2, \ldots, n)$.

By analogy with the discrete-time case, define

$$P^* = (\lambda_1^{-1}, \lambda_2^{-1}, \ldots, \lambda_n^{-1}) \Big/ \sum_{j=1}^{n} \lambda_j^{-1}.$$

From Theorem 8.8 it follows that if the hider plays P^* the searcher's optimal counter-strategy is

$$u_i(t) = \lambda_i^{-1} \Big/ \sum_{j=1}^{n} \lambda_j^{-1} \quad (i = 1, 2, \ldots, n; t \geqslant 0).$$

Call this strategy U^*. From 8.13 it follows that if the searcher plays U^*, the expected time until the object is found is

$$\int_0^\infty \exp\left\{ -t \bigg/ \sum_{j=1}^n \lambda_j^{-1} \right\} dt,$$

irrespective of the hider's strategy. Thus U^* is an equalizer strategy, the conditions of Theorem 8.3 are satisfied, and we have the following theorem.

Theorem 8.9. The minimax theorem holds for the continuous-time hide-and-seek game. P^* is maximin for the hider and U^* is minimax for the searcher.

As in the case of discrete time a second version of the continuous-time game can be defined. Again the hider hides the object in one of the boxes, but now the searcher can look only for an infinitesimal interval. The hider's aim is to minimize the probability that the object is found during this infinitesimal interval, and the searcher's aim is to maximize this probability. It is easy to show that P^* and U^* (for the infinitesimal interval only) are minimax and maximin, so that in continuous time the same policies are best for both the extended and instantaneous versions of the game. This is in contrast to the discrete-time case. The difference is that in continuous time the extended game may be regarded as a succession of plays of the instantaneous game, despite the hider's inability to move the object. The reason is that in continuous time the searcher's strategy U^* ensures that the location probability vector remains precisely at P^*, rather than only approximately so because of the overshoot discussed in the previous section.

EXERCISES

Exercises 8.1, 8.2 and 8.3 involve Version 1 of the discrete-time hide-and-seek game.

8.1. $n=2$, $q_1=0.2$, $q_2=1-(0.8)^{3/2}$ (so (8.12) holds). Evaluate the function $V(P)$ at the set of points S (see page 195), and hence determine the hider's maximin strategy and the searcher's minimax strategy.

8.2. If the hider plays P^* it follows from Theorem 8.1 and the transformation (8.5) that an optimal index counter-strategy for the searcher starts by searching the n boxes once each in any order. Show that the probability of finding the object on the rth of these first n searches, conditional on it not being found during the first $r-1$ searches, is $(\Sigma q_i^{-1} - r + 1)^{-1}$. Show also that if $P(=(p_1, p_2, \ldots, p_n))$ is such that

$$\max_i p_i q_i \leqslant \left(\sum_{j=1}^n q_j^{-1} - n + 1 \right)^{-1} \tag{8.17}$$

then the sequence of location probability vectors into which P is transformed under an optimal policy all satisfy this inequality. Hence show that the hider's maximin strategy must also satisfy (8.17).

You may find it easier to consider first the case $n = 2$.

8.3. Show that the functions $V_i(x, 1)$ and $V_i(x, 0)$, $(i = 1, 2)$, defined in the proof of Lemma 8.4 are right and left continuous respectively.

8.4. Deterministic policies for Problem 3 are defined by the sequence; with repetitions, in which the boxes are to be searched, up to the point when the object is found. Attention may be restricted to sequences in which each of the boxes occurs infinitely often, since otherwise there is a positive probability that the object is never found. Denote by (p, i) the pth search of box i, and by $(p, i)(q, j)$ part of a sequence in which (p, i) immediately procedes (q, j).

Lemma 8.10. If $(p, i)(q, j)$ occurs in a given sequence, and that sequence is better than the one obtained by replacing $(p, i)(q, j)$ by $(q, j)(p, i)$, then any sequence which includes $(p, i)(q, j)$ is better than the one derived from it by interchanging (p, i) and (q, j).

Denote the relationship postulated in Lemma 8.10 between (p, i) and (q, j) by

$$(p, i) \succ (q, j).$$

For simplicity assume that $(p, i) \succ (q, j)$ excludes the possibility that $(q, j) \succ (p, i)$.

Lemma 8.11. If $(p, i) \succ (q, j)$ and $(q, j) \succ (r, k)$ then $(p, i) \succ (r, k)$.

Lemma 8.12. If $(p, i) \succ (q, j)$ then $(p, i) \succ (q + 1, j)$.

Prove the above three lemmas, and hence show that the optimal policy for Problem 3 is given by the index $p_i q_i / c_i$.

Similar lemmas leading to an optimal policy in the form of an index may be established for a variety of other discrete search problems, not all of which may easily be formulated as FABPs. Lehnerdt (1982) gives an interesting account of this approach which extends, for example, to more than one hidden object, dangerous searches in which there is a possibility of the searcher being put out of action, different objective functions, and problems with precedence constraints.

CHAPTER 9

In Conclusion

9.1 INTRODUCTION

The theme of this book has been exponentially discounted bandit problems. More specifically it has been exponentially discounted bandit problems as I have known them and been closely involved with them. Until fairly recently this distinction would not have been worth making, but in the last few years there have been a number of substantial further contributions. This final chapter is a review of these developments. Three strands may usefully be distinguished, associated initially, and in chronological order, with the names of Whittle, Glazebrook and Bather.

Whittle's contributions centre on his proof (1980) of the index theorem for a simple family of alternative bandit processes. Some people have found this proof more accessible than the alternative interchange arguments in the form in which they were available at that time. Certainly it is an elegant proof and provides valuable insight. For example, Eick (1988) has adapted it to establish a limited index theorem for bandit reward sampling processes (§6.1) for which the rewards do not become known immediately. Its main and major importance, however, is probably the expression which it provides for the optimal payoff function. This has already been exploited by Whittle and other authors but not, I am sure, to its full potential. For example, Whittle was able to obtain the optimal priority system for an M/G/1 queue under the expected weighted flow-time criterion (the nonpreemptive version of Theorem 3.28) from the expression for the case with Poisson arrivals of further bandit processes. This work is described in Whittle's book (1982). The Whittle index theorem is stated and proved in §9.2.

In his book Whittle also examines the implications of his earlier work on sequential analysis and design for bandit problems. This leads to expressions for allocation indices based on a no-overshoot approximation for a *simple* bandit process, and for the normal with known variance, and Bernoulli, reward processes. Simple bandit process (or arm) is the term coined by Keener (1986) for a sampling process (see Chapter 6) for which there are just two possible underlying distributions. Keener has in fact gone one better by finding an exact expression for the index in this case.

Glazebrook has written more than thirty papers on exponentially discounted bandit problems, almost certainly considerably more than any other author. These are all listed in the references, though there are a few which are not specifically mentioned in the text. They range widely over most of the subject, though recently there have been two main themes. These are applications to scheduling, of industrial research in particular, and the investigation of different kinds of suboptimality. This work is summarized in §9.3.

The properties of the Brownian reward process listed as parts (i) and (iii) of Theorem 6.28 form part of the rich harvest yielded by the relationship between the Wiener process and the heat equation expounded by Doob (1955). The possibility of using this relationship in the solution of sequential decision problems was noted by Lindley (1960) and Chernoff (1961), and discussed in detail by Chernoff (1968) and Bather (1970). Chernoff was interested in the diffusion version of the finite-horizon undiscounted two-armed bandit problem, but it was Bather (1983) who turned the comparison procedure which he had developed (1962) for finding approximate solutions to the free boundary problems associated with optimal stopping into an effective instrument for analysing the discounted Brownian reward process. The resulting elegant argument is outlined in §9.4.

In the last three years or so there have been a number of further significant contributions, both in the theory of discounted bandit problems and in terms of algorithms for computing index values. These are reviewed in §9.5.

9.2 THE WHITTLE INDEX THEOREM

Following his example we shall for simplicity present Whittle's index theorem for a SFABP $\mathscr{F} = \{B_1, B_2, \ldots, B_n\}$ in the Markov case. He has shown (1981) that it extends without difficulty to the semi-Markov setting, and may also be extended to give an interesting alternative proof of the index theorem for a FABP with a state-independent stationary arrival process. The proof hinges on spotting the form of the solution to the dynamic programming functional equation (2.2) when the decision process D is the SFABP $\{\mathscr{F}, M\}$ formed by adding to $\mathscr{F}$ a standard bandit process which produces a total discounted reward M if it is selected at all times, and hence a reward $(1 - a)M$ if it is selected for one time unit. The solution is expressed in terms of the solutions to the simpler SFABPs $\{B_i, M\}(i = 1, 2, \ldots, n)$. Whittle calls the standard process a *retirement option*, which is appropriate since if it is ever optimal to choose this option it must remain optimal to do so permanently. First we need some notation.

Let $x = (x_1, x_2, \ldots, x_n) \in \Theta = \Theta_1 \times \Theta_2 \times \cdots \Theta_n$ be the state of $\mathscr{F}$ and $r_i(x_i)$ the reward if B_i is selected when $\mathscr{F}$ is in state x. Let $\phi_i(x_i, M, \Omega_i)$ be the expected total discounted reward from $\{B_i, M\}$ when retirement is chosen at the time τ_i at which B_i first reaches a state in $\Omega_i \subset \Theta_i$ starting from x_i. Let $\Phi_g(x, M, \Omega)$ be the expected total discounted reward from $\{\mathscr{F}, M\}$ when policy g is applied to $\mathscr{F}$ and retirement is chosen at the time τ at which $\mathscr{F}$ first reaches a state in $\Omega \subset \Theta$ starting from x. Thus, recalling the notation defined in §§2.3 and 3.4,

$$\phi_i(x_i, M, \Omega_i) = R_{\tau_i}(B_i, x_i) + \mathbf{E}a^{\tau_i}M \qquad (9.1)$$

and

$$\Phi_g(x, M, \Omega) = R_{g\tau}(\mathscr{F}, x) + \mathbf{E}a^\tau M \qquad (9.2)$$

Denote by $\phi_i(x_i, M)$, $\Phi(x, M)$, and $G(x)$, the maximal payoffs from $\{B_i, M\}$, $\{\mathscr{F}, M\}$, and $\mathscr{F}$, so that

$$\phi_i(x_i, M) = \sup_{\Omega_i} \phi_i(x_i, M, \Omega_i),$$

$$\Phi(x, M) = \sup_{g, \Omega} \Phi_g(x, M, \Omega),$$

and

$$G(x) = \Phi(x, -\infty).$$

The operator L_i is defined to be such that

$$L_i f(x_i) = r_i(x_i) + a\mathbf{E}[f(x_i(t+1))|x_i(t) = x_i],$$

provided the expectation exists. Thus for $D = \{B_i, M\}$ equation (2.2) becomes

$$\phi_i(x_i, M) = \max\{M, L_i\phi_i(x_i, M)\}.$$

From the calibration argument used in the proof of Theorem 3.1, and since we are dealing with the Markov case, for which the discrete-time correction factor applies (see §2.3), we also have

$$(1-a)^{-1}v(B_i, x_i) = \min\{M : \phi_i(x_i, M) = M\}.$$

Define

$$M_i = (1-a)^{-1}v(B_i),$$

$$\delta_i(M) = \phi_i(M) - L_i\phi_i(M),$$

dropping the state variable from the notation, as we shall throughout most of the remainder of this section since the focus is on the variable M. Thus

$$\delta_i(M) \geqslant 0, \quad \text{with equality for } M \leqslant M_i.$$

For $D = \{\mathscr{F}, M\}$ equation (2.2) becomes

$$\Phi(x, M) = \max\left\{M, \max_i L_i\Phi(x, M)\right\}, \qquad (9.3)$$

where

$$L_i\Phi(x, M) = r_i(x_i) + a\mathbf{E}[\Phi(x_1, x_2, \ldots, x_i(t+1), \ldots, x_n, M)|x_i(t) = x_i].$$

Theorem 9.1 (The Whittle Index Theorem). If the SFABP $\mathscr{F}$ has reward functions r_i such that

$$(1-a)A < r_i(x_i) < (1-a)B \qquad (x_i \in \Theta_i, 1 \leqslant i \leqslant n)$$

for suitably chosen finite A and B, then (i) index policies are optimal, and (ii) the optimal payoff function $\Phi(M)$ for the augmented SFABP $\{\mathscr{F}, M\}$ is a non-decreasing convex function of M such that

$$\Phi(M) = \begin{cases} G & (M \leqslant A) \\ M & (B \leqslant M) \end{cases},$$

and

$$\Phi(M) = B - \int_{M}^{B} \prod_{i} \frac{\partial \phi_i(M)}{\partial m} \, dm \qquad (M \in \mathbf{R}).$$

Proof. We first prove (ii). Since

$$\Phi(M) = \sup_{g, \Omega} \Phi_g(M, \Omega)$$

and $\Phi_g(M, \Omega)$ is a non-decreasing convex (in fact linear) function of M it follows that $\Phi(M)$ is also non-decreasing and convex. Next note that if any policy for $\{\mathscr{F}, M\}$ which never retires achieves with certainty a higher total reward than is obtainable by interrupting it at any point and retiring, then it can never be optimal to retire. It follows that

$$\Phi(M) = G \qquad (M < A).$$

Also if any policy for $\{\mathscr{F}, M\}$ leads with certainty to a total reward which is at most M then immediate retirement must be optimal. Thus

$$\Phi(M) = M \qquad (B < M).$$

The general form of $\Phi(M)$ is less easily established. Note first that by setting $n = 1$ it follows that $\phi_i(M)$, like $\Phi(M)$, is non-decreasing and convex, so that $\partial \phi_i / \partial M$ exists for almost all M. Now define

$$\hat{\Phi}(M) = B - \int_{M}^{B} \prod_{i} \frac{\partial \phi_i(m)}{\partial m} \, dm. \tag{9.4}$$

We proceed to show that $\hat{\Phi}$ satisfies the functional equation (9.3).

Integrating by parts gives

$$\hat{\Phi}(M) = \phi_i(M) P_i(M) + \int_{M}^{B} \phi_i(m) \, dP_i(m), \tag{9.5}$$

where

$$P_i(M) = \prod_{j \neq i} \frac{\partial \phi_j(M)}{\partial M}.$$

Since ϕ_i is a non-decreasing, convex function of M and $\phi_i(M) = M \ (M \geqslant M_i)$ it follows that P_i is a non-negative, non-decreasing function of M and

$$P_i(M) = 1 \qquad \left(M \geqslant \max_{j \neq i} M_j \right).$$

Thus the upper integration limit B in (9.5) may be replaced by ∞.

From (9.5) and the properties just cited of ϕ_i and P_i it follows that

$$\hat{\Phi}(M) \geqslant M, \quad \text{with equality if } M \geqslant \max M_j. \tag{9.6}$$

From (9.5) and since $P_i(\infty) = 1$ we have

$$\hat{\Phi}(M) - L_i \hat{\Phi}(M) = \delta_i(M) P_i(M) + \int_M^{\infty} \delta_i(m)\, dP_i(m). \tag{9.7}$$

Now $\delta_i(M) = 0 \ (M \leqslant M_i)$ and $dP_i(m) = 0 \ (m \geqslant \max_{j \neq i} M_j)$. Thus it follows from (9.7) that

$$\hat{\Phi}(M) - L_i \hat{\Phi}(M) = 0 \text{ if } M_i = \max; \ M_j \geqslant M. \tag{9.8}$$

Since δ_i, P_i and dP_i are all non-negative it also follows that

$$\hat{\Phi}(M) - L_i \hat{\Phi}(M) \geqslant 0 \qquad (M \in \mathbf{R}). \tag{9.9}$$

The relations (9.6), (9.8) and (9.9) together are equivalent to the assertion that $\hat{\Phi}$ satisfies (9.3), which therefore follows. Since (9.3) has a unique solution (Theorem 2.1(ii)) the functions $\hat{\Phi}$ and Φ must be identical, completing the proof of (ii).

Finally, a policy for $\mathcal{F}$ is optimal (Theorem 2.1 (iii)) if at each stage the bandit process B_i selected is such that $G(x) = L_i G(x)$. Since $\Phi(x, M) = G(x) \ (M \leqslant A)$ it follows from (9.8) that the index policy is optimal for $\mathcal{F}$. $\qquad \Box$

Whittle also provides the motivation for the conjectured form (9.4), which is far from immediately obvious. This involves the concept of a *write-off* policy for $\{\mathcal{F}, M\}$. A policy g for $\{\mathcal{F}, M\}$ is a write-off policy if there are sets $\Omega_i \subset \Theta_i$ $(i = 1, 2, \ldots, n)$ such that under g (i) bandit process B_i is written off (i.e. never selected again) when it reaches a state in Ω_i, and (ii) the retirement option is chosen as soon as, though not before, each bandit process B_i has reached its write-off set Ω_i.

Any stationary policy for $\{\mathcal{F}, M\}$ is equivalent to a corresponding policy for $\mathcal{F}$ together with a retirement set. Let g be a policy for $\mathcal{F}$ which together with the retirement set Ω defines a write-off policy for $\{\mathcal{F}, M\}$ with write-off sets Ω_i for $B_i (i = 1, 2, \ldots, n)$. It follows that, in the notation of equations (9.1) and (9.2),

$$\tau = \sum_{i=1}^{n} \tau_i,$$

and that the write-off times τ_i are independent. Thus differentiating (9.1) and (9.2) we have

$$\frac{\partial \Phi_g(M, \Omega)}{\partial M} = \prod_{i=1}^{n} \frac{\partial \phi_i(M, \Omega_i)}{\partial M}. \tag{9.10}$$

These differentiations are on the basis that g and Ω are held fixed and only M changes. Whittle was able to show that if g and Ω are varied so as to give an optimal M-dependent write-off policy then equation (9.10) still holds for almost all M, the differentiations now taking into account the variation of g and Ω. This follows from the fact that the supremum over all write-off policies of $\Phi_g(M, \Omega)$ is a convex function of M. On the assumption that the best write-off policy is optimal over the set of all policies this means that

$$\frac{\partial \Phi(M)}{\partial M} = \prod_{i=1}^{n} \frac{\partial \phi_i(M)}{\partial M}.$$

The expression (9.4) now follows by integration.

Whittle needed Theorem 2.1 to establish the conjecture as no direct proof was available that the set of write-off policies includes an overall optimal policy. Tsitsiklis (1986) has provided the required direct proof. This follows from a lemma involving the SFABP $\mathcal{F}^i$ obtained from $\mathcal{F}$ by removing bandit process B_i. We may assume $n \geqslant 2$, since for $n = 1$ the conjecture is trivial. Let x^i denote the state of $\mathcal{F}^i$, so that the state x of $\mathcal{F}$ may be written as (x_i, x^i), and let Φ^i denote the optimal payoff function for $\{\mathcal{F}^i, M\}$.

Lemma 9.2. $\Phi(x, M) \leqslant \phi_i(x_i, M) + \Phi^i(x^i, M) - M \quad (i = 1, 2, \ldots, n)$.

Let the variable s denote compulsory retirement not later than time s, so that with $s = 0$ we have

$$\Phi(x, M, 0) = \phi_i(x_i, M, 0) = \Phi^i(x^i, M, 0) = M,$$

and thus

$$\Phi(x, M, s) \leqslant \phi_i(x_i, M, s) + \Phi^i(x^i, M, s) - M \tag{9.11}$$

when $s = 0$. The proof proceeds by first establishing by induction that (9.11) holds for any integer $s \in \mathbf{Z}$ and then letting $s \to \infty$. The details are left as an exercise.

Lemma 9.3. There is an optimal policy for $\{\mathcal{F}, M\}$ which is a write-off policy.

Proof. Let $\Omega_i = \{x_i \in \Theta_i : \phi_i(x_i, M) = M\}$. If $(x_i, x^i) \in \Theta$ is such that $x_i \in \Omega_i$ it follows from Lemma 9.2 that $\Phi(x, M) \leqslant \Phi^i(x^i, M)$. This means that there is an optimal policy for $\{\mathcal{F}, M\}$ which never uses B_i. Conversely if $\exists i$ such that $x_i \notin \Omega_i$ then $\phi_i(x_i, M) > M$ and it is better to select B_i than to retire. It follows that there must be an optimal policy within the set of write-off policies defined by the write-off sets Ω_i $(i = 1, 2, \ldots, n)$. $\qquad \square$

9.3 PERMUTATION SCHEDULES AND SUB-OPTIMALITY

If the allocation indices for a set of jobs are increasing up to completion it follows from Theorem 3.6 that the optimal policy in the absence of arrivals or precedence constraints is to schedule the jobs in a particular fixed order. When there are precedence constraints the optimal policy is still of this form under a rather stronger monotonicity condition (Theorem 4.8). These permutation schedules have two important advantages. Firstly, in many practical situations they are more easily implemented than more complex policies, both administratively and in terms of technological constraints. Secondly, much less complex algorithms are typically needed to find the best permutation schedule than to find the best policy overall when we cannot be sure that this is a permutation schedule. These considerations have led Glazebrook to explore in some detail the questions of (a) the conditions under which the globally optimal policy is a permutation schedule, (b) how to find the best permutation, and (c) how big the penalty is from using the best permutation schedule when the optimal policy is not of this type. The relevant papers are Glazebrook (1980b and c, 1981a and b, 1982b and e, 1983b), Glazebrook and Gittins (1981), and Glazebrook and Fay (1987).

The possible sub-optimality of permutation schedules is one of the links between the two topics of this section. The other is the major contribution that Glazebrook has made to both of them. Theorem 3.9 is one of his results, showing that the penalty of not following an index policy for a SFABP is bounded by a suitable overall measure of the extent to which the indices of the bandit processes selected at different times are not maximal. This typifies the Glazebrook approach to suboptimality, which is to measure the consequences of breaking some condition which is necessary for optimality in terms of the extent to which the condition has been broken. In addition to the two cases already mentioned he has carried out this programme for myopic scheduling strategies, strategies designed without regard for processor breakdown when the processor actually does break down, jobs for which the rewards, service-time distributions, or discount parameter have been estimated incorrectly, non-index strategies for families of superprocesses satisfying Whittle's Condition D (see Theorem 3.15) and for families of Nash's (1980) generalized bandit processes, and finally for index strategies for families of superprocesses which do not satisfy Whittle's condition, and for which index strategies are therefore (since Glazebrook has shown the condition to be necessary as well as sufficient) not optimal. This impressive body of work is reported in Glazebrook (1982a, b, c, d, e; 1983a and b; (1984, 1985a, 1987), and Glazebrook and Fay (1988).

9.4 MORE ABOUT THE BROWNIAN REWARD PROCESS

The Wiener (or standard Brownian motion) process is a random process W defined on $\mathbf{R}^+$, having independent normal increments on disjoint intervals, and such that

$$W(0)=0, \quad \mathbf{E}W(t)=0 \qquad (t \geqslant 0),$$

and

$$\mathbf{E}(W(s)-W(t))^2 = s-t \qquad (0 \leqslant t \leqslant s).$$

Let Z be a random process defined on the non-negative real line $\mathbf{R}^+$ and taking values in $\mathbf{R}^2$, such that for some fixed $(u, v) \in \mathbf{R}^2$

$$Z(t)=(u+W(t), v-t) \qquad (0 \leqslant t).$$

Let τ be a stopping time for Z (see §2.3) defined by a closed stopping subset Θ of $\mathbf{R}^2$ and let r be a real-valued function defined on $\mathbf{R}^2$. Finally let

$$f(u, v) = \mathbf{E}[r(Z(\tau))|Z(0)=(u, v)].$$

It may be shown that the function f satisfies the heat equation

$$\frac{1}{2}\frac{\partial^2 f}{\partial u^2} = \frac{\partial f}{\partial v} \tag{9.12}$$

at points $(u, v) \notin \Theta$, and clearly

$$f(u, v) = r(u, v) \qquad ((u, v) \in \Theta). \tag{9.13}$$

The optimal stopping problem defined by the reward function r is to choose Θ so as to maximize the expected reward $f(u, v)$. It may be shown along the lines of the proof of Lemma 2.5 that this is achieved simultaneously for all $(u, v) \in \mathbf{R}^2$ when

$$\Theta = \left\{ (x, y) : \sup_{\Theta} f(x, y) = r(x, y) \right\}.$$

Under regularity conditions it may also be shown that at points on the boundary of Θ

$$\frac{\partial f}{\partial u} = \frac{\partial r}{\partial u} \quad \text{and} \quad \frac{\partial f}{\partial v} = \frac{\partial r}{\partial v}. \tag{9.14}$$

The problem of finding Θ so that (9.12), (9.13) and (9.14) are satisfied is known as a *free boundary problem*. In general no closed form of solution can be found, though there are exceptions, in particular when the maximal f corresponds to a separable solution of the heat equation. This observation led Bather to develop a comparison procedure for approximating the optimal Θ.

Consider two stopping problems distinguished by subscripts 1 and 2. Suppose that the point (u, v) and the reward functions r_1 and r_2 are such that

$$f_2(u, v) = r_2(u, v) = r_1(u, v) \tag{9.15}$$

and

$$r_2(x, y) \geqslant r_1(x, y) \quad (x \in \mathbf{R}, y < v). \tag{9.16}$$

Since it is optimal to stop at (u, v) for problem 2, and the reward function for problem 2 is at least as great as it is for problem 1 at every point which the process Z may reach starting from (u, v), it must also be optimal to stop at (u, v) for problem 1. Thus if problem 2 can be solved and the conditions (9.15) and (9.16) are satisfied the point (u, v) will have been shown to belong to the optimal stopping set Θ_2 for the possibly less immediately tractable problem 1. The comparison procedure for a given problem (1) is to build up as large as possible a subset of the optimal stopping set Θ_1 by finding a tractable auxiliary problem (2), for which conditions (9.15) and (9.16) are satisfied, for each point (u, v) in the subset.

By using suitable transformations it may be shown that finding the allocation index for the Brownian reward process is equivalent to solving the stopping problem defined by the reward function

$$r(u, v) = \max(u, D)\exp(v^{-1}). \tag{9.17}$$

One of the steps in establishing this equivalence is the interesting observation that the posterior mean $\bar{\Sigma}$ for the parameter θ of a Brownian process at time t may itself be regarded as a Brownian process with $-t^{-1}$ as the time variable. Bather (1983) was then able to establish Theorem 6.29(iii) by means of the comparison procedure, and part (ii) by a more direct argument.

Chang and Lai (1987) also consider the stopping problem defined by (9.17). They derive Theorem 6.29(i) by considering the asymptotic properties of a class of solutions of the heat equation, together with comparison theorems somewhat similar to the one used by Bather. They also extend this result to provide an approximation to the index for a discrete reward process governed by a distribution belonging to an exponential family. The approximation holds in the limit as the discount factor $a \to 1$.

9.5 MORE PROOFS, GENERALIZATIONS AND EXTENSIONS

The last few years have seen significant further development in discounted multi-armed bandit theory in addition to what has already been described.

Varaiya, Walrand and Buyukkoc (1985) give attractive proofs of Theorems 3.6 and 3.15 along fairly similar lines to the proof of Lemma 3.16. They show that these results may be set in the context of monotone sequences of sigma-fields, also known as *filtrations*, for each bandit process, rather than the Markov or semi-Markov setting to which most other work has been restricted. They also give a simple algorithm for computing indices for a bandit process with a finite number of states, and show how it may be applied in the assignment of a single server to a finite set of queues between which jobs transfer in a stochastic manner on completion. Mandelbaum (1986, 1987) shows that recent extensions of the theory of stopping problems in which time is replaced by a partially ordered set provide a convenient language for describing bandit problems, and uses this to provide some extensions to the work of Varaiya *et al.* These include an elegant proof of the index theorem for a simple family of alternative bandit processes for each of which the undiscounted reward to date is a continuous function of process time. The proof starts from the discrete-time case and uses limit arguments in similar fashion to the proof of Theorem 5.1.

Lai and Ying (1988) establish an algorithm for computing indices for a bandit process with a finite number of states with a state-independent stationary arrival process of further similar bandit processes. When there are no arrivals the algorithm reduces to the one proposed by Varaiya *et al.* The justification of the algorithm uses Whittle's extension of the expression (9.4) for the optimal payoff function with arrivals. By letting the discount parameter $\gamma \to 0$ they obtain an alternative derivation and a generalization of the priority indices discovered by Klimov (1974, 1978) for the queueing network problem (as discussed by Varaiya *et al.* but with Poisson arrivals).

The Varaiya *et al.* algorithm for computing indices for a bandit process with a finite number of states proceeds by calculation in decreasing order of magnitude. Other algorithms have been proposed by Beale (1979), using policy iteration, Chen and Katehakis (1986) and Kallenberg (1986), using linear programming, and by Katehakis and Veinott (1986). The last of these is based on the observation that the index for a bandit process B in state x is equal to the maximal expected time-averaged rate of accumulation of rewards obtainable from a SFABP consisting of an infinite number of copies of B, all initially in state x, the same idea as we have used in the Proof of Theorem 3.4, parts (ii) and (iii). This enables them to use standard dynamic programming value-iteration methods, which are particularly powerful when the number of possible transitions is small.

Karatzas (1984) considers bandit processes for which the state $x(\in \mathbf{R})$ at process time t is a diffusion process satifying the stochastic differential equation

$$dx = \mu(x)\,dt + \sigma(x)\,dw,$$

where w is a Wiener process, and rewards accrue at the rate $h(x)$ per unit time when the bandit process is selected, h being an increasing function. The general form of the stopping set for the stopping time τ which maximizes the expression

$$\mathbf{E}\left[\int_0^\tau h(x(t))e^{-\gamma t}\,dt + Me^{-\gamma \tau}\right]$$

may be obtained along the lines described by Bensoussan and Lions (1978), leading to equations from which the allocation index may be calculated. For the case when the functions μ and σ are constants Karatzas shows that the allocation index becomes

$$v(x) = \int_0^\infty h\left(x + \frac{z}{\beta}\right) e^{-z}\,dz,$$

where

$$\beta = ((\mu^2 + 2\gamma\sigma^2)^{1/2} - \mu)\sigma^{-2}.$$

Karatzas also establishes the technically far from trivial fact that realizations of a SFABP made up of diffusion bandit processes of this type may be described in a suitable probability space when an index policy is followed. By mimicking the proof of Theorem 9.1 he shows that the index policy is optimal.

Eplett (1986) generalizes the Karatzas assumptions by allowing $x(t)$ to be an arbitrary continuous-time Markov process and considers discrete-time approximations, establishing conditions under which both the indices and the optimal payoff functions for the approximations converge to those for the original set-up. As he points out, this analysis is of practical importance as there must be some limit to the possibility of reallocating resources at very short notice, and it is also sometimes easier to calculate indices for discrete-time approximations.

Kertz (1986) and Whittle (1988) consider families of alternative bandit processes subject to global constraints. In Kertz's paper the constraints are inequalities on the expectation of the integral over time of the discounted value of some function of the state. Whittle's constraint is on the time-averaged number of servers, and actually therefore amounts to a relaxation of our standard assumption that only one server, or processor, can operate at a time. Another difference is that he considers the undiscounted case. Both authors formally remove these constraints by Lagrangian methods. Kertz shows that for an SFABP subject to global constraints the solution may be expressed as an index policy with modified index values, and obtains a generalization of the expression (9.4) for the optimal payoff function together with some explicit results for Bernoulli reward processes. Whittle shows that under certain circumstances the Lagrange multiplier may be identified with the allocation index. He also gives reasons for believing that when there are large numbers of bandit processes of each of several types, a given fraction of which may be processed simultaneously, the index policy is nearly optimal, even if the state of a bandit process may change when it is not being processed. A proof of this conjecture seems likely to be available soon. This extension of index theory to restless bandit processes is an important step forward.

Finally, consider a sequence of independent identically distributed real-valued random variables in the Bayesian setting of Chapter 6, which may be sampled one at a time at a fixed cost per observation, and with the option of stopping at any time and collecting a reward equal to the mean of the distribution. If we also have a discount factor this forms a stoppable bandit process (see §3.9), but one for which the stopping option need not improve with sampling, so that we cannot use

Lemma 3.17 to deduce the optimality of index policies for a simple family of such processes.

The problem thus posed has been termed the *buyer's problem*, as stopping may be equated with deciding to buy one of the available batches of some commodity. It was considered in depth and in much greater generality than we have stated it here by Bergman (1981), using and extending the theoretical framework for stopping problems set out by Chow, Robbins and Siegmund (1971), each sampling process being represented by a filtration. Bergman showed that in general index policies are indeed not optimal, though they are optimal if initially there are an infinite number of populations available for sampling, each with the same prior information.

Bather (1983) has also considered the buyer's problem. He notes that Lemma 9.2 (which he must have derived independently) continues to hold if the B_i's are superprocesses, and confirms Bergman's conclusion that an index policy is optimal with an infinite number of initially indistinguishable populations. This follows from the argument of Lemma 9.3 by setting $(1-a)M$ equal to the initial common index value of each stoppable bandit process (see equation (3.24)). In fact it is not difficult to see that the argument holds for any infinitely replicated superprocess. Using a diffusion approximation to a normal sampling process with known variance along the lines described in §6.6 enables Bather to obtain an upper bound to the index value by means of his comparison procedure.

EXERCISE

9.1. Write down the dynamic programming recurrence equation (2.1) for the finite horizon optimal payoff function $\Phi(x, M, s)$ in the set-up of Lemma 9.2, writing out separate expressions on the right-hand side for the three cases $u=$ retire, $u=$ select B_1, and $u=$ select B_i ($i \neq 1$). Write down too the similar recurrences for the sub-problem optimal payoff functions $\phi_1(x_1, M, s)$ and $\Phi^1(x^1, M, s)$.

Explain why it is that, for any $x \in \Theta$ and $s \in \mathbf{Z}$,

$$0 \leqslant a(\phi_1(x_1, M, s) - M) \leqslant \phi_1(x_1, M, s+1) - M,$$

and

$$0 \leqslant a(\Phi^1(x^1, M, s) - M) \leqslant \Phi^1(x^1, M, s+1) - M.$$

Now use the inductive hypothesis together with the right-hand side of the recurrence for $\Phi(x, m, s)$ to obtain an inequality for that quantity, and using also the above inequalities and the recurrences for ϕ_1 and Φ^1 transform the right-hand side of the inequality for $\Phi(x, m, s)$ so as to carry through the induction on s, establishing (9.11) for all s and, without loss of generality, for $i=1$, completing the proof of the lemma.

Note that the proof works equally well if $\mathscr{F}$ is a simple family of alternative superprocesses $\{S_1, S_2, \ldots, S_n\}$.

REFERENCES

Ahlswede, R. and Wegener, I. (1987). *Search Problems*, Wiley.

Amaral, J.A.F.P. do (1985). *Aspects of Optimal Sequential Resource Allocation*, D. Phil. thesis, Oxford.

Barra, J.-R. (1981). *Mathematical Basis of Statistics*, Academic Press.

Bather, J.A. (1962). Bayes procedures for deciding the sign of a normal mean, *Proc. Cambridge Phil Soc.*, **58**, 599–620.

Bather, J.A. (1970). Optimal stopping problems for Brownian motion, *Adv. Appl. Prob.*, **2**, 259–86.

Bather, J.A. (1983). Optimal stopping of Browian motion: a comparison technique, in *Recent Advances in Statistics*, ed. D. Siegmund, *et al.*, Academic Press.

Beale, E.M.L. (1979). Contribution to the discussion of Gittins, *J. Roy. Statist. Soc. B*, **41**, 171–2.

Bellman, R.E. (1956). A problem in the sequential design of Experiments, *Sankhyā A*, **30**, 221–52.

Bellman, R.E. (1957). *Dynamic Programming*, Princeton University Press.

Bensoussan, A. and Lions, J.L. (1978). *Applications des Inéquations Variationelles en Controle Stochastique*, Dunod.

Bergman, S.W. (1981). *Acceptance Sampling: The Buyer's Problem*, Ph.D. Dissertation, Yale University.

Bergman, S.W. and Gittins, J.C. (1985). *Statistical Methods for Planning Pharmaceutical Research*, Marcel Dekker.

Berninghaus, S. (1984). *Das 'Multi-Armed-Bandit' Paradigma*, Verlagsgruppe Athenaum.

Berry, D.A. and Fristedt, B. (1985). *Bandit Problems*, Chapman & Hall.

Blackwell, D. (1965). Discounted dynamic programming, *Ann. Math. Statist*, **36**, 226–35.

Cassels, J.W.S. (1980). *The Gittins Index*, private communication.

Chang, F. and Lai, T.L. (1987). Optimal stopping and dynamic allocation, *Adv. Appl. Prob.* **19**, 829–53.

Chen, Y.R. and Katehakis, M.N. (1986). Linear programming for finite state multi-armed bandit problems, *Math. OR*, **11**, 178–83.

Chernoff, H. (1961). Sequential tests for the mean of a normal distribution, *Proc. Fourth Berkeley Symposium on Statistics*, University of California Press, **1**, 79–91.

Chernoff, H. (1968). Optimal stochastic control, *Sankhya A*, **30**, 221–52.

Chow, Y.S., Robbins, H. and Siegmund, D. (1971). *Great Expectations: the Theory of Optimal Stopping*, Houghton Mifflin.

Coffman, E.G. (1976). *Computer and Job-Shop Scheduling Theory*. Wiley.

Conway, R.W., Maxwell, W.L., and Miller, L.W. (1967) *Theory of Scheduling*, Addison-Wesley.

Cramér, H. (1946). *Mathematical Methods of Statistics*, Princeton University Press.

Dean, B.V. and Goldhar, J.L. (1980). *Management of Research and Innovation*, North-Holland.

Dempster, M.A.H., Lenstra, J.K., and Rinooy Kan, A.H.G. (editors). (1982). *Deterministic and Stochastic Scheduling*, Reidel.

212

Doob, J.L. (1955). A probability approach to the heat equation, *Trans. Amer. Math. Soc.*, **80**, 216–280.

Eick, S.G. (1988). Gittins procedures for bandits with delayed responses, *J.R. Statist. Soc. B*, **50**, 125–32.

Eplett, W.J.R. (1986). Continuous-time allocation indices and their discrete-time approximation, *Adv. Appl. Prob.*, **18**, 724–46.

Ferguson, T.S. (1967). *Mathematical Statistics: a Decision Theoretic Approach*, Academic Press.

Fox, L. and Parker, I.B. (1968). *Chebyshev Polynomials in Numerical Analysis*, Oxford Univ. Press.

French, S. (1982). *Sequencing and Scheduling: An Introduction to the Mathematics of the Job-Shop*, Ellis Horwood.

Gal, S. (1980). *Search Games*, Academic Press.

Gittins, J.C. (1975). The two-armed bandit problem: variations on a conjecture by H. Chernoff, *Sankhya A*, **37**, 287–91.

Gittins, J.C. (1979). Bandit processes and dynamic allocation indices, *J. Roy. Statist. Soc. Ser. B*, **41**, 148–77.

Gittins, J.C. (1980). *Sequential resource allocation (a Progress Report)*, circulated privately.

Gittins, J.C. (1981). Multiserver scheduling of jobs with increasing completion rates, *J. Appl. Prob.* **18**, 321–4.

Gittins, J.C. (1982). Forwards induction and dynamic allocation indices, in *Deterministic and Stochastic Scheduling*, editors M.A.H. Dempster *et al.*, NATO Advanced Study Institutes Series, Reidel, 125–56.

Gittins, J.C. (1983). Dynamic allocation indices for Bayesian bandits, in *Mathematical Learning Models–Theory and Algorithms*, eds. U. Herkenrath *et al.*, Springer, 50–67.

Gittins, J.C. and Glazebrook, K.D. (1977). On Bayesian models in stochastic scheduling, *J. Appl. Prob.*, **14**, 556–65.

Gittins, J.C. and Jones, D.M. (1974). A dynamic allocation index for the sequential design of experiments, Read at the 1972 European Meeting of Statisticians, Budapest, *Progress in Statistics*, (ed. J. Gani *et al.*), 241–66, North-Holland.

Gittins, J.C. and Jones, D.M. (1979). A dynamic allocation index for the discounted multi-armed bandit problem, *Biometrika*, **66**, 561–6.

Gittins, J.C. and Nash, P. (1977). Scheduling, queues and dynamic allocation indices, *Proc. 1974 European Meeting of Statisticians, Prague, Czech. Academy of Sciences*, 191–202.

Gittins, J.C. and Roberts, D.M. (1979). The search for an intelligent evader: strategies for searcher and evader in the n-region problem, *Naval Research Logistics Quarterly*, **26**, 651–66.

Gittins, J.C. and Roberts, D.M. (1981). RESPRO — an Interactive Planning Procedure for new Product Chemical Research, *R & D Management*, **11**, 139–48.

Glazebrook, K.D. (1976a). Stochastic scheduling with order constraints, *International Journal of Systems Science*, **7**, 657–66.

Glazebrook, K.D. (1976b). A profitability index for alternative research projects, *OMEGA (International Journal of Management Science)*, **4**, 79–83.

Glazebrook, K.D. (1978a). Some ranking formulae for alternative research projects, *OMEGA (International Journal of Management Science)*, **6**, 193–4.

Glazebrook, K.D. (1978b). On the optimal allocation of two or more treatments in a controlled clinical trial, *Biometrika*, **65**, 335–40.

Glazebrook, K.D. (1978c). On a class of non-Markov decision processes, *Journal of Applied Probability*, **15**, 689–98.

Glazebrook, K.D. (1979a). On the optimal control of a class of continuous-time non-Markov decision processes, *International Journal of Systems Science*, **10**, 135–44.

Glazebrook, K.D. (1979b). Decomposable jump decision processes, *Stochastic Processes and their Applications*, **9**, 19–33.

Glazebrook, K.D. (1979c). Stoppable families of alternative bandit processes, *Journal of Applied Probability*, **16**, 843–54.

Glazebrook, K.D. (1980a). On the design of efficient experiments for choosing between two Bernoulli populations, *Communications in Statistics A*, **9**, 255–64. (With T.F. Cox)

Glazebrook, K.D. (1980b). On single-machine sequencing with order constraints, *Naval Research Logistics Quartery*, **27**, 123–30.

Glazebrook, K.D. (1980c). On stochastic scheduling with precedence relations and switching costs, *Journal of Applied Probability*, **17**, 1016–24.

Glazebrook, K.D. (1980d). On randomized dynamic allocation indices for the sequential design of experiments, *Journal of the Royal Statistical Society B*, **42**, 342–6.

Glazebrook, K.D. (1981a). On non-preemptive strategies in stochastic scheduling, *Naval Research Logistic Quarterly*, **28**, 289–300.

Glazebrook, K.D. (1981b). On non-preemptive strategies for stochastic scheduling problems in continuous time, *International Journal of Systems Science*, **12**, 771–82.

Glazebrook, K.D. (1982a). On a sufficient condition for superprocesses due to Whittle, *Journal of Applied Probability*, **19**, 99–110.

Glazebrook, K.D. (1982b). On the evaluation of non-preemptive strategies in stochastic scheduling, from *Deterministic and Stochastic Scheduling*, M.A.H. Dempster *et al.*, NATO Advanced Study Institute Series, Reidel, 375–84.

Glazebrook, K.D. (1982c). On the evaluation of strategies for stochastic scheduling problems with order constraints, *International Journal of Systems Science*, **13**, 349–58.

Glazebrook, K.D. (1982d). On the evaluation of suboptimal strategies for families of alternative bandit processes, *Journal of Applied Probability*, **19**, 716–22.

Glazebrook, K.D. (1982e). On the evaluation of fixed permutations as strategies in stochastic scheduling, *Stochastic Processes and their Applications*, **13**, 171–87.

Glazebrook, K.D. (1982f). Myopic strategies for Bayesian models in stochastic scheduling, *Opsearch*, **19**, 160–70.

Glazebrook, K.D. (1983a). On the role of dynamic allocation indices in the evaluation of suboptimal strategies for families of alternative bandit processes, proceedings of *Mathematical Learning Models – Theory and Algorithms, Springer Lecture Notes in Statistics*, Vol. 20, U. Herkenrath, D. Kalin, W. Vogel, 68–77.

Glazebrook, K.D. (1983b). Methods for the evaluation of permutations as strategies in stochastic scheduling, *Management Science*, **29**, 10, 1142–55.

Glazebrook, K.D. (1983c). Optimal strategies for families of alternative bandit processes, *IEEE Transactions on Automatic Control*, AC-**28**, 8, 858–61.

Glazebrook, K.D. (1983d). Some reward-penalty rules for the multi-armed bandit problem which are symptotically optimal, *Advances in Applied Probability*, **15**, 221–2.

Glazebrook, K.D. (1984). Scheduling stochastic jobs on a single machine subject to breakdowns, *Naval Research Logistics Quarterly*, **31**, 251–64.

Glazebrook, K.D. (1985a). Methods for evaluating strategies for families of alternative bandit processes, *Journal of Organizational Behaviour and Statistics*, **2**, 1, 1–18.

Glazebrook, K.D. (1985b). On semi-Markov models for single machine stochastic scheduling problems, *International Journal of Systems Science*, **16**, 573–87.

Glazebrook, K.D. (1987a). Evaluating the effects of machine breakdowns in stochastic scheduling problems, *Naval Research Logistics*, **34**, 319–35.

Glazebrook, K.D. (1987b). Sensitivity analysis for stochastic scheduling problems, *Mathematics for Operations Research*, **12**, 205–223.

Glazebrook, K.D. (to appear). Evaluating strategies for Markov decision processes in parallel, *Mathematics for Operations Research*.

Glazebrook, K.D. (1988). On a reduction principle in dynamic programming, *Adv. Appl. Prob.*, **20**, 4.

Glazebrook, K.D. and Fay, N. (1987). On the scheduling of alternative stochastic jobs on a single machine, *Adv. Appl. Prob.*, **19**, 955–73.

Glazebrook, K.D. and Fay, N. (1988). Evaluating strategies for generalised bandit problems, *International J. of Syst. Sci.*, **19**, 1605–13.

Glazebrook, K.D. and Gittins, J.C. (1981). On single-machine scheduling with precedence relations and linear or discounted costs, *Oper. Res.*, **29**, 289–300.

Glazebrook, K.D. and Jones, D.M. (1983). Some best possible results for a discounted one-armed bandit, *Metrika*, **30**, 109–15.

Henrici, P. (1964). *Elements of Numerical Analysis*, Wiley.

Howard, R. (1960). *Dynamic Programming and Markov Processes*. MIT Press.

Howard, R. (1971). *Dynamic Probabilistic Systems, Vol 2, Semi-Markov Decision Process*, Wiley.

Jones, D.M. (1970). *A Sequential Method for Industrial Chemical Research*, M.Sc. thesis, U.C.W. Aberystwyth.

Jones, D.M. (1975). *Search Procedures for Industrial Chemical Research*, Ph.D. thesis, Cambridge.

Kadane, J.B. and Simon, H.A. (1977). Optimal strategies for a class of constrained sequential problems, *Ann. Stats.*, **5**, 237–55.

Kallenberg, L.C.M. (1986). A note on Katehakis and Y.-R. Chen's computation of the Gittins index, *Math. OR*, **11**, 184–6.

Karatzas, I. (1984). Gittins indices in the dynamic allocation problem for diffusion processes, *Ann. Prob.*, **12**, 173–92.

Katehakis, M.N. and Derman, C. (1986). Computing optimal sequential allocation rules in Clinical Trials. *Institute of Math. Stats. Lecture Note Series: Adoptive Statistical Procedures and Related Topics*, **8**, 29–39.

Katehakis, M.N. and Veinott, A.F. (1987). The Multi-armed bandit problem: decomposition and computation *Math. OR*, **12**, 262–8.

Keener, R.W. (1986). Multi-armed bandits with simple arms, *Adv. Appl. Maths.*, **7**, 199–204.

Kelly, F.P. (1979). Contribution to the discussion of Gittins, *J. Roy. Statist. Soc. B*, **41**, 167–8.

Kelly, F.P. (1981). Multi-armed bandits with discount factor near one: the Bernoulli case, *Ann. Statist*, **9**, 987–1001.

Kendall, D.G. (1951). Some problems in the theory of queues, *J.R. Statist. Soc. B*, **13**, 151–85.

Kertz, R.P. (1986). Decision Processes under Total Expected Concomitant Constraints with Applications to Bandit Processes. Private communication.

Klimov, G.P. (1974). Time-sharing service systems I, *Theory Prob. & Appl.*, **19**, 532–51.

Klimov, G.P. (1978). Time-sharing service systems II, *Theory Prob. & Appl.*, **23**, 314–21.

Koopman, B. O. (1979). An operational critique of detection laws, *Operations Res.*, **27**, 115–133.

Kumar, P.R. (1985). A Survey of Some Results in Stochastic Adaptive Control, *SIAM J. Control and Optimization*, **23**, 329–80.

Lai, T.L. and Ying, Z. (1988). Open bandit processes and optimal scheduling of queuing networks, *Adv. Appl. Prob.* **20**, 447–72.

Lehnerdt, M. (1982). On the Structure of Discrete Sequential Search Problems and of their Solutions, *Math Operationsforschung u. Statist.*, *Ser. Optimization*, **13**, 523–57.

Lindley, D.V. (1960). Dynamic programming and decision theory, *Appl. Statist.*, **10**, 39–51.

Lippman, S.A. and McCall, J.J. (1981). The economics of belated information, *Internat. Econ. Rev.*, **22**, 135–46.

Mandelbaum, A. (1986). Discrete multi-armed bandits and multi-parameter processes. *Probability Theory & Related Fields*, **71**, 129–47.

Mandelbaum, A. (1987). Continuous multi-armed bandits and multi-parameter processes, *Ann. Probab.*, **15**, 1527–56.

McCall, B.P. and McCall, J.J. (1981). Systematic search, belated information, and the Gittins index, *Economics Letters*, **8**, 327–33.

Matula, D. (1964). A periodic optimal search, *Amer. Math. Monthly*, **71**, 15–21.

Miller, R.A. (1984). Job matching and occupational choice, *J. Political Econ.*, **92**, 1086–1120.

Nash, P. (1973). *Optimal Allocation of Resources Between Research Projects*, Ph.D. Thesis, Cambridge.

Nash, P. (1980). A generalised bandit problem, *J.R. Statist. Soc. B*, **42**, 165–9.

Nash, P. and Gittins, J.C. (1977). A Hamiltonian approach to optimal stochastic resource allocation, *Adv. Appl. Prob.*, **9**, 55–68.

von Neumann, J. and Morgenstern, O. (1928). *Theory of Games and Economic Behavior*, Princeton University Press.

Norris, R.C. (1962). *Studies in Search for a Conscious Evader*, MIT Lincoln Lab. Tech. Rpt No. 279.

Olivier, von G. (1972). Cost-Minimum Priorities in Queueing Systems of Type M/G/1, *Elektron. Rechenanl.*, **14**, 262–71.

Papadimitriou, C.H. and Steiglitz, K. (1982). *Combinatorial Optimization: Algorithms and Complexity*, Prentice-Hall.

Prabhu, N.U. (1965). *Queues and Inventories*, Wiley.

Presman, E.L. and Sonin, I.M. (1982). *Sequential Control with Incomplete Information: The Bayesian Approach*, Nauka, Moscow.

Raiffa, H. and Schlaifer, R. (1961). *Applied Statistical Decision Theory*, Harvard Business School.

Roberts, D.M. and Gittins, J.C. (1978). The search for an intelligent evader: strategies for searcher and evader in the two-region problem, *Naval Research Logistics Quarterly*, **25**, 95–106.

Robinson, D.R. (1982). Algorithms for evaluating the dynamic allocation index, *Op. Res. Letters*, **1**, 72–4.

Ross, S.M. (1970). *Applied Probability Models with Optimization Applications*, Holden-Day.

Rothkopt, M.H. (1966). Scheduling with random service times, *Management Sci.*, **12**, 707–13.

Sidney, J.B. (1975). Decomposition algorithms for single-machine scheduling with precedence relations and deferral costs, *Operations Res.*, **23**, 283–93.

Stone, L.D. (1975). *Theory of Optimal Search*, Academic Press.

Strümpfer, J. (1980). *Search Theory Index*, TN-017-80, Institute for Maritime Technology, Simonstown, S. Africa.

Teugels, J.L. (1976). A bibliography on semi-Markov processes. *J. Comp. Appl. Maths.*, **2**, 125–44.

Thron, C. (1984). *The Multi-Armed Bandit Problem and Optimality of the Gittins Index Strategy*, private communication.

Tsitsiklis, J.N. (1986). A lemma on the MAB problem, *IEEE Trans. on Automatic Control*, **31**, 576–77.

Varaiya, P., Walrand, J.C. and Buyukkoc, C. (1985). Extensions of the multiarmed bandit problem: the discounted case, *IEEE Trans. Automat. Contr.*, AC-**30**, 426–39.

Wald, A. (1950). *Statistical Decision Functions*, Wiley.

Weber, R.R. (1982). Scheduling jobs with stochastic processing requirement on parallel machines to minimise makespan or flowtime, *J. Appl. Prob.*, **19**, 167–82.

Weitzman, M.L. (1979). Optimal search for the best alternative *Econometrica*, **47**, 641–54.

Whittle, P. (1980). Multi-armed bandits and the Gittins index. *J. Roy. Statist. Soc. Ser. B.*, **42**, 143–9.

Whittle, P. (1981). Arm-acquiring bandits, *Ann. Prob.*, **9**, 284–92.

Whittle, P. (1983). *Optimization over Time*, Wiley, Volume 1, 1982, Volume 2, 1983.

Whittle, P. (1988). Restless bandits: activity allocation in changing world. *Journal of Appl. Prob.*

Table 1. Normal Reward Process (Known Variance)
Values of $n(1-a)^{\frac{1}{2}}\nu(0,n,1,a)$, (§7.2)

n \ a	.5	.6	.7	.8	.9	.95	.99	.995
1	.14542	.17451	.20218	.22582	.23609	.22263	.15758	.12852
2	.17209	.20815	.24359	.27584	.29485	.28366	.20830	.17192
3	.18522	.22513	.26515	.30297	.32876	.32072	.24184	.20137
4	.19317	.23560	.27874	.32059	.35179	.34687	.26709	.22398
5	.19855	.24277	.28820	.33314	.36879	.36678	.28736	.24242
6	.20244	.24801	.29521	.34261	.38200	.38267	.30429	.25803
7	.20539	.25202	.30063	.35005	.39265	.39577	.31881	.27158
8	.20771	.25520	.30496	.35607	.40146	.40682	.33149	.28356
9	.20959	.25777	.30851	.36105	.40889	.41631	.34275	.29428
10	.21113	.25991	.31147	.36525	.41526	.42458	.35285	.30400
20	.21867	.27048	.32642	.38715	.45047	.47295	.41888	.36986
30	.22142	.27443	.33215	.39593	.46577	.49583	.45587	.40886
40	.22286	.27650	.33520	.40070	.47448	.50953	.48072	.43613
50	.22374	.27778	.33709	.40370	.48013	.51876	.49898	.45679
60	.22433	.27864	.33838	.40577	.48411	.52543	.51313	.47324
70	.22476	.27927	.33932	.40728	.48707	.53050	.52451	.48677
80	.22508	.27974	.34003	.40843	.48935	.53449	.53391	.49817
90	.22534	.28011	.34059	.40934	.49117	.53771	.54184	.50796
100	.22554	.28041	.34104	.41008	.49266	.54037	.54864	.51648
200	.22646	.28177	.34311	.41348	.49970	.55344	.58626	.56637
300	.22678	.28223	.34381	.41466	.50219	.55829	.60270	.59006
400	.22693	.28246	.34416	.41525	.50347	.56084	.61220	.60436
500	.22703	.28260	.34438	.41561	.50425	.56242	.61844	.61410
600	.22709	.28270	.34452	.41585	.50478	.56351	.62290	.62123
700	.22714	.28276	.34462	.41602	.50516	.56431	.62629	.62674
800	.22717	.28281	.34470	.41615	.50545	.56493	.62896	.63116
900	.22720	.28285	.34476	.41625	.50568	.56543	.63121	.63481
1000	.22722	.28288	.34480	.41633	.50587	.56583	.63308	.63789
∞	.22741	.28316	.34524	.41714	.5092	.583		

Table 1. Normal Reward Process (Known Variance)
Values of $n(1-a)^{\frac{1}{2}}\nu(0,n,1,a)$, (§7.2) *(continued)*

	$a = .95$				$a = .99$		
$n/100$		$n/100$		$n/100$		$n/100$	
10	.56583	10	.63308	37	.65345	64	.66348
11	.56618	11	.63468	38	.65390	65	.66379
12	.56648	12	.63611	39	.65435	66	.66409
13	.56674	13	.63737	40	.65478	67	.66440
14	.56698	14	.63852	41	.65521	68	.66470
15	.56719	15	.63956	42	.65562	69	.66499
16	.56738	16	.64054	43	.65604	70	.66529
17	.56756	17	.64144	44	.65644	71	.66558
18	.56772	18	.64229	45	.65684	72	.66586
19	.56787	19	.64309	46	.65724	73	.66615
20	.56802	20	.64385	47	.65763	74	.66643
21	.56815	21	.64458	48	.65801	75	.66671
22	.56828	22	.64527	49	.65838	76	.66698
23	.56840	23	.64594	50	.65875	77	.66726
24	.56851	24	.64658	51	.65912	78	.66753
25	.56862	25	.64720	52	.65948	79	.66780
26	.56872	26	.64780	53	.65984	80	.66806
27	.56882	27	.64838	54	.66019	81	.66833
28	.56892	28	.64895	55	.66053	82	.66859
29	.56901	29	.64949	56	.66088	83	.66885
30	.56910	30	.65003	57	.66122	84	.66911
31	.56918	31	.65055	58	.66155	85	.66936
32	.56926	32	.65106	59	.66188	86	.66962
33	.56934	33	.65156	60	.66221	87	.66987
34	.56942	34	.65205	61	.66253	88	.67012
35	.56950	35	.65253	62	.66285	89	.67036
36	.56957	36	.65299	63	.66317	90	.67061

$a = .995$

$n/100$		$n/100$	
10	.63789	16	.65020
11	.64055	17	.65167
12	.64288	18	.65304
13	.64492	19	.65433
14	.64689	20	.65554
15	.64861		

Table 2. Brownian Reward Process
Lower Bound for $Tu(T)$, (§§6.6 & 7.2)

T	$Tu(T) >$	T	$Tu(T) >$
0.005	0.12852	1	0.56637
0.01	0.17192	2	0.60436
0.02	0.22398	5	0.63789
0.05	0.30400	10	0.65554
0.1	0.36986	20	0.64385
0.2	0.43617	40	0.65478
0.5	0.51648	80	0.66806

Table 3. Normal Reward Process, Ratio of Indices for Cases of Unknown Variance and Known Variance (§7.3)

Values shown are for $\zeta\xi^{-1} - 1$ for $n = 2,3,4$

a / n	.5	.6	.7	.8	.9	.95	.99	.995
2	4.971	5.365	5.995	8.132	10.088	14.988	36.767	52.950
3	.789	.810	.847	.923	1.123	1.438	2.848	3.851
4	.389	.396	.408	.432	.496	.597	1.011	1.289

Values shown are for $100(\zeta\xi^{-1} - 1)$ for $n \geq 5$

	.5	.6	.7	.8	.9	.95	.99	.995
5	25.681	26.003	26.601	27.842	31.228	36.487	57.509	71.073
6	19.012	19.212	19.585	20.365	22.512	25.865	39.095	47.376
7	15.165	15.303	15.563	16.110	17.635	20.039	29.567	35.483
8	12.612	12.714	12.907	13.316	14.470	16.308	23.643	28.192
9	10.797	10.882	11.025	11.345	12.257	13.722	19.624	23.292
10	9.440	9.502	9.622	9.881	10.624	11.829	16.728	19.784
20	4.188	4.203	4.233	4.301	4.507	4.868	6.471	7.522
30	2.689	2.696	2.710	2.741	2.840	3.021	3.886	4.479
40	1.980	1.984	1.992	2.010	2.068	2.179	2.738	3.136
50	1.566	1.569	1.574	1.586	1.625	1.700	2.098	2.390
60	1.296	1.298	1.301	1.310	1.337	1.392	1.693	1.919
70	1.106	1.106	1.109	1.115	1.136	1.178	1.414	1.597
80	.963	.964	.966	.972	.987	1.020	1.212	1.364
90	.854	.854	.856	.860	.873	.899	1.059	1.188
100	.766	.767	.768	.771	.782	.804	.939	1.050
200	.380	.380	.379	.380	.383	.390	.435	.476
300	.253	.251	.252	.252	.254	.257	.286	.305
400	.188	.188	.189	.189	.190	.192	.207	.224
500	.151	.151	.151	.151	.152	.153	.165	.178
600	.126	.126	.125	.126	.126	.127	.137	.149
700	.108	.107	.108	.108	.108	.109	.119	.129
800	.094	.095	.094	.095	.095	.096	.110	.114
900	.083	.084	.084	.084	.084	.085	.093	.102
1000	.074	.076	.076	.076	.076	.078	.084	.094

Table 4. Bernoulli Reward Process, Ranges of Explicit Tabulation & Accuracy of Approximation (§7.4)

a	Tabulated Range	Accuracy in Untabulated Range
0.5, 0.6, 0.7	$(\alpha, \beta) \leq 20$	4-figure accuracy for $0.025 \leq \alpha/(\alpha + \beta) \leq 0.975$ endpoint errors at most -0.0001
0.8	$(\alpha, \beta) \leq 40$	4-figure accuracy for $0.025 \leq \alpha/(\alpha + \beta) \leq 0.975$ endpoint errors at most -0.0002
0.9	$(\alpha, \beta) \leq 40$	4-figure accuracy for $0.025 \leq \alpha/(\alpha + \beta) \leq 0.975$ endpoint errors at most -0.0003
0.95	$(\alpha, \beta) \leq 40$	errors up to 0.0002 for $40 \leq \alpha + \beta \leq 50$ and $0.025 \leq \alpha/(\alpha + \beta) \leq 0.975$ 4-figure accuracy for $\alpha + \beta > 50$ and $0.025 \leq \alpha/(\alpha + \beta) \leq 0.975$ endpoint errors at most -0.0005
0.99	$(\alpha, \beta) \leq 40$	errors up to 0.0002 for $40 \leq \alpha + \beta \leq 70$ and $0.025 \leq \alpha/(\alpha + \beta) \leq 0.975$ 4-figure accuracy for $\alpha + \beta > 70$ and $0.025 \leq \alpha/(\alpha + \beta) \leq 0.975$ endpoint errors at most -0.0009

Endpoint errors were nearly all negative. For that reason the bounds quoted are expressed as negative numbers.

Table 5. Bernoulli Reward Process
Index Values, $a = 0.5$

α / β	1	2	3	4	5	6	7	8	9	10
1	.5590	.7060	.7772	.8199	.8485	.8691	.8847	.8969	.9067	.9148
2	.3758	.5359	.6289	.6899	.7333	.7658	.7911	.8114	.8280	.8419
3	.2802	.4298	.5258	.5937	.6441	.6832	.7144	.7399	.7611	.7791
4	.2223	.3577	.4512	.5201	.5736	.6161	.6507	.6795	.7038	.7247
5	.1837	.3058	.3947	.4626	.5165	.5606	.5971	.6279	.6543	.6771
6	.1563	.2668	.3504	.4163	.4697	.5140	.5515	.5835	.6111	.6353
7	.1358	.2364	.3149	.3783	.4306	.4745	.5121	.5447	.5732	.5982
8	.1200	.2122	.2859	.3465	.3974	.4407	.4781	.5107	.5396	.5652
9	.1074	.1923	.2616	.3196	.3688	.4113	.4482	.4807	.5096	.5355
10	.0972	.1758	.2411	.2965	.3440	.3855	.4218	.4541	.4828	.5087
11	.0888	.1619	.2236	.2764	.3223	.3627	.3983	.4302	.4587	.4845
12	.0816	.1500	.2084	.2589	.3032	.3423	.3773	.4086	.4369	.4625
13	.0756	.1397	.1951	.2434	.2862	.3242	.3583	.3891	.4170	.4424
14	.0703	.1307	.1833	.2297	.2709	.3078	.3411	.3713	.3988	.4240
15	.0657	.1228	.1729	.2174	.2572	.2931	.3255	.3551	.3821	.4070
16	.0617	.1158	.1636	.2064	.2448	.2796	.3113	.3402	.3668	.3913
17	.0582	.1095	.1553	.1964	.2335	.2673	.2982	.3265	.3526	.3767
18	.0550	.1039	.1477	.1873	.2232	.2561	.2862	.3139	.3395	.3632
19	.0522	.0988	.1409	.1790	.2138	.2457	.2751	.3022	.3273	.3506
20	.0496	.0942	.1346	.1714	.2051	.2362	.2648	.2913	.3159	.3389

Table 5. Bernoulli Reward Process
Index Values, $a = 0.5$ (continued)

α / β	11	12	13	14	15	16	17	18	19	20
1	.9216	.9274	.9324	.9367	.9405	.9439	.9469	.9496	.9520	.9542
2	.8537	.8638	.8726	.8804	.8872	.8933	.8988	.9037	.9081	.9122
3	.7946	.8080	.8197	.8301	.8393	.8476	.8550	.8618	.8679	.8736
4	.7428	.7586	.7726	.7850	.7961	.8061	.8152	.8235	.8310	.8379
5	.6971	.7147	.7304	.7444	.7570	.7684	.7788	.7883	.7970	.8050
6	.6566	.6755	.6925	.7077	.7215	.7340	.7454	.7559	.7656	.7745
7	.6205	.6403	.6582	.6743	.6890	.7024	.7147	.7260	.7365	.7461
8	.5880	.6085	.6271	.6439	.6593	.6734	.6864	.6984	.7095	.7198
9	.5587	.5797	.5987	.6161	.6321	.6467	.6602	.6727	.6843	.6952
10	.5321	.5534	.5728	.5906	.6069	.6220	.6359	.6488	.6609	.6721
11	.5079	.5294	.5490	.5670	.5837	.5990	.6133	.6266	.6390	.6506
12	.4859	.5073	.5270	.5453	.5621	.5777	.5922	.6058	.6185	.6304
13	.4657	.4870	.5068	.5251	.5420	.5578	.5726	.5863	.5992	.6113
14	.4470	.4683	.4880	.5063	.5233	.5393	.5541	.5680	.5811	.5934
15	.4298	.4510	.4706	.4888	.5059	.5219	.5368	.5509	.5641	.5765
16	.4139	.4349	.4544	.4726	.4896	.5055	.5205	.5347	.5480	.5605
17	.3991	.4199	.4392	.4573	.4743	.4902	.5052	.5194	.5327	.5454
18	.3853	.4058	.4251	.4431	.4599	.4758	.4908	.5049	.5183	.5310
19	.3724	.3927	.4118	.4296	.4464	.4622	.4772	.4913	.5047	.5174
20	.3603	.3804	.3992	.4170	.4337	.4494	.4643	.4784	.4917	.5045

Table 6. Bernoulli Reward Process
Index Values, $a = 0.6$

α β	1	2	3	4	5	6	7	8	9	10
1	.5788	.7189	.7864	.8268	.8539	.8735	.8883	.8999	.9093	.9171
2	.3911	.5483	.6388	.6980	.7400	.7714	.7959	.8156	.8317	.8451
3	.2916	.4403	.5348	.6015	.6508	.6890	.7195	.7444	.7652	.7828
4	.2310	.3666	.4591	.5272	.5800	.6218	.6558	.6841	.7080	.7285
5	.1906	.3133	.4018	.4690	.5224	.5660	.6020	.6324	.6584	.6809
6	.1618	.2732	.3567	.4222	.4751	.5190	.5561	.5878	.6151	.6390
7	.1404	.2419	.3205	.3837	.4356	.4792	.5165	.5488	.5770	.6018
8	.1238	.2170	.2908	.3513	.4020	.4451	.4821	.5145	.5432	.5686
9	.1107	.1965	.2661	.3240	.3731	.4154	.4521	.4844	.5130	.5388
10	.1000	.1796	.2451	.3005	.3480	.3893	.4255	.4575	.4861	.5118
11	.0912	.1652	.2272	.2801	.3260	.3662	.4018	.4335	.4619	.4875
12	.0837	.1530	.2117	.2623	.3066	.3457	.3805	.4117	.4399	.4654
13	.0774	.1424	.1981	.2466	.2893	.3273	.3614	.3921	.4199	.4452
14	.0720	.1331	.1861	.2326	.2739	.3108	.3440	.3741	.4016	.4266
15	.0672	.1250	.1755	.2201	.2600	.2958	.3282	.3578	.3847	.4095
16	.0631	.1178	.1660	.2089	.2474	.2822	.3138	.3427	.3693	.3937
17	.0594	.1114	.1575	.1987	.2360	.2698	.3006	.3289	.3550	.3791
18	.0561	.1056	.1498	.1895	.2255	.2584	.2885	.3162	.3417	.3654
19	.0532	.1004	.1428	.1811	.2160	.2479	.2773	.3044	.3294	.3528
20	.0505	.0957	.1364	.1734	.2072	.2382	.2669	.2934	.3180	.3409

Table 6. Bernoulli Reward Process
Index Values, $a = 0.6$ (continued)

α β	11	12	13	14	15	16	17	18	19	20
1	.9236	.9291	.9339	.9381	.9417	.9450	.9479	.9505	.9529	.9550
2	.8566	.8664	.8750	.8825	.8891	.8951	.9004	.9052	.9095	.9135
3	.7979	.8110	.8225	.8326	.8416	.8497	.8570	.8636	.8697	.8752
4	.7462	.7618	.7755	.7877	.7986	.8085	.8174	.8255	.8329	.8397
5	.7006	.7180	.7334	.7472	.7596	.7709	.7811	.7904	.7990	.8069
6	.6600	.6787	.6955	.7105	.7241	.7365	.7478	.7581	.7677	.7765
7	.6238	.6435	.6611	.6771	.6917	.7049	.7171	.7283	.7386	.7482
8	.5913	.6116	.6300	.6467	.6619	.6759	.6888	.7006	.7116	.7218
9	.5618	.5827	.6016	.6188	.6346	.6491	.6625	.6749	.6864	.6972
10	.5351	.5563	.5756	.5932	.6094	.6244	.6382	.6510	.6630	.6741
11	.5108	.5321	.5517	.5696	.5861	.6014	.6156	.6288	.6411	.6526
12	.4886	.5099	.5296	.5477	.5645	.5800	.5944	.6079	.6205	.6323
13	.4683	.4896	.5092	.5274	.5443	.5601	.5747	.5884	.6012	.6133
14	.4496	.4708	.4904	.5086	.5256	.5414	.5562	.5701	.5831	.5953
15	.4323	.4534	.4729	.4910	.5080	.5239	.5388	.5528	.5660	.5784
16	.4163	.4372	.4566	.4747	.4916	.5075	.5225	.5366	.5498	.5623
17	.4014	.4221	.4414	.4594	.4763	.4922	.5071	.5212	.5345	.5471
18	.3875	.4080	.4271	.4451	.4619	.4777	.4926	.5067	.5201	.5327
19	.3745	.3948	.4138	.4316	.4483	.4641	.4790	.4930	.5064	.5191
20	.3623	.3824	.4012	.4189	.4355	.4512	.4661	.4801	.4934	.5061

Table 7. Bernoulli Reward Process
Index Values, $a = 0.7$

α / β	1	2	3	4	5	6	7	8	9	10
1	.6046	.7358	.7985	.8359	.8612	.8794	.8932	.9041	.9129	.9202
2	.4118	.5650	.6520	.7088	.7489	.7790	.8024	.8213	.8367	.8496
3	.3075	.4546	.5472	.6121	.6599	.6970	.7265	.7506	.7708	.7878
4	.2434	.3789	.4701	.5370	.5887	.6296	.6627	.6904	.7137	.7337
5	.2005	.3237	.4116	.4779	.5305	.5734	.6088	.6386	.6640	.6861
6	.1699	.2822	.3654	.4304	.4825	.5259	.5626	.5938	.6207	.6441
7	.1471	.2497	.3282	.3911	.4426	.4856	.5225	.5545	.5824	.6068
8	.1295	.2238	.2978	.3581	.4085	.4511	.4878	.5199	.5483	.5734
9	.1155	.2025	.2724	.3301	.3791	.4211	.4575	.4894	.5179	.5434
10	.1042	.1849	.2508	.3062	.3535	.3946	.4306	.4624	.4906	.5162
11	.0948	.1700	.2324	.2854	.3311	.3712	.4066	.4381	.4663	.4916
12	.0870	.1573	.2164	.2671	.3114	.3504	.3851	.4161	.4441	.4694
13	.0803	.1463	.2024	.2510	.2938	.3317	.3657	.3962	.4239	.4490
14	.0745	.1367	.1901	.2368	.2781	.3149	.3481	.3781	.4054	.4303
15	.0695	.1283	.1792	.2240	.2639	.2997	.3320	.3615	.3884	.4131
16	.0651	.1208	.1694	.2125	.2510	.2859	.3175	.3463	.3728	.3971
17	.0613	.1141	.1606	.2021	.2394	.2733	.3041	.3323	.3583	.3823
18	.0578	.1081	.1527	.1927	.2288	.2617	.2918	.3194	.3449	.3686
19	.0547	.1027	.1455	.1841	.2191	.2510	.2804	.3075	.3325	.3558
20	.0519	.0978	.1390	.1762	.2101	.2412	.2699	.2964	.3209	.3438

Table 7. Bernoulli Reward Process
Index Values, $a = 0.7$ (continued)

α / β	11	12	13	14	15	16	17	18	19	20
1	.9263	.9316	.9361	.9400	.9435	.9466	.9494	.9519	.9541	.9562
2	.8606	.8700	.8782	.8855	.8919	.8976	.9027	.9073	.9115	.9154
3	.8024	.8151	.8263	.8361	.8449	.8527	.8598	.8662	.8721	.8774
4	.7510	.7662	.7796	.7915	.8021	.8117	.8204	.8284	.8356	.8423
5	.7054	.7224	.7375	.7510	.7632	.7743	.7843	.7935	.8019	.8096
6	.6647	.6832	.6996	.7144	.7278	.7399	.7510	.7612	.7706	.7792
7	.6285	.6478	.6652	.6810	.6953	.7084	.7204	.7314	.7416	.7510
8	.5958	.6159	.6340	.6505	.6655	.6793	.6920	.7037	.7146	.7246
9	.5662	.5868	.6055	.6226	.6382	.6525	.6657	.6780	.6894	.7000
10	.5393	.5603	.5794	.5969	.6129	.6277	.6414	.6541	.6659	.6769
11	.5148	.5360	.5554	.5731	.5895	.6047	.6187	.6318	.6439	.6553
12	.4924	.5136	.5332	.5512	.5678	.5832	.5975	.6109	.6234	.6351
13	.4720	.4931	.5126	.5308	.5476	.5632	.5777	.5913	.6040	.6160
14	.4532	.4742	.4936	.5118	.5287	.5444	.5591	.5729	.5858	.5979
15	.4358	.4567	.4761	.4941	.5110	.5269	.5417	.5556	.5686	.5809
16	.4196	.4404	.4597	.4777	.4945	.5104	.5253	.5393	.5524	.5649
17	.4046	.4252	.4444	.4624	.4791	.4949	.5098	.5238	.5371	.5496
18	.3905	.4110	.4301	.4479	.4647	.4804	.4952	.5092	.5226	.5352
19	.3775	.3977	.4166	.4343	.4510	.4667	.4815	.4955	.5088	.5214
20	.3652	.3852	.4039	.4216	.4381	.4538	.4685	.4825	.4957	.5083

Table 8. Bernoulli Reward Process
Index Values, $a = 0.8$

α / β	1	2	3	4	5	6	7	8	9	10
1	.6413	.7596	.8157	.8492	.8718	.8881	.9006	.9104	.9184	.9251
2	.4430	.5898	.6715	.7248	.7623	.7904	.8123	.8299	.8444	.8565
3	.3320	.4761	.5659	.6280	.6735	.7090	.7372	.7601	.7793	.7955
4	.2629	.3977	.4866	.5521	.6020	.6414	.6733	.7000	.7225	.7417
5	.2163	.3398	.4267	.4914	.5431	.5848	.6192	.6481	.6727	.6942
6	.1830	.2963	.3790	.4430	.4940	.5367	.5726	.6030	.6292	.6520
7	.1581	.2620	.3403	.4026	.4534	.4955	.5320	.5634	.5906	.6146
8	.1389	.2347	.3087	.3687	.4186	.4606	.4965	.5284	.5563	.5809
9	.1236	.2122	.2823	.3398	.3885	.4300	.4659	.4972	.5255	.5506
10	.1113	.1936	.2598	.3151	.3622	.4030	.4386	.4700	.4978	.5231
11	.1011	.1778	.2406	.2936	.3391	.3791	.4142	.4453	.4731	.4981
12	.0925	.1643	.2239	.2748	.3189	.3578	.3923	.4230	.4507	.4757
13	.0852	.1527	.2094	.2581	.3009	.3386	.3725	.4028	.4302	.4552
14	.0790	.1426	.1965	.2434	.2847	.3214	.3545	.3844	.4115	.4362
15	.0736	.1337	.1851	.2302	.2701	.3059	.3381	.3675	.3942	.4187
16	.0688	.1258	.1749	.2183	.2569	.2918	.3232	.3520	.3783	.4025
17	.0646	.1187	.1658	.2075	.2449	.2788	.3096	.3377	.3637	.3875
18	.0609	.1124	.1575	.1978	.2340	.2670	.2970	.3246	.3500	.3736
19	.0575	.1067	.1500	.1889	.2240	.2560	.2854	.3124	.3373	.3606
20	.0545	.1015	.1432	.1807	.2148	.2459	.2746	.3011	.3256	.3484
21	.0518	.0969	.1370	.1732	.2063	.2366	.2646	.2906	.3146	.3370
22	.0494	.0926	.1312	.1663	.1984	.2280	.2553	.2807	.3044	.3264
23	.0471	.0886	.1259	.1599	.1911	.2199	.2466	.2715	.2947	.3164
24	.0451	.0850	.1211	.1540	.1843	.2124	.2385	.2629	.2857	.3070
25	.0432	.0817	.1165	.1485	.1780	.2054	.2309	.2548	.2771	.2981
26	.0415	.0786	.1123	.1434	.1721	.1988	.2238	.2471	.2691	.2898
27	.0399	.0757	.1084	.1386	.1665	.1926	.2171	.2399	.2615	.2818
28	.0384	.0730	.1047	.1341	.1613	.1868	.2107	.2332	.2543	.2743
29	.0370	.0705	.1013	.1299	.1565	.1814	.2047	.2267	.2475	.2672
30	.0357	.0682	.0981	.1259	.1518	.1762	.1991	.2207	.2411	.2604
31	.0345	.0660	.0951	.1222	.1475	.1713	.1937	.2149	.2349	.2539
32	.0334	.0640	.0923	.1186	.1434	.1667	.1886	.2094	.2291	.2478
33	.0323	.0620	.0896	.1153	.1395	.1623	.1838	.2042	.2236	.2419
34	.0313	.0602	.0870	.1122	.1358	.1581	.1792	.1992	.2183	.2364
35	.0304	.0585	.0846	.1092	.1323	.1542	.1749	.1945	.2132	.2310
36	.0295	.0569	.0824	.1064	.1290	.1504	.1707	.1900	.2084	.2259
37	.0287	.0553	.0802	.1037	.1258	.1468	.1667	.1857	.2038	.2211
38	.0279	.0539	.0782	.1011	.1228	.1434	.1629	.1816	.1994	.2164
39	.0272	.0525	.0762	.0987	.1199	.1401	.1593	.1776	.1951	.2119
40	.0265	.0512	.0744	.0964	.1172	.1370	.1558	.1739	.1911	.2076

TABLES

Table 8. Bernoulli Reward Process
Index Values, $a = 0.8$ (continued)

α / β	11	12	13	14	15	16	17	18	19	20
1	.9306	.9354	.9396	.9432	.9464	.9492	.9518	.9541	.9562	.9581
2	.8668	.8756	.8833	.8901	.8962	.9015	.9064	.9108	.9147	.9183
3	.8094	.8215	.8321	.8415	.8499	.8574	.8642	.8703	.8759	.8811
4	.7584	.7730	.7859	.7973	.8076	.8168	.8252	.8329	.8398	.8463
5	.7129	.7294	.7440	.7571	.7689	.7796	.7893	.7982	.8063	.8138
6	.6721	.6901	.7061	.7205	.7335	.7453	.7561	.7661	.7752	.7836
7	.6358	.6546	.6716	.6871	.7011	.7138	.7255	.7363	.7462	.7554
8	.6029	.6226	.6404	.6565	.6712	.6847	.6972	.7087	.7193	.7291
9	.5731	.5934	.6117	.6285	.6438	.6578	.6708	.6829	.6941	.7045
10	.5460	.5666	.5854	.6026	.6184	.6330	.6464	.6589	.6705	.6814
11	.5212	.5421	.5612	.5787	.5949	.6098	.6236	.6365	.6485	.6597
12	.4984	.5195	.5388	.5566	.5730	.5882	.6024	.6155	.6279	.6394
13	.4779	.4986	.5181	.5360	.5527	.5681	.5824	.5959	.6084	.6202
14	.4588	.4797	.4988	.5169	.5336	.5492	.5638	.5774	.5901	.6021
15	.4412	.4620	.4812	.4990	.5158	.5315	.5462	.5600	.5729	.5851
16	.4249	.4455	.4647	.4825	.4991	.5149	.5296	.5435	.5566	.5689
17	.4096	.4301	.4492	.4670	.4836	.4992	.5140	.5280	.5411	.5536
18	.3954	.4158	.4347	.4524	.4690	.4846	.4993	.5133	.5265	.5390
19	.3822	.4023	.4211	.4387	.4553	.4709	.4855	.4994	.5126	.5252
20	.3697	.3897	.4083	.4258	.4423	.4578	.4725	.4863	.4994	.5120
21	.3581	.3778	.3962	.4136	.4300	.4454	.4601	.4739	.4870	.4995
22	.3471	.3666	.3849	.4021	.4183	.4337	.4483	.4621	.4752	.4877
23	.3368	.3560	.3741	.3912	.4073	.4226	.4371	.4508	.4639	.4764
24	.3271	.3460	.3639	.3808	.3968	.4120	.4264	.4401	.4531	.4656
25	.3179	.3365	.3542	.3710	.3868	.4019	.4162	.4298	.4428	.4552
26	.3092	.3276	.3451	.3616	.3773	.3923	.4065	.4201	.4330	.4453
27	.3010	.3192	.3363	.3527	.3683	.3831	.3972	.4107	.4236	.4359
28	.2932	.3111	.3281	.3443	.3597	.3744	.3884	.4017	.4145	.4268
29	.2858	.3034	.3202	.3362	.3514	.3660	.3799	.3932	.4059	.4180
30	.2787	.2961	.3127	.3285	.3436	.3580	.3717	.3849	.3975	.4097
31	.2720	.2892	.3055	.3211	.3360	.3503	.3640	.3770	.3896	.4016
32	.2656	.2825	.2987	.3141	.3288	.3429	.3565	.3695	.3819	.3938
33	.2595	.2762	.2921	.3074	.3219	.3359	.3493	.3622	.3745	.3864
34	.2536	.2701	.2858	.3009	.3153	.3291	.3424	.3551	.3674	.3792
35	.2480	.2643	.2798	.2947	.3090	.3226	.3357	.3484	.3606	.3723
36	.2427	.2587	.2740	.2887	.3028	.3164	.3294	.3419	.3540	.3656
37	.2376	.2533	.2685	.2830	.2970	.3104	.3232	.3356	.3476	.3591
38	.2326	.2482	.2632	.2775	.2913	.3046	.3173	.3296	.3415	.3529
39	.2279	.2433	.2580	.2722	.2859	.2990	.3116	.3238	.3355	.3469
40	.2234	.2386	.2531	.2671	.2806	.2936	.3061	.3182	.3298	.3411

Table 8. Bernoulli Reward Process
Index Values, $a = 0.8$ (continued)

α / β	21	22	23	24	25	26	27	28	29	30
1	.9598	.9614	.9629	.9643	.9655	.9667	.9678	.9688	.9698	.9707
2	.9217	.9247	.9275	.9302	.9326	.9349	.9370	.9389	.9408	.9425
3	.8858	.8901	.8941	.8979	.9013	.9046	.9076	.9104	.9131	.9156
4	.8522	.8576	.8627	.8674	.8718	.8759	.8798	.8834	.8868	.8900
5	.8208	.8272	.8332	.8387	.8439	.8488	.8534	.8577	.8617	.8655
6	.7914	.7987	.8054	.8117	.8176	.8232	.8284	.8333	.8379	.8423
7	.7640	.7719	.7794	.7863	.7928	.7990	.8047	.8102	.8153	.8202
8	.7383	.7468	.7548	.7624	.7694	.7761	.7823	.7883	.7939	.7992
9	.7142	.7233	.7318	.7398	.7473	.7544	.7611	.7674	.7734	.7791
10	.6915	.7011	.7100	.7184	.7263	.7338	.7409	.7476	.7540	.7600
11	.6702	.6801	.6894	.6982	.7065	.7143	.7218	.7288	.7355	.7418
12	.6502	.6604	.6700	.6791	.6877	.6958	.7035	.7108	.7178	.7245
13	.6313	.6418	.6516	.6609	.6697	.6782	.6861	.6937	.7010	.7078
14	.6134	.6241	.6342	.6438	.6528	.6614	.6696	.6774	.6848	.6919
15	.5965	.6074	.6177	.6274	.6367	.6454	.6538	.6618	.6694	.6767
16	.5805	.5915	.6019	.6119	.6213	.6302	.6388	.6469	.6547	.6621
17	.5653	.5764	.5870	.5970	.6066	.6157	.6244	.6327	.6406	.6482
18	.5508	.5621	.5727	.5829	.5926	.6018	.6106	.6190	.6271	.6348
19	.5371	.5484	.5591	.5694	.5792	.5885	.5974	.6059	.6141	.6219
20	.5240	.5353	.5462	.5565	.5663	.5757	.5847	.5934	.6016	.6096
21	.5115	.5229	.5338	.5441	.5540	.5635	.5726	.5813	.5897	.5977
22	.4995	.5110	.5219	.5323	.5423	.5518	.5610	.5697	.5782	.5862
23	.4882	.4996	.5105	.5210	.5310	.5406	.5498	.5586	.5671	.5752
24	.4774	.4888	.4996	.5101	.5201	.5297	.5390	.5479	.5564	.5646
25	.4671	.4784	.4892	.4996	.5097	.5193	.5286	.5375	.5461	.5544
26	.4572	.4685	.4793	.4897	.4996	.5093	.5186	.5276	.5362	.5445
27	.4476	.4589	.4697	.4801	.4901	.4997	.5090	.5180	.5266	.5349
28	.4385	.4498	.4606	.4709	.4809	.4905	.4997	.5087	.5173	.5257
29	.4297	.4409	.4517	.4621	.4720	.4816	.4908	.4997	.5084	.5167
30	.4213	.4324	.4432	.4535	.4634	.4730	.4822	.4911	.4997	.5081
31	.4132	.4243	.4350	.4453	.4552	.4647	.4740	.4829	.4914	.4998
32	.4053	.4164	.4270	.4373	.4472	.4567	.4659	.4748	.4834	.4917
33	.3978	.4088	.4194	.4296	.4395	.4490	.4582	.4671	.4756	.4840
34	.3906	.4015	.4120	.4222	.4320	.4415	.4507	.4595	.4681	.4764
35	.3836	.3944	.4049	.4150	.4248	.4343	.4434	.4522	.4608	.4691
36	.3768	.3876	.3980	.4081	.4178	.4272	.4364	.4452	.4537	.4620
37	.3703	.3810	.3914	.4014	.4111	.4205	.4295	.4383	.4468	.4551
38	.3640	.3746	.3849	.3949	.4045	.4139	.4229	.4317	.4402	.4484
39	.3579	.3685	.3787	.3886	.3982	.4075	.4165	.4252	.4337	.4419
40	.3519	.3625	.3727	.3825	.3921	.4013	.4103	.4190	.4274	.4356

Table 8. Bernoulli Reward Process
Index Values, $a = 0.8$ (continued)

α β	31	32	33	34	35	36	37	38	39	40
1	.9715	.9723	.9731	.9738	.9745	.9751	.9758	.9763	.9769	.9774
2	.9442	.9457	.9472	.9486	.9499	.9511	.9523	.9535	.9545	.9556
3	.9180	.9202	.9223	.9243	.9262	.9281	.9298	.9314	.9330	.9345
4	.8930	.8958	.8986	.9011	.9036	.9059	.9081	.9102	.9122	.9141
5	.8692	.8726	.8758	.8789	.8818	.8846	.8873	.8898	.8923	.8946
6	.8465	.8504	.8541	.8577	.8610	.8643	.8673	.8703	.8731	.8757
7	.8248	.8292	.8334	.8373	.8411	.8447	.8482	.8515	.8546	.8576
8	.8042	.8090	.8136	.8179	.8221	.8260	.8298	.8334	.8369	.8402
9	.7846	.7897	.7946	.7993	.8038	.8081	.8122	.8161	.8198	.8235
10	.7658	.7713	.7765	.7815	.7863	.7909	.7952	.7994	.8034	.8073
11	.7479	.7537	.7592	.7645	.7695	.7743	.7790	.7834	.7877	.7918
12	.7308	.7368	.7426	.7481	.7534	.7585	.7633	.7680	.7725	.7768
13	.7144	.7207	.7267	.7324	.7379	.7432	.7483	.7531	.7578	.7624
14	.6987	.7052	.7114	.7174	.7231	.7285	.7338	.7389	.7437	.7484
15	.6837	.6904	.6968	.7029	.7088	.7144	.7199	.7251	.7301	.7350
16	.6692	.6761	.6827	.6890	.6950	.7008	.7064	.7118	.7170	.7220
17	.6555	.6624	.6691	.6756	.6818	.6877	.6935	.6990	.7043	.7095
18	.6422	.6493	.6561	.6627	.6690	.6751	.6810	.6866	.6921	.6974
19	.6295	.6367	.6436	.6503	.6567	.6629	.6689	.6747	.6803	.6856
20	.6172	.6245	.6316	.6384	.6449	.6512	.6573	.6631	.6688	.6743
21	.6054	.6128	.6200	.6268	.6335	.6399	.6460	.6520	.6577	.6633
22	.5940	.6015	.6087	.6157	.6224	.6289	.6352	.6412	.6470	.6527
23	.5831	.5906	.5979	.6049	.6117	.6183	.6246	.6308	.6367	.6424
24	.5725	.5801	.5875	.5946	.6014	.6080	.6144	.6206	.6266	.6324
25	.5623	.5700	.5774	.5845	.5914	.5981	.6046	.6108	.6169	.6228
26	.5525	.5602	.5676	.5748	.5818	.5885	.5950	.6013	.6074	.6134
27	.5429	.5507	.5582	.5654	.5724	.5792	.5858	.5921	.5983	.6042
28	.5337	.5415	.5490	.5563	.5633	.5702	.5768	.5832	.5894	.5954
29	.5248	.5326	.5402	.5475	.5546	.5614	.5680	.5745	.5807	.5868
30	.5162	.5240	.5316	.5389	.5460	.5529	.5596	.5661	.5723	.5784
31	.5079	.5157	.5233	.5306	.5378	.5447	.5514	.5579	.5642	.5703
32	.4998	.5076	.5152	.5226	.5297	.5367	.5434	.5499	.5562	.5624
33	.4920	.4998	.5074	.5148	.5219	.5289	.5356	.5422	.5485	.5547
34	.4845	.4922	.4998	.5072	.5143	.5213	.5281	.5346	.5410	.5472
35	.4771	.4849	.4925	.4998	.5070	.5139	.5207	.5273	.5337	.5399
36	.4700	.4778	.4854	.4927	.4998	.5068	.5136	.5202	.5266	.5328
37	.4631	.4709	.4784	.4858	.4929	.4998	.5066	.5132	.5196	.5259
38	.4564	.4642	.4717	.4791	.4862	.4931	.4998	.5064	.5129	.5191
39	.4499	.4577	.4652	.4725	.4796	.4865	.4933	.4998	.5063	.5125
40	.4436	.4513	.4588	.4661	.4732	.4802	.4869	.4935	.4998	.5061

Table 9. Bernoulli Reward Process
Index Values, $a = 0.9$

α β	1	2	3	4	5	6	7	8	9	10
1	.7029	.8001	.8452	.8723	.8905	.9039	.9141	.9221	.9287	.9342
2	.5001	.6346	.7072	.7539	.7869	.8115	.8307	.8461	.8588	.8695
3	.3796	.5163	.6010	.6579	.6996	.7318	.7573	.7782	.7956	.8103
4	.3021	.4342	.5184	.5809	.6276	.6642	.6940	.7187	.7396	.7573
5	.2488	.3720	.4561	.5179	.5676	.6071	.6395	.6666	.6899	.7101
6	.2103	.3245	.4058	.4677	.5168	.5581	.5923	.6212	.6461	.6677
7	.1815	.2871	.3647	.4257	.4748	.5156	.5510	.5811	.6071	.6300
8	.1591	.2569	.3308	.3900	.4387	.4795	.5144	.5454	.5723	.5960
9	.1413	.2323	.3025	.3595	.4073	.4479	.4828	.5134	.5409	.5652
10	.1269	.2116	.2784	.3332	.3799	.4200	.4548	.4853	.5125	.5373
11	.1149	.1942	.2575	.3106	.3558	.3951	.4296	.4601	.4871	.5116
12	.1049	.1793	.2396	.2906	.3343	.3729	.4069	.4371	.4642	.4886
13	.0965	.1664	.2239	.2729	.3155	.3529	.3864	.4163	.4432	.4677
14	.0892	.1552	.2100	.2571	.2985	.3348	.3678	.3972	.4240	.4483
15	.0828	.1453	.1977	.2431	.2831	.3187	.3507	.3798	.4062	.4303
16	.0773	.1366	.1867	.2304	.2692	.3039	.3351	.3638	.3898	.4137
17	.0724	.1288	.1768	.2190	.2565	.2904	.3210	.3490	.3747	.3983
18	.0681	.1218	.1679	.2086	.2450	.2780	.3079	.3352	.3606	.3840
19	.0642	.1155	.1598	.1991	.2344	.2665	.2958	.3227	.3475	.3706
20	.0608	.1098	.1524	.1904	.2247	.2559	.2846	.3110	.3353	.3581
21	.0576	.1046	.1456	.1824	.2157	.2462	.2742	.3001	.3240	.3463
22	.0548	.0998	.1394	.1750	.2074	.2371	.2645	.2899	.3134	.3353
23	.0522	.0955	.1337	.1682	.1997	.2287	.2554	.2803	.3035	.3250
24	.0498	.0915	.1284	.1619	.1926	.2208	.2470	.2714	.2941	.3154
25	.0477	.0878	.1236	.1560	.1859	.2134	.2390	.2629	.2853	.3062
26	.0457	.0844	.1190	.1506	.1796	.2065	.2316	.2550	.2770	.2976
27	.0438	.0812	.1148	.1455	.1738	.2000	.2246	.2475	.2691	.2894
28	.0421	.0783	.1108	.1407	.1683	.1940	.2180	.2405	.2617	.2816
29	.0406	.0756	.1072	.1362	.1631	.1882	.2117	.2338	.2546	.2743
30	.0391	.0730	.1037	.1320	.1583	.1828	.2058	.2275	.2479	.2673
31	.0377	.0706	.1004	.1280	.1537	.1777	.2002	.2215	.2416	.2606
32	.0364	.0683	.0974	.1243	.1493	.1728	.1949	.2158	.2356	.2543
33	.0352	.0662	.0945	.1207	.1452	.1682	.1899	.2104	.2298	.2482
34	.0341	.0642	.0918	.1174	.1413	.1639	.1851	.2053	.2244	.2425
35	.0330	.0623	.0892	.1142	.1377	.1597	.1806	.2003	.2191	.2370
36	.0320	.0606	.0868	.1112	.1341	.1558	.1762	.1957	.2141	.2317
37	.0311	.0589	.0845	.1084	.1308	.1520	.1721	.1912	.2094	.2267
38	.0302	.0573	.0823	.1056	.1276	.1484	.1681	.1869	.2048	.2218
39	.0294	.0558	.0802	.1031	.1246	.1450	.1644	.1828	.2004	.2172
40	.0286	.0543	.0782	.1006	.1217	.1417	.1608	.1789	.1962	.2128

Table 9. Bernoulli Reward Process
Index Values, $a = 0.9$ (continued)

α	11	12	13	14	15	16	17	18	19	20
β										
1	.9389	.9429	.9463	.9494	.9521	.9545	.9567	.9587	.9604	.9621
2	.8786	.8864	.8932	.8993	.9046	.9094	.9137	.9176	.9212	.9244
3	.8230	.8340	.8437	.8522	.8598	.8667	.8729	.8785	.8836	.8883
4	.7728	.7863	.7982	.8089	.8184	.8270	.8347	.8419	.8483	.8543
5	.7276	.7431	.7568	.7691	.7802	.7902	.7993	.8077	.8153	.8224
6	.6869	.7039	.7191	.7327	.7450	.7562	.7665	.7758	.7845	.7925
7	.6502	.6682	.6846	.6993	.7126	.7248	.7360	.7462	.7556	.7644
8	.6171	.6360	.6531	.6684	.6827	.6957	.7076	.7186	.7288	.7382
9	.5870	.6065	.6242	.6404	.6551	.6686	.6812	.6928	.7036	.7136
10	.5594	.5795	.5977	.6143	.6296	.6436	.6566	.6686	.6799	.6905
11	.5342	.5546	.5732	.5902	.6058	.6203	.6338	.6462	.6578	.6686
12	.5109	.5316	.5505	.5678	.5838	.5985	.6123	.6251	.6371	.6483
13	.4898	.5103	.5294	.5469	.5631	.5782	.5922	.6053	.6175	.6290
14	.4705	.4908	.5097	.5275	.5439	.5591	.5733	.5866	.5991	.6108
15	.4525	.4728	.4916	.5092	.5258	.5412	.5555	.5690	.5817	.5935
16	.4357	.4561	.4749	.4923	.5087	.5243	.5388	.5524	.5652	.5772
17	.4201	.4404	.4591	.4766	.4929	.5083	.5229	.5367	.5496	.5618
18	.4056	.4257	.4444	.4618	.4781	.4934	.5079	.5217	.5348	.5470
19	.3920	.4119	.4305	.4479	.4642	.4795	.4939	.5076	.5207	.5330
20	.3792	.3989	.4174	.4347	.4509	.4662	.4807	.4943	.5073	.5197
21	.3672	.3868	.4050	.4222	.4384	.4537	.4681	.4817	.4946	.5070
22	.3560	.3753	.3934	.4104	.4265	.4417	.4561	.4697	.4827	.4949
23	.3453	.3644	.3824	.3993	.4153	.4304	.4447	.4583	.4712	.4835
24	.3353	.3542	.3720	.3887	.4046	.4196	.4339	.4474	.4603	.4726
25	.3259	.3445	.3621	.3787	.3944	.4093	.4235	.4370	.4499	.4621
26	.3170	.3352	.3527	.3691	.3847	.3995	.4136	.4271	.4399	.4521
27	.3085	.3266	.3437	.3600	.3755	.3902	.4042	.4175	.4303	.4425
28	.3005	.3183	.3352	.3513	.3667	.3813	.3952	.4084	.4211	.4333
29	.2929	.3105	.3271	.3431	.3583	.3727	.3865	.3997	.4123	.4244
30	.2856	.3030	.3195	.3351	.3502	.3645	.3782	.3913	.4039	.4159
31	.2787	.2958	.3121	.3276	.3425	.3567	.3703	.3833	.3957	.4077
32	.2721	.2890	.3051	.3205	.3351	.3492	.3627	.3756	.3879	.3998
33	.2658	.2825	.2984	.3136	.3280	.3420	.3553	.3682	.3804	.3922
34	.2597	.2762	.2919	.3070	.3213	.3350	.3483	.3610	.3732	.3849
35	.2540	.2702	.2858	.3006	.3148	.3284	.3415	.3541	.3662	.3779
36	.2485	.2645	.2798	.2945	.3086	.3221	.3350	.3475	.3595	.3711
37	.2432	.2590	.2742	.2887	.3026	.3159	.3287	.3411	.3530	.3645
38	.2381	.2537	.2687	.2830	.2968	.3100	.3227	.3349	.3468	.3582
39	.2333	.2486	.2634	.2776	.2912	.3043	.3169	.3290	.3407	.3521
40	.2286	.2438	.2584	.2724	.2859	.2988	.3113	.3233	.3349	.3461

Table 9. Bernoulli Reward Process
Index Values, $a = 0.9$ (continued)

α / β	21	22	23	24	25	26	27	28	29	30
1	.9636	.9649	.9662	.9674	.9685	.9695	.9705	.9714	.9722	.9730
2	.9274	.9301	.9327	.9350	.9372	.9392	.9411	.9429	.9446	.9462
3	.8926	.8966	.9003	.9038	.9069	.9099	.9127	.9153	.9178	.9201
4	.8598	.8649	.8696	.8740	.8781	.8819	.8855	.8889	.8921	.8951
5	.8289	.8349	.8405	.8458	.8507	.8553	.8596	.8636	.8675	.8711
6	.7998	.8067	.8131	.8191	.8247	.8300	.8349	.8396	.8440	.8481
7	.7726	.7802	.7873	.7939	.8001	.8060	.8115	.8167	.8216	.8263
8	.7470	.7552	.7629	.7701	.7769	.7832	.7892	.7949	.8003	.8054
9	.7229	.7317	.7398	.7475	.7548	.7616	.7681	.7742	.7800	.7855
10	.7003	.7095	.7181	.7262	.7339	.7411	.7479	.7544	.7606	.7664
11	.6789	.6885	.6975	.7060	.7140	.7216	.7288	.7356	.7421	.7482
12	.6587	.6686	.6780	.6868	.6952	.7031	.7106	.7177	.7244	.7309
13	.6398	.6499	.6595	.6686	.6772	.6854	.6932	.7006	.7076	.7143
14	.6218	.6322	.6421	.6514	.6602	.6685	.6765	.6842	.6914	.6984
15	.6048	.6154	.6254	.6349	.6440	.6525	.6607	.6685	.6759	.6831
16	.5886	.5994	.6096	.6193	.6285	.6373	.6456	.6536	.6612	.6684
17	.5733	.5842	.5945	.6044	.6137	.6226	.6311	.6393	.6470	.6544
18	.5587	.5697	.5802	.5901	.5996	.6086	.6173	.6255	.6335	.6410
19	.5448	.5559	.5664	.5765	.5861	.5952	.6040	.6124	.6204	.6281
20	.5315	.5427	.5533	.5635	.5732	.5824	.5913	.5997	.6079	.6157
21	.5188	.5301	.5408	.5510	.5608	.5701	.5790	.5876	.5958	.6037
22	.5067	.5180	.5288	.5391	.5489	.5583	.5673	.5759	.5842	.5922
23	.4952	.5064	.5173	.5276	.5375	.5469	.5560	.5647	.5730	.5811
24	.4843	.4954	.5062	.5166	.5265	.5360	.5451	.5539	.5623	.5704
25	.4738	.4850	.4957	.5060	.5160	.5255	.5347	.5435	.5519	.5601
26	.4638	.4750	.4856	.4959	.5058	.5154	.5246	.5334	.5419	.5501
27	.4541	.4653	.4760	.4862	.4961	.5056	.5148	.5237	.5322	.5405
28	.4449	.4560	.4667	.4769	.4868	.4962	.5054	.5143	.5229	.5311
29	.4360	.4471	.4577	.4680	.4778	.4873	.4964	.5052	.5138	.5221
30	.4274	.4385	.4491	.4593	.4692	.4786	.4878	.4965	.5051	.5134
31	.4192	.4302	.4408	.4510	.4608	.4703	.4794	.4882	.4967	.5049
32	.4112	.4222	.4328	.4429	.4527	.4622	.4713	.4801	.4886	.4968
33	.4036	.4145	.4250	.4352	.4449	.4544	.4635	.4723	.4808	.4890
34	.3962	.4071	.4175	.4276	.4374	.4468	.4559	.4647	.4732	.4814
35	.3891	.3999	.4103	.4204	.4301	.4395	.4485	.4573	.4658	.4740
36	.3822	.3930	.4033	.4133	.4230	.4324	.4414	.4502	.4586	.4668
37	.3756	.3863	.3966	.4065	.4162	.4255	.4345	.4432	.4517	.4599
38	.3692	.3798	.3900	.3999	.4095	.4188	.4278	.4365	.4449	.4531
39	.3630	.3735	.3837	.3936	.4031	.4124	.4213	.4300	.4384	.4466
40	.3570	.3675	.3776	.3874	.3969	.4061	.4150	.4236	.4320	.4402

Table 9. Bernoulli Reward Process
Index Values, $a = 0.9$ (continued)

α / β	31	32	33	34	35	36	37	38	39	40
1	.9738	.9745	.9751	.9758	.9764	.9769	.9775	.9780	.9785	.9789
2	.9477	.9491	.9504	.9517	.9529	.9540	.9551	.9561	.9571	.9580
3	.9223	.9244	.9263	.9282	.9300	.9316	.9332	.9348	.9362	.9376
4	.8979	.9006	.9031	.9055	.9078	.9100	.9120	.9140	.9159	.9177
5	.8745	.8777	.8808	.8837	.8864	.8891	.8916	.8940	.8963	.8985
6	.8521	.8558	.8593	.8627	.8659	.8690	.8719	.8747	.8774	.8799
7	.8307	.8348	.8388	.8426	.8462	.8497	.8530	.8561	.8591	.8620
8	.8102	.8148	.8192	.8234	.8273	.8311	.8348	.8383	.8416	.8448
9	.7907	.7956	.8004	.8049	.8092	.8133	.8173	.8210	.8247	.8281
10	.7720	.7773	.7823	.7872	.7918	.7962	.8004	.8045	.8084	.8121
11	.7541	.7597	.7651	.7702	.7751	.7797	.7842	.7885	.7927	.7966
12	.7370	.7429	.7485	.7538	.7590	.7639	.7686	.7732	.7775	.7817
13	.7207	.7268	.7326	.7382	.7435	.7486	.7536	.7583	.7629	.7673
14	.7050	.7113	.7173	.7231	.7287	.7340	.7391	.7441	.7488	.7534
15	.6899	.6964	.7027	.7086	.7144	.7199	.7252	.7303	.7352	.7400
16	.6754	.6821	.6885	.6947	.7006	.7063	.7118	.7170	.7221	.7270
17	.6615	.6684	.6750	.6813	.6873	.6932	.6988	.7042	.7094	.7145
18	.6483	.6552	.6619	.6683	.6745	.6805	.6863	.6918	.6972	.7023
19	.6355	.6426	.6494	.6559	.6622	.6683	.6741	.6798	.6853	.6906
20	.6232	.6303	.6373	.6439	.6503	.6565	.6625	.6682	.6738	.6792
21	.6113	.6186	.6256	.6324	.6389	.6451	.6512	.6570	.6627	.6682
22	.5998	.6072	.6143	.6212	.6278	.6341	.6403	.6462	.6520	.6575
23	.5888	.5963	.6034	.6104	.6170	.6235	.6297	.6358	.6416	.6472
24	.5782	.5857	.5929	.5999	.6067	.6132	.6195	.6256	.6315	.6372
25	.5679	.5755	.5828	.5898	.5966	.6032	.6096	.6157	.6217	.6275
26	.5580	.5656	.5729	.5800	.5869	.5935	.6000	.6062	.6122	.6181
27	.5484	.5560	.5634	.5706	.5775	.5842	.5907	.5969	.6030	.6089
28	.5391	.5468	.5542	.5614	.5684	.5751	.5816	.5879	.5941	.6000
29	.5301	.5378	.5453	.5525	.5595	.5663	.5728	.5792	.5854	.5914
30	.5214	.5292	.5367	.5439	.5509	.5577	.5643	.5707	.5769	.5830
31	.5130	.5208	.5283	.5356	.5426	.5494	.5561	.5625	.5687	.5748
32	.5048	.5126	.5201	.5274	.5345	.5414	.5480	.5545	.5607	.5668
33	.4969	.5047	.5122	.5195	.5266	.5335	.5402	.5467	.5530	.5591
34	.4893	.4970	.5045	.5119	.5190	.5259	.5326	.5391	.5454	.5516
35	.4820	.4897	.4971	.5044	.5116	.5185	.5252	.5317	.5380	.5442
36	.4748	.4825	.4900	.4972	.5043	.5112	.5180	.5245	.5309	.5371
37	.4678	.4755	.4830	.4903	.4973	.5042	.5110	.5175	.5239	.5301
38	.4611	.4688	.4762	.4835	.4905	.4974	.5041	.5107	.5171	.5233
39	.4545	.4622	.4696	.4769	.4840	.4908	.4975	.5040	.5104	.5166
40	.4481	.4558	.4632	.4705	.4775	.4844	.4911	.4975	.5039	.5102

Table 10. Bernoulli Reward Process
Index Values, $a = 0.95$

α / β	1	2	3	4	5	6	7	8	9	10
1	.7614	.8381	.8736	.8948	.9092	.9197	.9278	.9343	.9396	.9440
2	.5601	.6810	.7443	.7845	.8128	.8340	.8505	.8637	.8746	.8838
3	.4334	.5621	.6392	.6903	.7281	.7568	.7797	.7984	.8139	.8271
4	.3477	.4753	.5556	.6133	.6563	.6899	.7174	.7400	.7589	.7753
5	.2877	.4094	.4898	.5493	.5957	.6326	.6628	.6881	.7099	.7285
6	.2439	.3576	.4372	.4964	.5440	.5830	.6152	.6425	.6658	.6863
7	.2106	.3172	.3937	.4528	.4999	.5397	.5733	.6019	.6267	.6484
8	.1847	.2842	.3573	.4154	.4627	.5018	.5361	.5657	.5914	.6140
9	.1640	.2568	.3270	.3834	.4301	.4694	.5029	.5331	.5595	.5829
10	.1471	.2341	.3011	.3554	.4014	.4405	.4743	.5036	.5305	.5544
11	.1331	.2147	.2787	.3312	.3761	.4146	.4483	.4780	.5041	.5284
12	.1214	.1982	.2591	.3101	.3535	.3915	.4248	.4544	.4808	.5043
13	.1114	.1839	.2421	.2912	.3334	.3706	.4035	.4329	.4592	.4830
14	.1029	.1713	.2270	.2744	.3155	.3517	.3842	.4132	.4394	.4632
15	.0954	.1603	.2136	.2592	.2993	.3346	.3665	.3951	.4211	.4448
16	.0889	.1505	.2016	.2457	.2846	.3191	.3502	.3785	.4041	.4277
17	.0832	.1419	.1909	.2335	.2712	.3049	.3352	.3631	.3885	.4118
18	.0781	.1340	.1812	.2223	.2589	.2918	.3217	.3489	.3740	.3970
19	.0735	.1270	.1723	.2121	.2476	.2798	.3090	.3356	.3604	.3832
20	.0695	.1206	.1643	.2027	.2373	.2687	.2973	.3235	.3477	.3703
21	.0658	.1148	.1569	.1942	.2278	.2583	.2864	.3121	.3358	.3581
22	.0625	.1095	.1501	.1863	.2190	.2488	.2762	.3015	.3249	.3467
23	.0594	.1047	.1439	.1790	.2108	.2399	.2667	.2915	.3146	.3360
24	.0566	.1002	.1382	.1722	.2031	.2316	.2578	.2822	.3049	.3260
25	.0541	.0961	.1328	.1659	.1960	.2238	.2494	.2734	.2957	.3166
26	.0518	.0923	.1279	.1600	.1894	.2165	.2417	.2651	.2870	.3076
27	.0496	.0888	.1233	.1545	.1832	.2097	.2343	.2573	.2789	.2991
28	.0476	.0855	.1190	.1494	.1773	.2032	.2274	.2499	.2711	.2911
29	.0458	.0824	.1149	.1445	.1718	.1971	.2208	.2430	.2638	.2834
30	.0441	.0796	.1112	.1400	.1666	.1914	.2146	.2364	.2568	.2762
31	.0425	.0769	.1076	.1357	.1618	.1860	.2087	.2301	.2502	.2693
32	.0410	.0744	.1043	.1317	.1572	.1809	.2031	.2241	.2439	.2627
33	.0396	.0720	.1012	.1279	.1528	.1760	.1979	.2185	.2380	.2564
34	.0383	.0698	.0982	.1243	.1486	.1714	.1929	.2131	.2323	.2504
35	.0370	.0677	.0954	.1209	.1447	.1670	.1881	.2080	.2268	.2447
36	.0359	.0657	.0927	.1177	.1410	.1629	.1835	.2030	.2216	.2392
37	.0348	.0639	.0902	.1146	.1374	.1589	.1792	.1984	.2166	.2340
38	.0337	.0621	.0878	.1117	.1341	.1551	.1750	.1939	.2119	.2290
39	.0328	.0604	.0856	.1089	.1308	.1515	.1711	.1896	.2073	.2242
40	.0318	.0588	.0834	.1063	.1278	.1480	.1672	.1855	.2029	.2196

Table 10. Bernoulli Reward Process
Index Values, $a = 0.95$ (continued)

α / β	11	12	13	14	15	16	17	18	19	20
1	.9477	.9510	.9538	.9563	.9585	.9605	.9623	.9639	.9654	.9667
2	.8916	.8984	.9043	.9096	.9142	.9184	.9221	.9255	.9286	.9314
3	.8383	.8482	.8568	.8645	.8713	.8775	.8830	.8881	.8927	.8969
4	.7894	.8017	.8126	.8223	.8310	.8388	.8460	.8525	.8584	.8639
5	.7448	.7590	.7718	.7832	.7935	.8027	.8112	.8190	.8261	.8326
6	.7043	.7202	.7344	.7471	.7586	.7692	.7788	.7876	.7957	.8031
7	.6674	.6846	.7000	.7139	.7265	.7380	.7485	.7581	.7671	.7754
8	.6342	.6521	.6683	.6831	.6966	.7090	.7203	.7307	.7403	.7493
9	.6036	.6224	.6394	.6548	.6688	.6818	.6939	.7050	.7152	.7248
10	.5758	.5951	.6126	.6286	.6432	.6567	.6691	.6808	.6916	.7017
11	.5502	.5698	.5878	.6042	.6193	.6333	.6462	.6581	.6692	.6798
12	.5265	.5465	.5648	.5815	.5970	.6113	.6246	.6369	.6485	.6593
13	.5045	.5248	.5434	.5604	.5761	.5907	.6043	.6170	.6288	.6399
14	.4848	.5045	.5234	.5406	.5566	.5714	.5852	.5981	.6102	.6216
15	.4664	.4863	.5046	.5221	.5382	.5532	.5672	.5803	.5926	.6042
16	.4493	.4692	.4876	.5046	.5209	.5361	.5502	.5635	.5760	.5877
17	.4333	.4531	.4716	.4886	.5045	.5199	.5342	.5476	.5602	.5721
18	.4183	.4381	.4565	.4736	.4896	.5045	.5189	.5325	.5452	.5572
19	.4043	.4239	.4422	.4593	.4753	.4903	.5044	.5181	.5309	.5430
20	.3912	.4106	.4288	.4459	.4619	.4769	.4910	.5044	.5173	.5295
21	.3788	.3981	.4162	.4331	.4491	.4641	.4783	.4916	.5043	.5166
22	.3672	.3863	.4042	.4211	.4370	.4519	.4661	.4795	.4922	.5042
23	.3562	.3752	.3929	.4096	.4254	.4404	.4545	.4679	.4806	.4926
24	.3459	.3646	.3822	.3988	.4145	.4293	.4434	.4568	.4695	.4816
25	.3360	.3546	.3721	.3885	.4041	.4188	.4329	.4462	.4589	.4710
26	.3269	.3451	.3624	.3787	.3942	.4088	.4228	.4360	.4487	.4608
27	.3182	.3360	.3532	.3694	.3847	.3992	.4131	.4263	.4389	.4510
28	.3099	.3276	.3444	.3605	.3757	.3901	.4039	.4170	.4296	.4416
29	.3020	.3195	.3360	.3520	.3670	.3814	.3951	.4081	.4206	.4326
30	.2945	.3118	.3282	.3438	.3588	.3730	.3866	.3996	.4120	.4239
31	.2873	.3044	.3206	.3360	.3509	.3650	.3785	.3914	.4037	.4156
32	.2805	.2974	.3134	.3287	.3433	.3573	.3707	.3835	.3957	.4075
33	.2739	.2906	.3065	.3216	.3360	.3499	.3632	.3759	.3881	.3998
34	.2677	.2842	.2998	.3148	.3291	.3428	.3560	.3686	.3807	.3923
35	.2617	.2780	.2935	.3083	.3224	.3359	.3490	.3616	.3736	.3852
36	.2560	.2721	.2874	.3020	.3160	.3294	.3423	.3548	.3667	.3782
37	.2505	.2664	.2815	.2960	.3099	.3232	.3359	.3483	.3601	.3715
38	.2453	.2609	.2759	.2902	.3039	.3171	.3297	.3419	.3537	.3651
39	.2403	.2557	.2705	.2846	.2982	.3113	.3238	.3358	.3476	.3588
40	.2355	.2506	.2652	.2793	.2927	.3056	.3181	.3300	.3416	.3528

Table 10. Bernoulli Reward Process
Index Values, $a = 0.95$ (continued)

β \ α	21	22	23	24	25	26	27	28	29	30
1	.9680	.9691	.9702	.9711	.9721	.9729	.9737	.9745	.9752	.9759
2	.9341	.9365	.9387	.9408	.9427	.9445	.9462	.9477	.9492	.9506
3	.9008	.9044	.9077	.9108	.9137	.9164	.9189	.9213	.9235	.9256
4	.8689	.8736	.8779	.8819	.8857	.8892	.8925	.8956	.8985	.9013
5	.8386	.8442	.8495	.8543	.8589	.8631	.8671	.8709	.8745	.8778
6	.8100	.8165	.8225	.8281	.8333	.8383	.8429	.8473	.8514	.8553
7	.7831	.7902	.7969	.8032	.8091	.8146	.8198	.8247	.8293	.8337
8	.7576	.7654	.7727	.7796	.7860	.7920	.7977	.8031	.8082	.8131
9	.7337	.7420	.7498	.7571	.7641	.7706	.7767	.7826	.7881	.7933
10	.7111	.7199	.7281	.7359	.7432	.7501	.7567	.7629	.7688	.7744
11	.6897	.6989	.7076	.7157	.7234	.7307	.7376	.7441	.7503	.7563
12	.6693	.6790	.6880	.6965	.7046	.7122	.7194	.7262	.7328	.7390
13	.6504	.6602	.6694	.6782	.6866	.6945	.7020	.7091	.7159	.7224
14	.6323	.6424	.6519	.6609	.6694	.6776	.6853	.6927	.6997	.7064
15	.6151	.6254	.6352	.6444	.6532	.6615	.6694	.6770	.6842	.6912
16	.5988	.6093	.6192	.6287	.6376	.6461	.6542	.6620	.6693	.6765
17	.5833	.5939	.6040	.6136	.6228	.6314	.6397	.6476	.6552	.6624
18	.5685	.5793	.5895	.5992	.6085	.6174	.6258	.6338	.6415	.6489
19	.5545	.5654	.5757	.5855	.5949	.6038	.6124	.6206	.6284	.6359
20	.5411	.5520	.5624	.5724	.5819	.5909	.5996	.6079	.6158	.6234
21	.5282	.5393	.5498	.5598	.5694	.5785	.5872	.5956	.6037	.6114
22	.5159	.5270	.5376	.5477	.5573	.5666	.5754	.5839	.5920	.5998
23	.5041	.5153	.5260	.5361	.5458	.5551	.5640	.5725	.5807	.5886
24	.4931	.5040	.5148	.5250	.5347	.5441	.5530	.5616	.5699	.5778
25	.4825	.4934	.5040	.5142	.5240	.5334	.5424	.5511	.5594	.5674
26	.4723	.4833	.4938	.5039	.5137	.5232	.5322	.5409	.5493	.5573
27	.4625	.4735	.4840	.4941	.5038	.5133	.5224	.5311	.5395	.5476
28	.4531	.4641	.4746	.4847	.4944	.5037	.5129	.5216	.5301	.5382
29	.4440	.4550	.4655	.4756	.4853	.4946	.5036	.5125	.5209	.5291
30	.4353	.4462	.4568	.4669	.4766	.4859	.4949	.5036	.5121	.5203
31	.4269	.4378	.4483	.4584	.4681	.4774	.4864	.4951	.5035	.5117
32	.4188	.4297	.4402	.4502	.4599	.4693	.4783	.4869	.4953	.5034
33	.4110	.4219	.4323	.4423	.4520	.4613	.4703	.4790	.4874	.4955
34	.4035	.4143	.4247	.4347	.4443	.4536	.4626	.4713	.4797	.4878
35	.3963	.4070	.4173	.4273	.4369	.4462	.4552	.4639	.4723	.4804
36	.3893	.3999	.4102	.4202	.4297	.4390	.4480	.4566	.4650	.4731
37	.3825	.3931	.4034	.4132	.4228	.4320	.4410	.4496	.4580	.4661
38	.3760	.3865	.3967	.4065	.4161	.4253	.4342	.4428	.4512	.4593
39	.3697	.3802	.3903	.4001	.4095	.4187	.4276	.4362	.4445	.4526
40	.3636	.3740	.3840	.3938	.4032	.4123	.4212	.4297	.4381	.4461

Table 10. Bernoulli Reward Process
Index Values, $a = 0.95$ (continued)

β \ α	31	32	33	34	35	36	37	38	39	40
1	.9765	.9771	.9777	.9782	.9787	.9792	.9797	.9801	.9805	.9809
2	.9519	.9532	.9544	.9555	.9566	.9576	.9585	.9595	.9603	.9612
3	.9276	.9295	.9312	.9329	.9345	.9361	.9375	.9389	.9402	.9415
4	.9039	.9063	.9087	.9109	.9130	.9150	.9169	.9188	.9205	.9222
5	.8810	.8840	.8869	.8896	.8922	.8946	.8970	.8992	.9014	.9034
6	.8590	.8625	.8658	.8690	.8720	.8749	.8776	.8803	.8828	.8852
7	.8379	.8419	.8456	.8492	.8526	.8559	.8590	.8620	.8648	.8676
8	.8177	.8220	.8262	.8302	.8339	.8375	.8410	.8443	.8475	.8505
9	.7983	.8030	.8075	.8119	.8160	.8199	.8237	.8273	.8307	.8340
10	.7797	.7848	.7896	.7943	.7987	.8029	.8070	.8108	.8146	.8182
11	.7619	.7673	.7725	.7774	.7821	.7866	.7909	.7950	.7989	.8028
12	.7449	.7505	.7559	.7611	.7660	.7708	.7753	.7797	.7839	.7879
13	.7285	.7344	.7401	.7454	.7506	.7556	.7604	.7649	.7694	.7736
14	.7128	.7190	.7248	.7304	.7358	.7410	.7459	.7507	.7553	.7597
15	.6978	.7041	.7101	.7159	.7215	.7269	.7320	.7370	.7417	.7463
16	.6833	.6898	.6960	.7020	.7077	.7133	.7186	.7237	.7286	.7334
17	.6693	.6760	.6824	.6886	.6945	.7001	.7056	.7109	.7159	.7208
18	.6560	.6627	.6692	.6756	.6816	.6874	.6931	.6985	.7037	.7087
19	.6431	.6500	.6567	.6630	.6692	.6752	.6809	.6864	.6918	.6969
20	.6308	.6378	.6445	.6510	.6573	.6633	.6691	.6748	.6803	.6855
21	.6188	.6259	.6328	.6394	.6458	.6519	.6578	.6636	.6691	.6745
22	.6073	.6145	.6215	.6282	.6347	.6409	.6469	.6527	.6583	.6638
23	.5962	.6035	.6105	.6173	.6239	.6302	.6363	.6422	.6479	.6534
24	.5855	.5928	.5999	.6068	.6134	.6198	.6260	.6320	.6378	.6434
25	.5751	.5826	.5897	.5966	.6033	.6098	.6160	.6221	.6280	.6336
26	.5651	.5726	.5798	.5868	.5935	.6001	.6064	.6125	.6184	.6242
27	.5554	.5630	.5702	.5773	.5841	.5906	.5970	.6032	.6092	.6150
28	.5461	.5536	.5610	.5680	.5749	.5815	.5879	.5941	.6001	.6060
29	.5370	.5446	.5520	.5591	.5659	.5726	.5791	.5853	.5914	.5973
30	.5282	.5358	.5432	.5504	.5573	.5640	.5705	.5768	.5829	.5888
31	.5197	.5274	.5348	.5419	.5489	.5556	.5622	.5685	.5746	.5806
32	.5114	.5191	.5265	.5337	.5407	.5475	.5540	.5604	.5666	.5726
33	.5034	.5111	.5186	.5258	.5328	.5396	.5462	.5526	.5588	.5648
34	.4957	.5033	.5108	.5180	.5251	.5319	.5385	.5449	.5511	.5572
35	.4882	.4958	.5032	.5105	.5176	.5244	.5310	.5375	.5437	.5498
36	.4810	.4886	.4960	.5032	.5102	.5171	.5238	.5302	.5365	.5426
37	.4740	.4816	.4890	.4961	.5031	.5100	.5167	.5231	.5294	.5356
38	.4671	.4747	.4821	.4893	.4963	.5031	.5098	.5162	.5226	.5287
39	.4605	.4681	.4755	.4826	.4896	.4964	.5030	.5095	.5159	.5220
40	.4540	.4616	.4690	.4762	.4831	.4899	.4965	.5029	.5093	.5155

Table 11. Bernoulli Reward Process
Index Values, $a = 0.99$

β \ α	1	2	3	4	5	6	7	8	9	10
1	.8699	.9102	.9285	.9395	.9470	.9525	.9568	.9603	.9631	.9655
2	.7005	.7844	.8268	.8533	.8719	.8857	.8964	.9051	.9122	.9183
3	.5671	.6726	.7308	.7696	.7973	.8184	.8350	.8485	.8598	.8693
4	.4701	.5806	.6490	.6952	.7295	.7561	.7773	.7949	.8097	.8222
5	.3969	.5093	.5798	.6311	.6697	.6998	.7249	.7456	.7631	.7781
6	.3415	.4509	.5225	.5756	.6172	.6504	.6776	.7004	.7203	.7373
7	.2979	.4029	.4747	.5277	.5710	.6061	.6352	.6599	.6811	.6997
8	.2632	.3633	.4337	.4876	.5300	.5665	.5970	.6230	.6456	.6653
9	.2350	.3303	.3986	.4520	.4952	.5308	.5625	.5895	.6130	.6337
10	.2117	.3020	.3679	.4208	.4640	.5002	.5310	.5589	.5831	.6045
11	.1922	.2778	.3418	.3932	.4359	.4722	.5034	.5307	.5556	.5776
12	.1756	.2571	.3187	.3685	.4108	.4469	.4782	.5057	.5302	.5527
13	.1614	.2388	.2982	.3468	.3882	.4239	.4551	.4827	.5072	.5295
14	.1491	.2228	.2799	.3274	.3677	.4030	.4340	.4615	.4862	.5083
15	.1384	.2086	.2637	.3097	.3491	.3839	.4145	.4420	.4666	.4889
16	.1290	.1960	.2491	.2938	.3324	.3663	.3967	.4238	.4484	.4707
17	.1206	.1847	.2359	.2792	.3170	.3501	.3801	.4070	.4314	.4537
18	.1132	.1746	.2239	.2659	.3028	.3355	.3648	.3914	.4156	.4378
19	.1066	.1654	.2130	.2539	.2898	.3218	.3505	.3769	.4009	.4228
20	.1006	.1570	.2031	.2428	.2778	.3092	.3374	.3633	.3870	.4088
21	.0952	.1494	.1940	.2325	.2666	.2974	.3252	.3505	.3741	.3957
22	.0903	.1425	.1856	.2231	.2564	.2864	.3138	.3387	.3619	.3833
23	.0858	.1361	.1778	.2142	.2468	.2762	.3031	.3277	.3503	.3716
24	.0817	.1302	.1707	.2061	.2379	.2666	.2930	.3173	.3396	.3605
25	.0780	.1248	.1640	.1985	.2295	.2577	.2836	.3074	.3295	.3500
26	.0745	.1198	.1578	.1914	.2217	.2493	.2747	.2981	.3199	.3402
27	.0713	.1151	.1521	.1848	.2143	.2414	.2663	.2894	.3109	.3309
28	.0684	.1107	.1467	.1786	.2075	.2340	.2583	.2811	.3023	.3221
29	.0656	.1067	.1416	.1727	.2010	.2269	.2509	.2732	.2941	.3136
30	.0631	.1029	.1369	.1673	.1949	.2203	.2439	.2658	.2864	.3056
31	.0607	.0994	.1325	.1621	.1891	.2140	.2372	.2587	.2790	.2980
32	.0585	.0960	.1283	.1572	.1837	.2080	.2308	.2520	.2719	.2907
33	.0564	.0929	.1244	.1526	.1785	.2024	.2248	.2456	.2652	.2838
34	.0545	.0900	.1206	.1483	.1736	.1971	.2190	.2395	.2588	.2771
35	.0526	.0872	.1171	.1441	.1690	.1920	.2135	.2337	.2527	.2708
36	.0509	.0846	.1138	.1402	.1645	.1871	.2083	.2282	.2469	.2647
37	.0493	.0821	.1106	.1365	.1603	.1825	.2033	.2229	.2414	.2588
38	.0478	.0797	.1076	.1329	.1563	.1781	.1985	.2178	.2360	.2533
39	.0463	.0775	.1048	.1296	.1525	.1739	.1940	.2129	.2309	.2479
40	.0449	.0754	.1021	.1264	.1488	.1699	.1896	.2083	.2260	.2428

Table 11. Bernoulli Reward Process
Index Values, $a = 0.99$ (continued)

β \ α	11	12	13	14	15	16	17	18	19	20
1	.9675	.9693	.9709	.9722	.9735	.9746	.9756	.9765	.9774	.9781
2	.9234	.9278	.9318	.9352	.9383	.9411	.9436	.9458	.9479	.9498
3	.8775	.8846	.8909	.8964	.9014	.9059	.9099	.9136	.9169	.9200
4	.8331	.8426	.8509	.8584	.8651	.8711	.8766	.8816	.8861	.8903
5	.7912	.8027	.8129	.8220	.8302	.8377	.8444	.8506	.8563	.8615
6	.7522	.7653	.7771	.7877	.7972	.8059	.8137	.8210	.8276	.8338
7	.7161	.7307	.7437	.7554	.7660	.7758	.7846	.7928	.8003	.8072
8	.6826	.6984	.7125	.7253	.7369	.7474	.7571	.7660	.7743	.7820
9	.6521	.6685	.6833	.6970	.7094	.7208	.7312	.7408	.7497	.7579
10	.6236	.6408	.6564	.6706	.6835	.6956	.7067	.7169	.7264	.7352
11	.5973	.6150	.6312	.6460	.6595	.6719	.6834	.6942	.7043	.7136
12	.5728	.5911	.6077	.6229	.6369	.6498	.6617	.6728	.6832	.6930
13	.5501	.5687	.5856	.6012	.6155	.6289	.6412	.6527	.6634	.6734
14	.5288	.5477	.5650	.5808	.5955	.6091	.6217	.6336	.6446	.6550
15	.5091	.5280	.5456	.5617	.5766	.5904	.6033	.6154	.6267	.6373
16	.4911	.5096	.5273	.5436	.5587	.5728	.5859	.5982	.6097	.6205
17	.4741	.4929	.5100	.5265	.5418	.5560	.5693	.5818	.5935	.6045
18	.4581	.4769	.4943	.5103	.5258	.5402	.5536	.5662	.5781	.5893
19	.4431	.4619	.4793	.4955	.5105	.5251	.5387	.5514	.5634	.5747
20	.4290	.4477	.4651	.4814	.4965	.5106	.5244	.5372	.5494	.5608
21	.4157	.4343	.4517	.4679	.4831	.4974	.5107	.5237	.5359	.5475
22	.4031	.4216	.4389	.4551	.4703	.4846	.4981	.5107	.5231	.5347
23	.3913	.4096	.4268	.4430	.4582	.4725	.4860	.4987	.5107	.5225
24	.3801	.3983	.4153	.4314	.4466	.4609	.4744	.4871	.4992	.5107
25	.3694	.3875	.4044	.4204	.4355	.4498	.4632	.4760	.4882	.4997
26	.3593	.3772	.3941	.4099	.4249	.4391	.4526	.4654	.4775	.4891
27	.3497	.3675	.3842	.3999	.4148	.4290	.4424	.4552	.4673	.4789
28	.3405	.3582	.3747	.3904	.4052	.4193	.4326	.4454	.4575	.4691
29	.3320	.3493	.3657	.3813	.3960	.4099	.4233	.4359	.4480	.4596
30	.3238	.3408	.3571	.3725	.3872	.4010	.4143	.4269	.4390	.4505
31	.3159	.3328	.3489	.3642	.3787	.3925	.4056	.4182	.4302	.4417
32	.3084	.3252	.3409	.3561	.3706	.3843	.3974	.4098	.4218	.4333
33	.3013	.3178	.3335	.3484	.3628	.3764	.3894	.4018	.4137	.4251
34	.2944	.3108	.3263	.3410	.3553	.3688	.3817	.3941	.4059	.4172
35	.2878	.3040	.3194	.3340	.3480	.3615	.3743	.3866	.3984	.4096
36	.2815	.2975	.3128	.3273	.3410	.3544	.3672	.3794	.3911	.4023
37	.2755	.2913	.3064	.3208	.3344	.3476	.3603	.3724	.3841	.3952
38	.2697	.2853	.3002	.3145	.3281	.3410	.3537	.3657	.3773	.3884
39	.2641	.2796	.2943	.3084	.3219	.3348	.3472	.3592	.3707	.3818
40	.2587	.2741	.2886	.3026	.3160	.3288	.3410	.3529	.3644	.3754

Table 11. Bernoulli Reward Process
Index Values, $a = 0.99$ (continued)

α/β	21	22	23	24	25	26	27	28	29	30
1	.9788	.9795	.9801	.9807	.9812	.9817	.9822	.9827	.9831	.9835
2	.9516	.9532	.9547	.9561	.9574	.9587	.9598	.9609	.9619	.9629
3	.9228	.9255	.9279	.9302	.9323	.9343	.9362	.9379	.9396	.9411
4	.8942	.8978	.9011	.9043	.9072	.9099	.9124	.9148	.9171	.9192
5	.8663	.8708	.8750	.8789	.8825	.8859	.8891	.8921	.8950	.8977
6	.8394	.8447	.8497	.8543	.8586	.8626	.8664	.8700	.8734	.8766
7	.8137	.8197	.8253	.8306	.8355	.8401	.8445	.8486	.8525	.8562
8	.7891	.7958	.8020	.8078	.8133	.8185	.8234	.8280	.8323	.8364
9	.7657	.7730	.7797	.7861	.7921	.7977	.8030	.8080	.8128	.8174
10	.7435	.7512	.7584	.7652	.7717	.7778	.7835	.7889	.7941	.7990
11	.7223	.7305	.7382	.7454	.7522	.7586	.7647	.7705	.7761	.7813
12	.7021	.7107	.7188	.7264	.7336	.7403	.7468	.7529	.7586	.7642
13	.6829	.6918	.7003	.7082	.7157	.7228	.7295	.7359	.7420	.7478
14	.6647	.6738	.6825	.6908	.6986	.7059	.7130	.7196	.7260	.7320
15	.6473	.6568	.6657	.6740	.6821	.6898	.6970	.7039	.7105	.7168
16	.6308	.6404	.6496	.6582	.6664	.6742	.6817	.6888	.6957	.7021
17	.6150	.6248	.6342	.6430	.6514	.6594	.6670	.6743	.6813	.6880
18	.5999	.6099	.6194	.6284	.6370	.6452	.6530	.6604	.6675	.6743
19	.5854	.5956	.6052	.6144	.6232	.6315	.6395	.6471	.6543	.6613
20	.5717	.5819	.5917	.6010	.6098	.6183	.6264	.6342	.6416	.6487
21	.5584	.5688	.5787	.5881	.5971	.6056	.6139	.6217	.6292	.6365
22	.5458	.5562	.5662	.5757	.5848	.5935	.6018	.6097	.6174	.6247
23	.5336	.5442	.5542	.5638	.5730	.5817	.5901	.5982	.6059	.6133
24	.5219	.5326	.5427	.5523	.5616	.5704	.5789	.5870	.5948	.6023
25	.5106	.5214	.5316	.5413	.5506	.5595	.5680	.5762	.5841	.5917
26	.5001	.5106	.5209	.5306	.5400	.5490	.5576	.5658	.5737	.5814
27	.4899	.5004	.5105	.5204	.5298	.5388	.5474	.5557	.5637	.5714
28	.4801	.4906	.5007	.5104	.5199	.5290	.5376	.5460	.5540	.5618
29	.4706	.4812	.4913	.5009	.5103	.5194	.5282	.5366	.5447	.5524
30	.4615	.4721	.4822	.4919	.5012	.5102	.5190	.5274	.5356	.5434
31	.4527	.4633	.4734	.4831	.4924	.5014	.5101	.5186	.5267	.5346
32	.4442	.4548	.4649	.4746	.4840	.4929	.5015	.5100	.5182	.5261
33	.4360	.4466	.4567	.4664	.4757	.4847	.4934	.5017	.5098	.5178
34	.4282	.4386	.4487	.4584	.4678	.4768	.4854	.4938	.5018	.5097
35	.4205	.4310	.4410	.4507	.4600	.4690	.4777	.4861	.4942	.5019
36	.4131	.4235	.4336	.4432	.4525	.4615	.4702	.4786	.4867	.4945
37	.4060	.4164	.4264	.4360	.4453	.4543	.4629	.4713	.4794	.4873
38	.3991	.4094	.4194	.4290	.4382	.4472	.4559	.4642	.4723	.4802
39	.3924	.4027	.4126	.4222	.4314	.4403	.4490	.4573	.4654	.4733
40	.3860	.3962	.4060	.4156	.4248	.4337	.4423	.4507	.4587	.4666

Table 11. Bernoulli Reward Process
Index Values, $a = 0.99$ (continued)

α / β	31	32	33	34	35	36	37	38	39	40
1	.9839	.9842	.9846	.9849	.9852	.9855	.9858	.9860	.9863	.9866
2	.9638	.9646	.9655	.9662	.9670	.9677	.9684	.9690	.9696	.9702
3	.9426	.9440	.9453	.9466	.9478	.9489	.9500	.9510	.9520	.9530
4	.9213	.9232	.9250	.9267	.9284	.9300	.9315	.9329	.9343	.9356
5	.9002	.9027	.9049	.9071	.9092	.9112	.9131	.9149	.9166	.9183
6	.8797	.8826	.8853	.8879	.8904	.8928	.8950	.8972	.8993	.9013
7	.8597	.8630	.8662	.8692	.8721	.8748	.8774	.8799	.8823	.8846
8	.8403	.8441	.8476	.8510	.8542	.8573	.8603	.8631	.8658	.8684
9	.8217	.8258	.8297	.8334	.8369	.8403	.8436	.8467	.8497	.8526
10	.8036	.8081	.8123	.8164	.8202	.8239	.8275	.8308	.8341	.8372
11	.7863	.7910	.7955	.7999	.8040	.8080	.8118	.8154	.8189	.8223
12	.7695	.7745	.7793	.7839	.7883	.7926	.7966	.8005	.8042	.8078
13	.7533	.7586	.7637	.7685	.7732	.7776	.7819	.7860	.7900	.7938
14	.7378	.7433	.7486	.7536	.7585	.7632	.7677	.7720	.7761	.7801
15	.7228	.7286	.7341	.7393	.7444	.7492	.7539	.7584	.7627	.7669
16	.7084	.7143	.7200	.7254	.7307	.7357	.7406	.7452	.7497	.7540
17	.6944	.7005	.7064	.7120	.7174	.7226	.7276	.7325	.7371	.7416
18	.6809	.6872	.6932	.6990	.7046	.7100	.7151	.7201	.7249	.7295
19	.6679	.6743	.6805	.6865	.6922	.6977	.7030	.7081	.7130	.7178
20	.6555	.6620	.6682	.6743	.6801	.6858	.6912	.6965	.7015	.7064
21	.6434	.6500	.6564	.6626	.6685	.6742	.6798	.6852	.6903	.6953
22	.6317	.6385	.6450	.6513	.6573	.6631	.6687	.6741	.6794	.6846
23	.6204	.6273	.6339	.6403	.6464	.6523	.6581	.6636	.6689	.6741
24	.6095	.6165	.6232	.6296	.6358	.6419	.6477	.6533	.6587	.6640
25	.5990	.6060	.6128	.6193	.6256	.6317	.6376	.6433	.6488	.6541
26	.5887	.5958	.6027	.6093	.6157	.6218	.6278	.6336	.6391	.6446
27	.5788	.5860	.5929	.5996	.6060	.6122	.6183	.6241	.6298	.6352
28	.5693	.5764	.5834	.5901	.5966	.6029	.6090	.6149	.6206	.6262
29	.5600	.5672	.5742	.5810	.5875	.5939	.6000	.6060	.6117	.6173
30	.5509	.5582	.5653	.5721	.5787	.5851	.5913	.5973	.6031	.6087
31	.5422	.5495	.5566	.5635	.5701	.5765	.5828	.5888	.5947	.6004
32	.5337	.5411	.5482	.5551	.5618	.5682	.5745	.5806	.5865	.5922
33	.5255	.5329	.5400	.5469	.5536	.5601	.5664	.5726	.5785	.5843
34	.5174	.5249	.5320	.5390	.5457	.5523	.5586	.5647	.5707	.5765
35	.5096	.5171	.5243	.5313	.5380	.5446	.5510	.5571	.5631	.5690
36	.5020	.5095	.5167	.5237	.5305	.5371	.5435	.5497	.5557	.5616
37	.4948	.5021	.5094	.5164	.5232	.5298	.5362	.5425	.5485	.5544
38	.4878	.4951	.5022	.5092	.5161	.5227	.5292	.5354	.5415	.5474
39	.4809	.4883	.4954	.5023	.5091	.5158	.5223	.5285	.5346	.5406
40	.4742	.4816	.4887	.4957	.5024	.5090	.5155	.5218	.5279	.5339

Table 12. Bernoulli Reward Process
Fitted A-Values (Equation (7.16))

a r/n	0.5	0.6	0.7	0.8	0.9	0.95	0.99
0.025	9.2681	7.0096	5.2973	3.9360	2.7944	2.6787	6.1145
0.05	6.8028	5.1840	4.0111	3.0983	2.6018	3.1825	10.3081
0.1	5.1912	4.0473	3.2564	2.7799	2.9147	4.4136	13.4238
0.2	4.2665	3.4833	3.0181	2.8967	3.6013	5.8120	17.1924
0.3	4.2057	3.6679	3.4095	3.5452	4.8016	7.5217	20.2994
0.4	4.3423	3.8339	3.6279	3.8171	5.0397	7.8905	21.5139
0.5	4.2710	3.7488	3.5316	3.7004	5.0189	8.0621	22.9575
0.6	5.0689	4.6046	4.4692	4.7972	6.3305	9.6717	24.9623
0.7	5.9458	5.4645	5.3591	5.7447	7.4003	10.9401	27.2915
0.8	7.6929	7.0607	6.8503	7.1881	9.0770	12.9649	31.0320
0.9	11.9057	11.1842	10.9371	11.3723	13.4908	17.9812	38.4591
0.95	17.9304	17.0536	17.4127	18.2349	21.1040	26.1323	49.5095
0.975	24.1764	28.0856	27.4593	26.7203	33.5284	40.8216	65.9106

Table 13. Bernoulli Reward Process
Fitted B-Values (Equation (7.16))

a r/n	0.5	0.6	0.7	0.8	0.9	0.95	0.99
0.025	9.2848	7.0047	5.3042	3.9386	2.7539	2.1869	1.7222
0.05	6.8134	5.1994	4.0089	3.0752	2.2996	1.9403	1.6483
0.1	5.1849	4.0386	3.2071	2.5684	2.0446	1.7979	1.6080
0.2	4.2244	3.3829	2.7739	2.3034	1.9105	1.7207	1.5820
0.3	3.9207	3.1798	2.6402	2.2198	1.8645	1.6918	1.5709
0.4	3.8095	3.1074	2.5938	2.1922	1.8525	1.6865	1.5716
0.5	3.8637	3.1624	2.6451	2.2363	1.8838	1.7079	1.5795
0.6	3.8097	3.1074	2.5940	2.1924	1.8530	1.6872	1.5745
0.7	3.9210	3.1798	2.6402	2.2200	1.8675	1.6967	1.5796
0.8	4.2241	3.3829	2.7740	2.3041	1.9114	1.7225	1.5892
0.9	5.1853	4.0388	3.2047	2.5692	2.0467	1.8009	1.6200
0.95	6.8137	5.2011	4.0082	3.0757	2.3015	1.9440	1.6793
0.975	9.3011	6.9875	5.2973	3.9503	2.7577	2.1901	1.7798

Table 14. Bernoulli Reward Process
Fitted C-Values (Equation (7.16))

r/n \ a	0.5	0.6	0.7	0.8	0.9	0.95	0.99
0.025	0.5004	-0.0603	0.1432	0.0912	-0.8275	-9.8562	36.7720
0.05	0.1667	0.2466	-0.0355	-0.2108	-2.8142	-11.7535	-119.1096
0.1	-0.0331	-0.0169	-0.1904	-0.9422	-4.0195	-19.3552	-122.1469
0.2	-0.0286	-0.0080	-0.3423	-1.2125	-4.7172	-22.6347	-139.2893
0.3	-0.4630	-0.9010	-1.3409	-2.8740	-10.6097	-34.4107	-168.5174
0.4	-0.6325	-0.8255	-1.2365	-2.5030	-7.0787	-27.4223	-152.4617
0.5	-0.0068	-0.1195	-0.3624	-1.0104	-4.9064	-23.2428	-155.0706
0.6	-0.8394	-1.0776	-1.5097	-2.8257	-7.3069	-28.5828	-150.1690
0.7	-1.1022	-1.4691	-2.2118	-3.6767	-7.7188	-28.2543	-148.8783
0.8	-2.3833	-2.2668	-2.4562	-3.2635	-7.9008	-27.6156	-151.9041
0.9	-5.8849	-5.9008	-5.9116	-6.7079	-9.4755	-28.1731	-126.5893
0.95	-15.3026	-13.0777	-19.9554	-20.7245	-23.3043	-30.9933	-79.0757
0.975	21.7826	-109.5406	-83.8863	-6.3995	-68.3475	-82.6998	36.7720

Table 15. Bernoulli Reward Process
Ranges of n-values used to fit $A, B \& C$ (Equation (7.16))

λ \ a	0.5, 0.6, 0.7	0.8, 0.9	0.95	0.99
.025 & .975	40 to 160	40 to 160	40 to 160	40 to 160
.05 & .95	20 to 180	20 to 160	20 to 140	40 to 160
.1 to .9	10 to 180	10 to 160	20 to 140	30 to 150

Table 16. Exponential Reward Process
Values of $(n - 1 - \Sigma_n)(1 - a)^{\frac{1}{2}}$, (§7.5)

a	0.5	0.6	0.7	0.8	0.9	0.95	0.99
n							
2	.07460	.10921	.14797	.18418	.19738	.17342	.09412
3	.08883	.13189	.18309	.23769	.27612	.26250	.16397
4	.09476	.14145	.19831	.26214	.31598	.31249	.21213
5	.09797	.14666	.20675	.27609	.34010	.34452	.24689
6	.09998	.14993	.21209	.28513	.35638	.36705	.27340
7	.10135	.15217	.21578	.29148	.36821	.38395	.29456
8	.10235	.15379	.21848	.29619	.37725	.39721	.31203
9	.10310	.15502	.22054	.29983	.38441	.40797	.32684
10	.10369	.15599	.22216	.30273	.39025	.41693	.33966
20	.10620	.16012	.22921	.31582	.41849	.46337	.41557
30	.10699	.16143	.23148	.32025	.42913	.48279	.45428
40	.10737	.16206	.23261	.32251	.43486	.49390	.47938
50	.10760	.16244	.23328	.32387	.43847	.50119	.49747
60	.10775	.16269	.23373	.32479	.44095	.50638	.51135
70	.10786	.16287	.23404	.32545	.44277	.51028	.52242
80	.10794	.16300	.23428	.32595	.44417	.51332	.53153
90	.10800	.16310	.23447	.32634	.44527	.51577	.53920
100	.10805	.16319	.23462	.32665	.44617	.51778	.54575
200	.10827	.16355	.23528	.32807	.45035	.52757	.58211
300	.10834	.16367	.23550	.32855	.45181	.53119	.59815
400	.10838	.16374	.23561	.32878	.45256	.53309	.60745
500	.10840	.16377	.23568	.32893	.45301	.53427	.61371
600	.10842	.16380	.23572	.32902	.45331	.53507	.61814
700	.10842	.16381	.23575	.32909	.45353	.53567	.62163
800	.10843	.16382	.23578	.32914	.45370	.53615	.62426
900	.10843	.16384	.23579	.32918	.45383	.53650	.62643

Table 17. Exponential Target Process
Values of $\Sigma(\nu, n)$, (§7.6)

n \ ν	0.3	0.25	0.2	0.15	0.1	0.05
2	1.2075	0.9905	0.7911	0.6031	0.4191	0.2288
3	2.0211	1.6911	1.3863	1.0963	0.8073	0.4940
4	2.8430	2.4017	1.9938	1.6049	1.2154	0.7863
5	3.6683	3.1164	2.6065	2.1202	1.6324	1.0916
6	4.4954	3.8333	3.2218	2.6390	2.0543	1.4043
7	5.3235	4.5514	3.8388	3.1599	2.4792	1.7216
8	6.1521	5.2702	4.4568	3.6822	2.9061	2.0421
9	6.9812	5.9896	5.0755	4.2055	3.3343	2.3648
10	7.8106	6.7094	5.6946	4.7295	3.7636	2.6893
20	16.1110	13.9157	11.8979	9.9854	8.0812	5.9766
30	24.4150	21.1267	18.1077	15.2510	12.4143	9.2927
40	32.7200	28.3390	24.3192	20.5192	16.7519	12.6182
50	41.0252	35.5517	30.5314	25.7886	21.0915	15.9480
60	49.3307	42.7647	36.7440	31.0585	25.4321	19.2802
70	57.6363	49.9778	42.9569	36.3288	29.7734	22.6138
80	65.9419	57.1910	49.1698	41.5992	34.1150	25.9485
90	74.2476	64.4042	55.3828	46.8699	38.4570	29.2838
100	82.5533	71.6175	61.5959	52.1406	42.7991	32.6196
200	165.6111	143.7515	123.7283	104.8502	86.2247	65.9896
300	248.6693	215.8860	185.8614	157.5610	129.6528	99.3665
400	331.7275	288.0206	247.9947	210.2721	173.0816	132.7453
500	414.7858	360.1553	310.1281	262.9834	216.5107	166.1248
600	497.8442	432.2900	372.2615	315.6948	259.9398	199.5048
700	580.9025	504.4247	434.3949	368.4061	303.3691	232.8850
800	663.9608	576.5595	496.5284	421.1176	346.7984	266.2654

Table 17. Exponential Target Process
Values of $\Sigma(\nu,n)$, (§7.6) *(continued)*

ν	0.04	0.03	0.025	0.02	0.015	0.01
n						
2	0.1884	0.1466	0.1250	0.1029	0.0801	0.0565
3	0.4233	0.3473	0.3064	0.2630	0.2162	0.1644
4	0.6873	0.5789	0.5197	0.4558	0.3853	0.3047
5	0.9655	0.8264	0.7497	0.6661	0.5728	0.4642
6	1.2518	1.0830	0.9894	0.8870	0.7717	0.6360
7	1.5434	1.3455	1.2355	1.1147	0.9781	0.8161
8	1.8385	1.6121	1.4861	1.3473	1.1898	1.0020
9	2.1362	1.8817	1.7398	1.5833	1.4054	1.1924
10	2.4358	2.1535	1.9959	1.8220	1.6239	1.3861
20	5.4815	4.9298	4.6214	4.2801	3.8895	3.4163
30	8.5610	7.7466	7.2919	6.7888	6.2131	5.5147
40	11.6517	10.5775	9.9782	9.3156	8.5579	7.6389
50	14.7477	13.4151	12.6722	11.8514	10.9133	9.7762
60	17.8468	16.2565	15.3706	14.3924	13.2751	11.9216
70	20.9477	19.1004	18.0719	16.9368	15.6411	14.0724
80	24.0498	21.9459	20.7752	19.4836	18.0099	16.2269
90	27.1529	24.7926	23.4798	22.0321	20.3809	18.3841
100	30.2565	27.6402	26.1855	24.5818	22.7534	20.5435
200	61.3086	56.1371	53.2680	50.1107	46.5198	42.1931
300	92.3699	84.6472	80.3665	75.6600	70.3132	63.8809
400	123.4339	113.1610	107.4698	101.2153	94.1149	85.5812
500	154.4989	141.6764	134.5750	126.7734	117.9203	107.2870
600	185.5645	170.1927	161.6814	152.3327	141.7275	128.9958
700	216.6304	198.7094	188.7883	177.8929	165.5360	150.7064
800	247.6965	227.2265	215.8956	203.4537	189.3457	172.4182

Table 17. Exponential Target Process
Values of $\Sigma(\nu,n)$, (§7.6) *(continued)*

	$\nu = .01$				$\nu = .001$		
n		n		n		n	
2	0.0565	100	20.5435	2	0.0090	100	12.6465
3	0.1644	200	42.1931	3	0.0388	200	26.8361
4	0.3047	300	63.8809	4	0.0882	300	41.1591
5	0.4642	400	85.5812	5	0.1527	400	55.5365
6	0.6360	500	107.2870	6	0.2287	500	69.9429
7	0.8161	600	128.9958	7	0.3135	600	84.3670
8	1.0020	700	150.7064	8	0.4052	700	98.8030
9	1.1924	800	172.4182	9	0.5025	800	113.2474
10	1.3861	900	194.1306	10	0.6042	900	127.6979
20	3.4163	1000	215.8435	20	1.7609	1000	142.1530
30	5.5147	1100	237.5568	30	3.0338	1100	156.6118
40	7.6389	1200	259.2704	40	4.3571	1200	171.0734
50	9.7762	1300	280.9843	50	5.7079	1300	185.5373
60	11.9216	1400	302.6983	60	7.0764	1400	200.0032
70	14.0724			70	8.4572	2000	286.8229
80	16.2269			80	9.8471	3000	431.5699
90	18.3841			90	11.2440	4000	576.3392
100	20.5435			100	12.6465	5000	721.1166

Table 18. Exponential Target Process
Values of $n(\nu - \Sigma^n(1+\Sigma)^{-n})\nu^{-\frac{1}{2}}(-log_e\nu)^{-1}$,
where $\nu = \nu(\Sigma, n)$, (§7.6)

ν	0.3	0.25	0.2	0.15	0.1	0.05
n						
2	.00241	.00684	.01365	.02308	.03510	.04577
3	.00270	.00801	.01639	.02845	.04477	.06200
4	.00291	.00880	.01827	.03219	.05170	.07436
5	.00307	.00937	.01963	.03496	.05697	.08417
6	.00318	.00980	.02068	.03710	.06114	.09222
7	.00328	.01014	.02150	.03882	.06453	.09896
8	.00335	.01042	.02217	.04022	.06735	.10472
9	.00341	.01064	.02273	.04140	.06974	.10970
10	.00347	.01083	.02319	.04239	.07179	.11408
20	.00373	.01180	.02560	.04766	.08311	.13998
30	.00383	.01217	.02654	.04980	.08797	.15228
40	.00388	.01236	.02704	.05096	.09070	.15963
50	.00392	.01248	.02735	.05170	.09245	.16454
60	.00394	.01256	.02756	.05220	.09368	.16809
70	.00395	.01262	.02772	.05257	.09458	.17076
80	.00397	.01267	.02784	.05285	.09528	.17286
90	.00398	.01270	.02793	.05307	.09583	.17455
100	.00398	.01273	.02800	.05325	.09628	.17595
200	.00402	.01286	.02835	.05408	.09838	.18272
300	.00403	.01291	.02846	.05436	.09911	.18518
400	.00404	.01293	.02852	.05450	.09949	.18647
500	.00404	.01294	.02856	.05459	.09971	.18725
600	.00404	.01295	.02858	.05465	.09986	.18778
700	.00404	.01296	.02860	.05469	.09997	.18816
800	.00404	.01296	.02861	.05472	.10006	.18845

Table 18. Exponential Target Process
Values of $n(\nu - \Sigma^n(1+\Sigma)^{-n})\nu^{-\frac{1}{2}}(-\log_e\nu)^{-1}$,
where $\nu = \nu(\Sigma, n)$, (§7.6) (continued)

ν	0.04	0.03	0.025	0.02	0.015	0.01
n						
2	.04620	.04498	.04339	.04084	.03693	.03100
3	.06380	.06360	.06222	.05948	.05473	.04682
4	.07750	.07855	.07763	.07512	.07012	.06103
5	.08855	.09086	.09052	.08842	.08352	.07379
6	.09772	.10124	.10149	.09992	.09531	.08531
7	.10548	.11015	.11099	.10997	.10578	.09577
8	.11216	.11791	.11933	.11888	.11517	.10531
9	.11799	.12474	.12672	.12683	.12364	.11406
10	.12314	.13082	.13333	.13400	.13134	.12211
20	.15432	.16875	.17532	.18059	.18296	.17860
30	.16961	.18811	.19731	.20574	.21197	.21219
40	.17891	.20019	.21124	.22198	.23115	.23516
50	.18524	.20854	.22099	.23350	.24500	.25214
60	.18984	.21471	.22825	.24217	.25556	.26534
70	.19334	.21947	.23389	.24896	.26393	.27595
80	.19611	.22326	.23841	.25444	.27074	.28469
90	.19836	.22635	.24211	.25897	.27641	.29204
100	.20022	.22893	.24521	.26277	.28120	.29833
200	.20937	.24187	.26097	.28239	.30649	.33248
300	.21277	.24678	.26703	.29007	.31663	.34669
400	.21454	.24937	.27024	.29415	.32206	.35440
500	.21563	.25097	.27222	.29667	.32542	.35918
600	.21637	.25205	.27357	.29837	.32767	.36236
700	.21691	.25284	.27454	.29960	.32927	.36460
800	.21731	.25343	.27526	.30047	.33017	.36625

Table 18. Exponential Target Process
Values of $n(\nu - \Sigma^n(1+\Sigma)^{-n})\nu^{-\frac{1}{2}}(-\log_e\nu)^{-1}$,
where $\nu = \nu(\Sigma, n)$, (§7.6) (continued)

$\nu = .01$				$\nu = .001$			
n		n		n		n	
2	.03100	100	.29833	2	.00843	100	.23103
3	.04682	200	.33248	3	.01302	200	.30758
4	.06103	300	.34669	4	.01752	300	.34972
5	.07379	400	.35440	5	.02196	400	.37717
6	.08531	500	.35918	6	.02632	500	.39668
7	.09577	600	.36236	7	.03063	600	.41130
8	.10531	700	.36460	8	.03487	700	.42265
9	.11406	800	.36625	9	.03904	800	.43169
10	.12211	900	.36754	10	.04315	900	.43904
20	.17860	1000	.36854	20	.08020	1000	.44512
30	.21219	1100	.36933	30	.11067	1100	.45019
40	.23516	1200	.36995	40	.13597	1200	.45449
50	.25214	1300	.37045	50	.15742	1300	.45814
60	.26534	1400	.37086	60	.17594	1400	.46127
70	.27595			70	.19215	2000	.47315
80	.28469			80	.20653	3000	.47990
90	.29204			90	.21940	4000	.48046
100	.29833			100	.23103	5000	.47864

Table 19. Exponential Target Process
Fitted Values of $A, B, C, D \& E$ (§7.6)

ν	A	B	C	$1000D$	E
0.3	0.0089	2.221	-1.0410	-0.12462	-1.4441
0.25	0.0360	2.666	-1.0479	-0.04785	-1.5699
0.2	0.1032	3.242	-1.0491	-0.03422	-1.7848
0.15	0.2687	4.189	-1.0497	-0.06770	-2.1413
0.1	0.7311	5.834	-1.0417	-0.07456	-2.6759
0.05	2.5341	8.038	-0.9805	-0.17755	-3.2358
0.04	3.5126	8.601	-0.9559	-0.19105	-3.3707
0.03	5.1426	8.283	-0.8998	-0.29122	-3.2936
0.025	6.4394	8.581	-0.8792	-0.30000	-3.3755
0.02	8.3339	8.606	-0.8471	-0.33041	-3.3970
0.015	11.3539	8.200	-0.7967	-0.37843	-3.3255
0.01	17.1229	10.119	-0.7932	-0.23122	-3.8292

Index

249